# Perspectives in Ring Theory

# NATO ASI Series

## Advanced Science Institutes Series

*A Series presenting the results of activities sponsored by the NATO Science Committee, which aims at the dissemination of advanced scientific and technological knowledge, with a view to strengthening links between scientific communities.*

The Series is published by an international board of publishers in conjunction with the NATO Scientific Affairs Division

| | |
|---|---|
| A **Life Sciences**<br>B **Physics** | Plenum Publishing Corporation<br>London and New York |
| C **Mathematical and Physical Sciences**<br>D **Behavioural and Social Sciences**<br>E **Applied Sciences** | Kluwer Academic Publishers<br>Dordrecht, Boston and London |
| F **Computer and Systems Sciences**<br>G **Ecological Sciences**<br>H **Cell Biology** | Springer-Verlag<br>Berlin, Heidelberg, New York, London, Paris and Tokyo |

Series C: Mathematical and Physical Sciences - Vol. 233

# Perspectives in Ring Theory

edited by

F. van Oystaeyen

and

Lieven Le Bruyn

Department of Mathematics,
University of Antwerp, Belgium

**Kluwer Academic Publishers**

Dordrecht / Boston / London

Published in cooperation with NATO Scientific Affairs Division

Proceedings of the NATO Advanced Research Workshop on
Perspectives in Ring Theory,
Antwerp, Belgium
July 19–29, 1987

**Library of Congress Cataloging in Publication Data**

NATO Advanced Research Workshop on Perspectives in Ring Theory (1987 : Antwerp, Belgium)
Perspectives in ring theory / editor, F. van Oystaeyen ; co-editor, Lieven Le Bruyn.
p. cm. -- (NATO ASI series. Series C, Mathematical and physical sciences ; vol. 233)
"Proceedings of the NATO Advanced Research Workshop on Perspectives in Ring Theory, Antwerp, Belgium, July 19-29, 1987"--T.p. verso.
Published in cooperation with NATO Scientific Affairs Division.
Includes bibliographies and indexes.
ISBN 9027727368
1. Rings (Algebra)--Congresses. I. Oystaeyen, F. van, 1947- . II. Le Bruyn, Lieven, 1958- . III. North Atlantic Treaty Organization. Scientific Affairs Division. IV. Title. V. Series: NATO ASI series. Series C, Mathematical and physical sciences ; no. 233.
QA247.N366 1987
512'.4--dc19 88-6608
CIP

**ISBN-13: 978-94-010-7841-2** **e-ISBN-13: 978-94-009-2985-2**
**DOI: 10.107/978-94-009-2985-2**

---

Published by Kluwer Academic Publishers,
P.O. Box 17, 3300 AA Dordrecht, The Netherlands.

Kluwer Academic Publishers incorporates the publishing programmes of
D. Reidel, Martinus Nijhoff, Dr W. Junk, and MTP Press.

Sold and distributed in the U.S.A. and Canada
by Kluwer Academic Publishers,
101 Philip Drive, Norwell, MA 02061, U.S.A.

In all other countries, sold and distributed
by Kluwer Academic Publishers Group,
P.O. Box 322, 3300 AH Dordrecht, The Netherlands.

---

# Contents

# Preface.

This proceedings is composed of the papers resulting from the NATO work-shop "Perspectives in Ring Theory" and the work-shop "Geometry and Invariant Theory of Representations of Quivers".

Three reports on problem sessions have been induced in the part corresponding to the work-shop where they belonged. One more report on a problem session, the "lost" problem session, will be published elsewhere eventually.

# Acknowledgement

The meeting became possible by the financial support of the Scientific Affairs Division of NATO. The people at this division have been very helpful in the organization of the meeting, in particular we commemorate Dr. Mario di Lullo, who died unexpectedly last year, but who has been very helpful with the organization of earlier meetings in Ring Theory.

For additional financial support we thank the national foundation for scientific research (NFWO), the rector of the University of Antwerp, UIA, and the Belgian Ministry of Education. We also gladly acknowledge support from the Belgian Friends of the Hebrew University and the chairman Prof. P. Van Remoortere who honored Prof. S. Amitsur for his continuous contributions to the mathematical activities at the University of Antwerp.

I thank the authors who contributed their paper(s) to this proceedings and the lecturers for their undisposable contributions towards the success of the work-shop. Finally I thank Danielle for allowing me to spoil another holiday period in favor of a congress.

F. Van Oystaeyen

# A Global View of the Meeting Perspectives in Ring Theory

by F. Van Oystaeyen

In recent years the field of Ring Theory has lost more and more of its coherence because some rapidly developing subdivisions are obtaining the status of (almost) independent topics in their own right. However one of the most important general observations concerning the development of Ring Theory that became evident during the work-shop is the emergence of new interactions or reciprocal applications between topics seemingly drifting further apart. As examples of such new "bridges" we just mention : the possibility to use the theory of rings of differential operators in problems connected to the famous Jacobian conjecture; application of the abstract theory of derived categories in the representation theory of algebras and quivers, in particular in tilting theory; representation theory and quivers in turn, show up in the classification of orders and maximal orders of low global dimension in combination with methods from graded ring theory; the use of Hopf algebras and smash-products in the study of Brauer Groups of fields; analytic properties of series associated to graded rings; interrelations between representations of quivers, invariant theory, generic matrices and bundles over projective space; Lie algebras and groups appearing in the representation theory of algebras and regular modules; etc...

The force-lines for future research made visible by the work-shop are mainly situated in these new "hybrid" topics some of which are still in a preliminary stage of development. In a call for synthesis one of the main problems raised at the meeting asks for the determination of the class of noncommutative rings that should be considered as the class representing the basic object for study of Ring Theory. It is also noteworty that some very classical ideas are still very much alive today e.g. : the problem of studying central simple algebras up to ring isomorphism and not merely up to algebra isomorphism over the centre; different notions of valuation rings in skewfields are further investigated after a recent succes for valuation theory proving another proof for the existence of non-crossed product skew fields.

The large number of participants that wanted to be part of the work-shop, even as observers, shows that the field of non-commutative algebra is perhaps more alive than ever before.

# Main Lectures

Amitsur, S., Hebrew University of Jerusalem, *An Example in Division Algebras.*

Anick, D., MIT, Boston, USA, *Graded Algebras and Modules, Homological Properties and Hilbert's Theorem.*

Artin, M., MIT, Boston, USA, *Graded Algebras of Global Dimension 3* and *Automorphisms of Elliptic Curves.*

Auslander, M., Brandeis University, USA, *Cohen-Macauley Approximation.*

Bass, H., Columbia University, USA, *Rings of Differential Operators and the Jacobian Conjecture.*

Bautista, R., Universidad Ciudad Mexico, Mexico City, Mexico, *Algebras of Infinite Representation Type.*

De Meyer, F., University of Colorado, Fort Collins, USA, *The Brauer Long Group and Smash-Products.*

Formanek, Penn. State University, USA, *A Note on the Jacobian Conjecture.*

Happel, D., University Bielefeld, BRD, *The Derived Categories for Tame Algebras.*

Lam, T., University of Berkeley, USA, *Algebraic Conjugacy Classes and Skew Polynomial Rings.*

Lorenz, M., Max Planck University, BRD, *Computation of* $G_0$ *for some Noetherian Algebras.*

Malliavin, M.P., Université Paris VI, Paris, France, *Local Cohomology* and *Lie Algebra Cohomology.*

McConnell, J., University of Leeds, UK, *GK, Dimension* and *Hilbert-Samuel Polynomials.*

Michler, G., University of Essen, BRD, *Amitsur-Levitski Theory for Determining the Simplicity of a Modular Group Representation.*

Montgomery, S., University of California, Los Angeles, USA, *Crossed Products of Hopf Algebras.*

Reiten, I., University of Trondheim, Norway, and M. Van den Bergh, UIA, Antwerp, Belgium, *Two Dimensional Orders of Finite Representation Type, (Part I, Part II).*

Ringel, C., University Bielefeld, BRD, *Seven Polynomials in Representation Theory.*

Robson, J. C., University of Leeds, UK, *Somewhat Commutative Rings and the Nulstellensatz.*

Salberger, P., University of Stockholm, Sweden, *Class Groups of Maximal Orders.*

Saltman, D., University of Austin, USA, *Linear Groups in Division Algebras.*

Schelter, W., University of Austin, USA, *Graded Algebras of Finite Global Dimension.*

Schofield, A., University College, London, UK, *Perpendicular Categories.*

Small, L., University of California, San Diego, USA, *Algebras Embeddable in Matrix Rings.*

Stafford, T., University of Leeds, UK, *Rings of Differential Operators.*

Tachikawa, H., University of Tsukuba, *Reflexive Auslander-Reiten Sequences.*

Van Oystaeyen, F., University of Antwerp, UIA, Antwerp, Belgium, *New Algebraic Approach to Microlocalization.*

Wadsworth, A., University of California, San Diego, USA, *Dubrovin Valuation Rings.*

| Name | Address |
|---|---|
| Alev, J. | U. De Paris VI, 4 Pl. Jusieu, 75230, Paris, France |
| Amitsur, S. | Hebrew U. Jerusalem, Israel |
| Anick, D. | M.I.T. Cambridge Mass. 02139, USA |
| Artin, M. | M.I.T. Cambridge Mass. 02139, USA |
| Auslander M. | Brandeis U., Waltham, Mass. 02154, USA |
| Awami, M. | U. of Antwerp, UIA, Wilrijk 2610, Belgium |
| Bacella, G. | U. de L'Aquila, via Roma 33, 67100 L'Aquila, It |
| Bakke, Ø | U. Trondhheim, Dragvoll, Trondheim 7055, Norway |
| Bartijn, J. | U. Utrecht, Boedapestlaan 56, Utrecht, Holland |
| Bass, H. | Columbia U., New York NY 10027 USA |
| Bautista, R. | U. Autonoma, Ciudad Univ., Mexico DFCP 04510 |
| Beattie, M. | Mt. Allison U. Sackville, New Brunswick, Ca. EO A3 Co |
| Brungs, H. | U. Alberta, Edmonton, 632 CA Building, Canada TCG 2611 |
| Bueso, J.L. | U. de Granada, Granada, Spain |
| Busque, C. | U. Autonoma de Barcelona, 08193 Barcelona, Spain |
| Caenepeel, S. | Free U. Brussels, Oefenpleinln. 2, Brussels, Belgium |
| Camps, R. | U. Autonoma de Barcelona, 08193 Barcelona, Spain |
| Cedo, F. | U. Autonoma de Barcelona, 08193 Barcelona, Spain |
| Childs, L. | State Univ. of N.Y. Albany NY12222, USA |
| Dabbour, A. | Ain Shams Univ., Cairo, Egypt |
| Dean, C. | U. of Leeds, Leeds LS 29JT, U.K. |
| De Meyer F. | Colorado St. U., Fort Collins, Colorado 80523 USA |
| Facchini, A. | U. di Udine, Via Zanon 6, 33100 Udine, Italy |
| Ford, T. | Florida Atlantic University, Boca Raton, Florida 33431-0991, USA |
| Formanek E. | Penn. St. U. 215 Mc Allister Bld.,Penn. 16802, USA |
| Gomez Pardo J.L. | Dpto. Algebra, Univ. de Murcia, Murcia, Spain |
| Hacque, J. | Univ. Lyon I, 43 bld. 11 Novembre 1918, 69622 Villeurbanne, Fr. |
| Happel, D. | U. Bielefeld Postf. 8640 4800 Bielefed BRD |
| Holland, M. | U. of Leeds, Leeds LS29JT, UK |
| Iqusa, Y. | North Eastern U., Mass. USA |
| Jordan, D. | U. Sheffield, Sheffield S37RH, U.K |
| Jøndrup, S. | U. Kobenhavn, Universitetsparken 5, 2100 Kobenhavn, Danmark |
| Jun Lun, L. | Beijing Normal U., Beijing, P.R. China |
| Kersten, I. | U. Wuppertal, Gauszstr. 20, 5600 Wuppertal BRD, |
| Kirkman, E. | U. Southern California, University Park, Los Angeles, California 900 89-1113, USA |

| **Name** | **Address** |
|---|---|
| Klein, A. | Bar Ilan Univ., Tel Aviv, Israel |
| Krause, G. | U. Manitoba, Canada |
| Le Bruyn, L. | U. Antwerp, UIA, 2610 Wilrijk, Belgium |
| Lorenz | Max-Planck Inst., Gottfried-Clarenstr., 3 Bonn 5300, BRD |
| Malliavin, M.P. | U. Paris VI, 4, Pl. Jusieu, 75230 Paris, France |
| Matczuck, J. | U. Warsaw, PKIN 009001, Warszawa, Poland |
| McConnell, J. | U. Leeds, Leeds LS 29JT, UK |
| Menini, C. | U. de l'Aquila, Via Roma 33, 67100 L'Aquila, Italy |
| Michler, G. | U. Essen, Postfach 6843, 4300 Essen, BRD |
| Misso, P. | U. de Palermo, Via Archirafi 34, 90123 Palermo, Italy, |
| Montgomery, S. | U. California, U. Park, Los Angeles, CA 90089-1113 |
| Müller, B. | McMaster U., 1280 Main str. W., Hamilton, L8S4K1 Canada |
| Nauwelaerts, E. | UCL, 3610 Diepenbeek, Univ. Campus, Belgium |
| Nelis, P. | U. Antwerp UIA, Universiteitsplein 1, 2610 Wilrijk, Belgium |
| Okninski, J. | U. Warsaw PKIN 00901, Warszawa, Poland |
| Ooms, A. | UCL, Univ. Campus, 3610 Diepenbeek, Belgium |
| Orsatti, A. | U. di Padova, V. Belzoni 7, 35131 Padova, Italy |
| Pittaluga, M. | Univ. degli Studi di Roma, P. Aldo Moro 2, Roma, Italy |
| Reiten, I. | Univ. of Trondheim, Dragvoll 7055 Norway |
| Rickard, J. | University College, London, UK |
| Riedtman, C. | Université de Grenoble, Inst. Fourier, 38402 St. Martin d'Heres, France |
| Ringel, C. | U. Bielefeld, Postfach 8640, 4800 Bielefeld BRD |
| Robson, J. C. | U. of Leeds, Leeds LS29JT, UK |
| Rohnes, B. | U. Trondheim, Dragvoll, 7055 Trondheim, Norway |
| Salberger, P. | U. Stockholm, Stockholm, Sweden |
| Saltman, D. | U. Austin, Texas 78712 USA |
| Sanchez Palacio, J. | Universidad de Barcelona, Gran Via 585, 08007 Barcelona, Spain |
| Sato, M. | Yamanishi Univ., Kofu 400, Japan |
| Schofield, A. | University College, London U., London, UK |
| Shaoxue, L. | Beijing Normal University, Beijing, P.R. China |
| Small, L. | University California, San Diego, Ca 92037, USA |
| Solberg, Ø | University of Trondheim, Dragvoll 7055, Norway |
| Stafford, T. | University of Leeds, Leeds LS 29JT, UK |

| Name | Address |
| --- | --- |
| Strooker, J. | Universiteit Utrecht, Budapestlaan 56, Utrecht, Nederland |
| Tachikawa, H. | Tsukuba U., Sakura Mura, Nihari-gun Ibaraki 305, Japan |
| Teranishi, Y. | University of Nagoya, Nagoya, Japan |
| Tignol, J.P. | Univ. de Louvain-la-Neuve, 2 Avenue du Cyclotron 1348 Louvain-la-Neuve, Belgium |
| Todorov, G. | Northeastern University, Boston, Mass., USA |
| Van den Bergh, M. | Univ. Antwerp, UIA, 2610 Wilrijk, Belgium |
| Van Oystaeyen, F. | Univ. Antwerp UIA, 2610 Wilrijk, Belgium |
| Verschoren, A. | Univ. of Antwerp RUCA, Antwerp, Belgium |
| Wadsworth, A. | University of California, San Diego, Ca 92037, USA |
| Zacharia, D. | St. U. North. Virginia, 4400 U. Drive, Fairfax Virginia 2203, USA |

# Deformations of Tilting Modules

Dieter Happel*
Fakultät für Mathematik
Universität Bielefeld
4800 Bielefeld, FDR

Mary Schaps
Department of Mathematics
Bar-Ilan University
Ramat-Gan, Israel

Abstract

We show that if a finite dimensional algebra $A$ has both a deformation $\tilde{A}$ and a tilting $B = \mathrm{End}_A(M)$, then we can deform the tilting module $M$ to $\tilde{M}$ in order to obtain a deformation $\tilde{B}$ of $B$ such that $\tilde{B} = \mathrm{End}_{\tilde{A}}(\tilde{M})$ is a tilting of $\tilde{A}$.

## 0. Introduction

Let $A$ be a finite-dimensional (associative, with unit) algebra over an algebraically-closed field $k$. By mod $A$ we denote the category of finite-dimensional left $A$-modules. The composition of morphisms $f: X \to Y$ and $g: Y \to Z$ in mod $A$ is denoted by $fg$.

The aim of this article is to combine two subjects which have been studied in recent years. On the geometric side there is the deformation theory of algebras and modules (see [S1] for further references on this topic)

---

*Part of the research was completed while the first author was visiting Bar-Ilan University. He wishes to express his gratitude for their warm hospitality.



F. van Oystaeyen and L. Le Bruyn (eds.), Perspectives in Ring Theory, 1–20.

while on the algebraic side there is the notion of a tilting module introduced in [HR] which is defined in homological terms. We refer to section 1 for more details. We will show here that tilting modules behave nicely with respect to deformation. More precisely, we will prove the following:

Theorem: Let $A$ be a finite-dimensional, basic $k$-algebra and let $\tilde{A}$ be a deformation of $A$. Let ${}_A M$ be a tilting module with endomorphism algebra $\mathrm{End}_A M = B$. Then there exists a tilting module ${}_{\tilde{A}}\tilde{M}$ such that $\tilde{B} = \mathrm{End}_{\tilde{A}}\tilde{M}$ is an algebra deformation of $B$.

We point out that even in the case where ${}_A M$ is multiplicity-free (i.e. each indecomposable direct summand of ${}_A M$ occurs with multiplicity one), the deformed tilting module ${}_{\tilde{A}}\tilde{M}$ will usually not be multiplicity-free. In other words, $\tilde{B}$ will usually not be a basic $k$-algebra. Examples for this will be given in section 3.

In section 1 we recall the relevant definitions and results from deformation theory and tilting theory. The proof of the theorem is given in section 2, while section 3 will contain several examples.

## 1. Preliminaries

### 1.1 Deformation theory

Let $\mathrm{Alg}_d$ be the affine variety of algebra structures (associative, with unit) on a vector space $V$ of dimension $d$ over the field $k$. This variety is the set of bilinear maps from $V \times V$ to $V$ that are associative and have a unit; for a formal definition and some of its properties the reader is referred to [G]. To a point $p$ of $Alg_d$ we associate the finite-dimensional k-algebra $A_p$. Let $G = GL(V)$. The canonical linear operation of $G$ on $V$ induces an operation of $G$ on $Alg_d$. For $p \in Alg_d$ we denote by $O_p$ the orbit of $p$ under $G$. Let $p,q \in Alg_d$. We say that $A_q$ deforms to $A_p$ (or $A_p$ degenerates to $A_q$) if $q \in \bar{O}_p$, the Zariski-closure of $O_p$. $A_p$ is then called an algebra deformation of $A_q$.

In this case there exists an irreducible pointed curve $(C,q)$ in $Alg_d$ with $C = \operatorname{Spec} R$, and a flat R-algebra $\mathbb{A}$ such that $A_q \simeq \mathbb{A} \otimes (R/m_q)$, $A_p \simeq \mathbb{A} \otimes (R/m_t)$, $t \neq p$, where $m_t$ denotes the maximal ideal in $R$ corresponding to a closed point $t \in C$.

Let $Gr(m, V^m)$ be the Grassmannian variety formed by the subspaces of $V^m$ of codimension $m$. Let $Alg\,mod_{d,m} \subseteq Gr(m, V^m) \times Alg_d$ be the subset of pairs $(U,p)$ such that $U$ is an $A_p$-submodule of $V^m$ where $V^m$ is given the structure of an $A_p$-module induced by the algebra structure $A_p$ on $V$; then $Alg\,mod_{d,m}$ is closed in $Gr(m, V^m) \times Alg_d$, and thus the natural projection from $Alg\,mod_{d,m}$ to $Alg_d$ is a proper map. To each point $(U,p)$ of $Alg\,mod_{d,m}$ we may associate the algebra $A_p$ and the $A_p$-module $X = V^m/U$. Since all modules of dimension $m$ occur as a quotient of $V^m$, every $A_p$-module of dimension $m$ is associated to some point in the fiber over $p$.

Let $p,q \in \mathrm{Alg}_d$ and assume that $A_q$ deforms to $A_p$. Then an $A_p$-module $\tilde{X}$ is called a deformation of an $A_q$-module $X$ if $X$ (considered an element in $\mathrm{Alg\,mod}_{d,m}$) is in the closure of the orbit of $\tilde{X}$ considered as an element in $\mathrm{Alg\,mod}_{d,m}$.

In section 2 we will need the following special case of a result in [S2].

<u>Lemma 1</u>: Let $p,q \in \mathrm{Alg}_d$ and assume that $A = A_q$ deforms to $\tilde{A} = A_p$. Let $X$ be an $A$-module satisfying $\mathrm{Ext}_A{}^1(X,X) = 0$. Then any deformation $\tilde{X}$ of $X$ satisfies $\mathrm{Ext}_{\tilde{A}}{}^1(\tilde{X},\tilde{X}) = 0$.

An easy extension of this also gives the following result. For the convenience of the reader we will include a proof.

<u>Lemma 2</u>: Let $p,q \in \mathrm{Alg}_d$ and assume that $A = A_q$ deforms to $\tilde{A} = A_p$. Let $X$ be an $A$-module such that $\mathrm{proj\,dim}_A X \leq s$. Then any deformation $\tilde{X}$ of $X$ satisfies $\mathrm{proj\,dim}_A \tilde{X} \leq s$.

<u>Proof</u>: The arguments given in [S2] show that the function from $\mathrm{Alg\,mod}_{d,m} \times \mathrm{Alg\,mod}_{d,\ell}$ to the integers sending a pair $(U,p)\times(U',p)$ to the dimension over $k$ of $\mathrm{Ext}_A{}^t(V^m/U, V^\ell/U')$ is upper semicontinuous.

The $A$-module $X$ satisfies $\mathrm{proj\,dim}_A X \leq s$ if and only if $\mathrm{Ext}_A{}^t(X,Y) = 0$ for all $t \geq s$ and all $A$-modules $Y$. Since $\mathrm{Gr}(\ell, V^\ell)$ is complete, any $\tilde{A}$-module $Z$ is a deformation of an $A$-module $Z'$. In particular, let $t \geq s$. Then $\dim_k \mathrm{Ext}_{\tilde{A}}{}^t(\tilde{X},Z) \leq \dim_k \mathrm{Ext}_A{}^t(X,Z') = 0$. Thus proj $\dim_{\tilde{A}}\tilde{X} \leq s$, completing the proof of the lemma.

Now assume that $A = A_q$ is a basic k-algebra and $\tilde{A} = A_p$ is a deformation of $A$. Let $e_1,\ldots,e_n$ be a complete set of primitive orthogonal idempotents. By Proposition 1 in [S1] we may assume that all primitive idempotents of $A$ deform to idempotents in $\tilde{A}$.

More precisely, there is a continuous idempotent section $\bar{e}_i = C \to \mathcal{A}$ such that $\bar{e}_i(q) = e_i$ and $\bar{e}_i(p) = \tilde{e}_i$. Let $_AP$ be an indecomposable projective $A$ module, say $_AP = Ae_i$. Let $\mathcal{P}$ be the $\mathcal{A}$-module $\mathcal{A}\bar{e}_i$. Then $_{\mathcal{A}}\mathcal{P}$ is a projective $\mathcal{A}$-module such that $_{\mathcal{A}}\mathcal{P}(q) = {}_AP$ and $_{\mathcal{A}}\mathcal{P}(p) = \tilde{A}\tilde{e}_i$ is also a projective $\tilde{A}$ module $_{\tilde{A}}\tilde{P}$, but need not be indecomposable, since $\tilde{e}_i$ might be nonprimitive. More generally, let $_AQ$ be a projective $A$-module, then $_{\tilde{A}}\tilde{Q}$ will denote the deformed $\tilde{A}$-projective module. In fact, if $_AQ = \bigoplus_{i=1}^{n} (Ae_i)^{n_i}$ then $_{\tilde{A}}\tilde{Q} = \bigoplus_{i=1}^{n} (\tilde{A}\tilde{e}_i)^{n_i}$.

<u>Lemma 3</u>: If $\alpha\colon {}_AP_1 \to {}_AP_0$ is a homomorphism of projective modules, then there is an $\mathcal{A}$-module homomorphism $\bar{\alpha}\colon {}_{\mathcal{A}}\mathcal{P}_1 \to {}_{\mathcal{A}}\mathcal{P}_0$ which reduces to $\alpha$ over $q$. If $\alpha$ is a monomorphism, then a general fiber $\tilde{\alpha} = \bar{\alpha}(p)$ is also a monomorphism.

<u>Proof</u>: We fix a deformation of the primitive idempotents of $A$ to a complete orthogonal set of idempotents sections for $\mathcal{A}$. We say that a basis

respects an idempotent set if the idempotents are basis elements and each element lies in a component of a Peirce decomposition of the algebra. We deform a basis of $A$ respecting the original idempotent set to a basis of $\mathcal{A}$ which respects the deformed idempotent set. This implies that basis elements lying in a particular indecomposable projective $_AP$ deform to elements of the deformed family $_{\mathcal{A}}\mathcal{P}$. For the first part of the lemma it suffices to show that we can construct $\bar{\alpha}$ when $P_1$ and $P_0$ are indecomposable projectives $P_1 = Ae_i$ and $P_0 = Ae_j$. Then $\alpha(e_i) \in Ae_j$ and $e_i\alpha(e_i) = \alpha(e_i \cdot e_i) = \alpha(e_i)$, so $\alpha(e_i) \in e_iAe_j$. Writing $\alpha(e_i)$ with respect to the basis of $e_iAe_j$ and taking an arbitrary lifting of the coefficients to elements of $R$, we define an element of $\bar{e}_i\mathcal{A}\bar{e}_j$ which we will take to be $\bar{\alpha}(\bar{e}_i)$. For general $P_1$ and $P_0$, we repeat this procedure for each pair consisting of a summand of $P_1$ and a summand of $P_0$, giving a total $\mathcal{A}$-homomorphism

$$\bar{\alpha}\colon {}_{\mathcal{A}}\mathcal{P}_1 \to {}_{\mathcal{A}}\mathcal{P}_0.$$

As for the second part of the lemma, the dimension of the kernel of an $\mathcal{A}$-homomorphism is an upper semicontinous function on the base curve $C$, and thus if it is zero at the special fiber $q$, it is zero at the general fiber $p$.

Using this lemma, we may define deformations of $A$-modules $X$ satisfying $\operatorname{proj\,dim}_A X \leq 1$ in a very explicit way, as the cokernels of such $\bar{\alpha}$ over a general point $p$. Over $p$ we have an $\tilde{A}$-algebra monomorphism

$\tilde{\alpha}: \tilde{P}_1 \to \tilde{P}_0$, and we denote its cokernel by $\tilde{X}$. As a corollary to the following lemma, we will show that when $\mathrm{Ext}^1(X,X) = 0$, the deformation $\tilde{X}$ is uniquely determined up to isomorphism.

<u>Lemma 4</u>: Let $\tilde{A}$ be a deformation of $A$ and let $X,Y$ be $A$-modules satisfying $\mathrm{proj\,dim}_A X \leq 1$, $\mathrm{proj\,dim}_A Y \leq 1$ and $\mathrm{Ext}_A{}^1(X,Y) = 0$. Let $\tilde{X},\tilde{Y}$ be deformations of $X$ and $Y$ obtained as above by deforming projective resolutions of $X$ and $Y$. Then $\mathrm{Hom}_A(X,Y)$ deforms to $\mathrm{Hom}_{\tilde{A}}(\tilde{X},\tilde{Y})$, monomorphisms deform to monomorphisms, and epimorphisms deform to epimorphisms. In particular, if $X$ and $Y$ are isomorphic, so are $\tilde{X}$ and $\tilde{Y}$.

<u>Proof</u>: Any $A$-linear map $f \in \mathrm{Hom}_A(X,Y)$ yields a commutative diagram of short exact sequences

$$\begin{array}{ccccccccc} 0 & \to & P_1 & \xrightarrow{u} & P_0 & \to & X & \to & 0 \\ & & \alpha\downarrow & & \beta\downarrow & & \downarrow f & & \\ 0 & \to & Q_1 & \xrightarrow{v} & Q_0 & \to & Y & \to & 0 \end{array}$$

As in lemma 3, all the $A$-homomorphisms between projective modules $u,v,\alpha,\beta$ can be lifted to $A$-homomorphisms $\bar{u},\bar{v},\bar{\alpha}$, and $\bar{\beta}$, and in addition $\bar{u}$ and $\bar{v}$ will still be monomorphisms. Letting $\mathcal{X}$ and $\mathcal{Y}$ denote the cokernels of $\bar{u}$ and $\bar{v}$ we get a (generally noncommutative) diagram with exact rows

$$\begin{array}{ccccccccc} 0 & \to & \mathcal{P}_1 & \xrightarrow{\bar{u}} & \mathcal{P}_0 & \to & \mathcal{X} & \to & 0 \\ & & \bar{\alpha}\downarrow & & \bar{\beta}\downarrow & & & & \\ 0 & \to & \mathcal{Q}_1 & \xrightarrow{\bar{v}} & \mathcal{Q}_0 & \to & \mathcal{Y} & \to & 0 \end{array}$$

Let $\bar{\gamma} = \bar{u}\bar{\beta} - \bar{\alpha}\bar{v}$ be the homomorphism which measures the noncommutativity of this diagram. It induces a morphism of complexes

$$\begin{array}{ccccccc} 0 \to & P_1 & \xrightarrow{\bar{u}} & P_0 & \to 0 & & \\ & \bar{\gamma}\downarrow & & & & & \\ 0 \to Q_1 & \xrightarrow{\bar{v}} & Q_0 & \to 0. & & & \end{array}$$

By the upper semicontinuity of $\mathrm{Ext}^1$, and the fact that $\mathrm{Ext}_A^1(X,Y) = 0$, the family $\mathrm{Ext}_\Lambda^1(\mathcal{X},\mathcal{Y}) = 0$ has zero fibers almost everywhere. Discarding the bad points of $C$ at which the fiber is nonzero, we may assume that $\mathrm{Ext}_\Lambda^1(\mathcal{X},\mathcal{Y}) = 0$. Then the homomorphism above must be homotopic to zero, that is, there exist

$$\gamma_0: P_0 \to Q_0$$

$$\gamma_1: P_1 \to Q_1$$

such that $\bar{\gamma} = \bar{u}\gamma_0 + \gamma_1\bar{v}$. Thus $\bar{u}(\bar{\beta}-\gamma_0) = (\bar{\alpha}+\gamma_1)\bar{v}$, so we obtain a commutative diagram of short exact sequences of $\Lambda$ modules and an induced $\Lambda$-homomorphism $\bar{f} \in \mathrm{Hom}_\Lambda(\mathcal{X},\mathcal{Y})$.

$$\begin{array}{ccccccccc} 0 \to & P_1 & \xrightarrow{\bar{u}} & P_0 & \to & \mathcal{X} & \to 0 \\ & \bar{\alpha}+\gamma_1\downarrow & & \bar{\beta}-\gamma_0\downarrow & & \vdots\downarrow\bar{f} & \\ 0 \to & Q_1 & \xrightarrow{\bar{v}} & Q_0 & \to & \mathcal{Y} & \to 0 \end{array}$$

Applying this procedure to a basis of $\mathrm{Hom}_A(X,Y)$ will produce a set of sections of $\mathrm{Hom}_{\mathcal{A}}(\mathcal{X},\mathcal{Y})$ which, since they are linearly independent in the special fiber, must be linearly independent in the general fiber, and thus provide a linearly independent subset of $\mathrm{Hom}_{\tilde{A}}(\tilde{X},\tilde{Y})$. This implies that $\dim \mathrm{Hom}_A(X,Y) \leq \dim \mathrm{Hom}_{\tilde{A}}(\tilde{X},\tilde{Y})$. On the other hand, $\mathrm{Hom}_{\mathcal{A}}(\mathcal{X},\mathcal{Y})$ is an algebraic subset of $\mathrm{Hom}_{\mathcal{R}}(\mathcal{X},\mathcal{Y})$, and thus the dimension of the fiber over C is upper semicontinuous showing that $\dim \mathrm{Hom}_A(X,Y) \geq \dim \mathrm{Hom}_{\tilde{A}}(\tilde{X},\tilde{Y})$. We conclude that the fibers of $\mathrm{Hom}_{\mathcal{A}}(\mathcal{X},\mathcal{Y})$ determine a flat family of $\mathcal{R}$-modules deforming $\mathrm{Hom}_A(X,Y)$ to $\mathrm{Hom}_{\tilde{A}}(\tilde{X},\tilde{Y})$ and that we have deformation of a basis of $\mathrm{Hom}_A(X,Y)$ to a basis of $\mathrm{Hom}_{\tilde{A}}(\tilde{X},\tilde{Y})$.

The dimension of the kernel of an $\mathcal{A}$-homomorphism is upper semicontinuous over C. Thus monomorphisms deform to monomorphisms.

We then note that for $\tilde{f} \in \mathrm{Hom}_{\tilde{A}}(\tilde{X},\tilde{Y})$,

$$\begin{aligned} \dim \ker \tilde{f} - \dim \operatorname{coker} \tilde{f} &= \dim \tilde{Y} - \dim \tilde{X} \\ &= \dim Y - \dim X \\ &= \dim \ker f - \dim \operatorname{coker} f \end{aligned}$$

Since the dimension of the kernel is upper semicontinuous, so is the dimension of the cokernel, and thus epimorphisms deform to epimorphisms. Therefore, if there is an isomorphism in $\mathrm{Hom}_A(X,Y)$, it lifts to an isomorphism between $\tilde{X}$ and $\tilde{Y}$.

Corollary: Suppose $M$ is an $A$-module satisfying proj $\dim_A M \leq 1$ and $\mathrm{Ext}^1(M,M) = 0$. Up to isomorphism, there is one and only one deformation of $M$ to a module $\tilde{M}$ over $\tilde{A}$.

Proof: Immediate from lemma 4.

## 1.2 Tilting theory

Let $A$ be a finite-dimensional $k$-algebra. Following [HR] an $A$-module ${}_AM$ is called a tilting module if we have (i) proj $\dim_A M \leq 1$, (ii) $\mathrm{Ext}_A{}^1(M,M) = 0$ and (iii) there exists a short exact sequence $0 \to {}_AA \to M^0 \to M^1 \to 0$ where $M^0, M^1$ are direct sums of directs summands of $M$. The interest in these modules comes from the fact that there is a close relationship between mod $A$ and mod $B$ where $B = \mathrm{End}_A M$ (see [HR] for details). An $A$-module ${}_AM$ satisfying proj $\dim_A M \leq 1$, $\mathrm{Ext}_A{}^1(M,M) = 0$ is called a partial tilting module.

Let ${}_AM$ be a tilting module and $B = \mathrm{End}_A M$. Then $M$ can be considered as a right $B$-module and $M_B$ is a right $B$-tilting module. Moreover $A \simeq \mathrm{End}_B(M_B)$, where the isomorphism is given by the canonical homomorphism from ${}_AA$ to $\mathrm{End}_B(M_B)$ given by the $A$-module structure on $M$.

## 2. Proof of the theorem

Theorem: Let $A$ be a finite dimensional, basic $k$-algebra and let $\tilde{A}$ be a deformation of $A$. Let ${}_A M$ be a tilting module with endomorphism algebra $\mathrm{End}_A M = B$. Then there exists a tilting module ${}_{\tilde{A}}\tilde{M}$ such that $\tilde{B} = \mathrm{End}_{\tilde{A}}\tilde{M}$ is an algebra deformation of $B$.

Proof: We will first construct the deformation ${}_{\tilde{A}}\tilde{M}$ of ${}_A M$ to a partial tilting module, then construct the deformation $\tilde{B}$ of $B$, and finally we will prove that ${}_{\tilde{A}}\tilde{M}$ is unique and is in fact a tilting module for $\tilde{A}$.

(a) The construction of ${}_{\tilde{A}}\tilde{M}$: Let $0 \to P_1 \xrightarrow{\alpha} P_0 \to M \to 0$ be a short projective resolution of $M$. Then by lemma 3 we can lift $\alpha$ to an $\mathcal{A}$-homomorphism $\bar{\alpha}$, which will be a monomorphism over the general point $p$ of $C$. We let $\tilde{M}$ be the fiber over $p$ of the cokernel $\mathcal{M}$ of $\bar{\alpha}$. By lemmas 1 and 2, ${}_{\tilde{A}}\tilde{M}$ is a partial tilting module.

(b) $\tilde{B}$ is an algebra deformation of $B$: We apply lemma 4 with $X = Y = M$. $\mathrm{End}_{\mathcal{A}}(\mathcal{M}) = \mathrm{Hom}_{\mathcal{A}}(\mathcal{M},\mathcal{M})$ is thus a flat deformation of modules of $\mathrm{End}_A(M)$ into $\mathrm{End}_{\tilde{A}}(\tilde{M})$ and a $k$-basis of $B=\mathrm{End}_A(M)$ deforms to a set of elements $h_1,\ldots,h_d$ of $\mathrm{End}_{\mathcal{A}}(\mathcal{M})$ which are linearly independent in each fiber. Lifting a $k$-basis of $M$ to $\mathcal{M}$ gives a set which is linearly independent almost everywhere. After discarding the bad points from $C$, we may assume that $\mathcal{M}$ is a free $R$-module on the given basis, and thus, since $\mathrm{End}_{\mathcal{A}}(\mathcal{M}) \subseteq \mathrm{End}_R(\mathcal{M})$, we obtain a matrix representation of $\mathrm{End}_{\mathcal{A}}(\mathcal{M})$ with entries in $R$.

Since $C$ was irreducible, $R$ is an integral domain. Let $K$ be its quotient field. Since the elements $h_1,\ldots,h_{d'}$ were linearly independent in each fiber, they form a basis of the $K$-vector space generated by $\mathrm{End}_{\mathcal{A}}(\mathcal{M})$.

Since the composition of two elements of $\mathrm{End}_{\mathcal{A}}(\mathcal{M})$ lies in $\mathrm{End}_{\mathcal{A}}(\mathcal{M})$, we conclude that the structure constants of the multiplication are elements of the quotient field $K$ of $R$ which have nonvanishing denominator over the distinguished point $q$. After we discard any bad points at which the denominators in the structure constants vanish, the structure constants give a mapping of $C$ onto a curve $(C',q')$ in $\mathrm{Alg}_{d'}$, with $A_{q'} \xrightarrow{\sim} B$ and $A_{p'} \xrightarrow{\sim} \tilde{B}$ for $p' \neq q'$. Thus $\tilde{B} = \mathrm{End}_{\tilde{A}}(\tilde{M})$ is an algebra deformation of $B = \mathrm{End}_A(M)$.

(c) <u>The isomorphism of</u> ${}_{\tilde{A}}\tilde{M}$ <u>and</u> $(\tilde{M}_B)_{\tilde{B}}$: The left $\tilde{A}$-module $\tilde{M}$ is also a right $\tilde{B} = \mathrm{End}_{\tilde{A}}(\tilde{M})$ module, and as such is a deformation of the right $B$-module $M_B$, which is a partial tilting module for $B$. Thus by the corollary to lemma 4, ${}_{\tilde{A}}\tilde{M}_{\tilde{B}}$ is the unique $\tilde{B}$ module $(\tilde{M}_B)_{\tilde{B}}$ obtained by deforming a projective $B$-resolution of $M_B$. Henceforward we will denote this $\tilde{A}$-$\tilde{B}$ bimodule by $\tilde{M}$.

(d) <u>The construction of a short sequence for</u> $\mathrm{End}_{\tilde{B}}(\tilde{M})$:

Our goal is to construct for $\tilde{A}$ a sequence as in (iii) of the definition of a tilting module in section 1.2. We will first construct it for $\mathrm{End}_{\tilde{B}}(\tilde{M})$ and then prove that $\tilde{A} \xrightarrow{\sim} \mathrm{End}_{\tilde{B}}(\tilde{M})$.

As shown in [HR], by taking the sequence

$$0 \to {}_AA \to M' \to M'' \to 0$$

in the definition of a tilting module and applying the functor $\mathrm{Hom}_A(\cdot, {}_AM)$, we get a new short exact sequence

$$0 \to R'' \to R' \to M \to 0$$

which is a short projective resolution of $M$ as a $B$-module. As in the corollary to lemma 4, this sequence has a deformation which is unique up to isomorphism of complexes

$$0 \to \tilde{R}'' \to \tilde{R}' \to \tilde{M} \to 0$$

Following [HR] again, we apply the functor $\mathrm{Hom}_{\tilde{B}}(\cdot, \tilde{M}_{\tilde{B}})$ to get a sequence

$$0 \to \mathrm{Hom}_{\tilde{B}}(\tilde{M},\tilde{M}) \to \mathrm{Hom}_{\tilde{B}}(\tilde{R}',\tilde{M}) \to \mathrm{Hom}_{\tilde{B}}(\tilde{R}'',\tilde{M})$$

The final map is surjective, since $\mathrm{Ext}_B{}^1(M,M) = 0$ and $\mathrm{Ext}^1$ is upper semicontinuous.

The projective $B$ modules $R'$ and $R''$ were generated by idempotents of $B = \mathrm{End}_A(M)$ which corresponded to the projections onto the various indecomposable direct summands of $M'$ and $M''$ respectively. Passing to an etale covering of $C$ and applying proposition 1 of [S1], we can deform these to idempotents of $\tilde{B}$, which are no longer necessarily primitive. These new idempotents correspond to projections onto direct summands of the deformed module $\tilde{M}$, which are not necessarily indecomposable. The choice of idempotents $f_i$ for $\tilde{B}$ corresponds to choosing a direct sum decomposition of $\tilde{M}$, each component $\tilde{M}_i$ of which is a flat deformation of an indecomposable summand $M_i$ of $M$. Then $\mathrm{Hom}_{\tilde{B}}(\tilde{f}_i\tilde{B},\tilde{M}) \xrightarrow{\sim} \tilde{M}_i$, since for any $\Theta \in \mathrm{Hom}_{\tilde{B}}(\tilde{f}_i\tilde{B},\tilde{M})$, $\Theta(\tilde{f}_i) = \Theta(\tilde{f}_i \cdot \tilde{f}_i) = \Theta(\tilde{f}_i)\tilde{f}_i \in \tilde{M}_i$ and conversely any choice $\Theta(f_i) \in \tilde{M}_i$ determines a $\tilde{B}$ homomorphism.

Thus

$$\mathrm{Hom}_{\tilde{B}}(\tilde{R}',\tilde{M}) \xrightarrow{\sim} \tilde{M}'$$

and

$$\mathrm{Hom}_{\tilde{B}}(\tilde{R}'',\tilde{M}) \xrightarrow{\sim} \tilde{M}''$$

Thus we have a short exact sequence

$$0 \to \mathrm{End}_{\tilde{A}}(\tilde{M}) \to \tilde{M}' \to \tilde{M}'' \to 0.$$

(e) <u>The isomorphism of</u> $\tilde{A}$ <u>and</u> $\mathrm{End}_{\tilde{B}}(\tilde{M})$: There is a natural homomorphism of $\tilde{A}$ into $\mathrm{End}_{\tilde{B}}(\tilde{M})$ determined by the $\tilde{A}$-module structure. Since this homomorphism is a monomorphism in the special fiber, it is a monomorphism everywhere. However

$$\begin{aligned}\dim_k\tilde{A} &= \dim_k A \\ &= \dim_k M' - \dim_k M'' \\ &= \dim_k\tilde{M}' - \dim_k\tilde{M}'' \\ &= \dim_k \mathrm{End}_{\tilde{B}}(\tilde{M})\end{aligned}$$

We conclude that $\tilde{A} \xrightarrow{\sim} \mathrm{End}_{B}(\tilde{M})$ and thus that $\tilde{M}$ is a tilting module.

## 3. Examples

If we diagram the d-dimensional algebras as a partially ordered set with the degenerations of an algebra lying below it, then there is a unique minimal point corresponding to the algebra $k[x_1,\ldots,x_{d-1}]/(x_1,\ldots,x_{d-1})^2$. As one goes up the diagram the representation theory gets progressively simpler. The maximal points, which are the rigid algebras without deformations, include the quiver algebras of global homological dimension 1. The algebras of finite representation type are also concentrated at the top of the chart [G].

Interesting examples of the tilting deformation theorem can be found in a band at the top of the diagram, far enough down to have nontrivial deformations, and far enough up to have nontrivial tilting modules.

Before we actually give the examples, let us describe the pictures we will use to illustrate them. The basis graph $Q_A$ of an algebra $A$ is a directed graph with one vertex for each idempotent in a complete set $\{e_1,\ldots e_n\}$ of primitive orthogonal idempotents, and

$$n_{ij} = \dim e_iAe_j$$

arrows from $e_i$ to $e_j$. If we let these arrows correspond to elements of a basis filtered by powers of the radical, then we will draw $s$ barbs on the arrow corresponding to an element of $J^s$. Nonradical arrows, which can be taken to correspond to matrix units, will be represented by arrows with solid barbs.

To simplify the verification that a given module is indeed a tilting module, it is convenient to work with algebras which are of finite representation type or tame. The algebras we consider all have a distributive ideal lattice, and thus by [S1] their deformations have the same property. Experience has shown that such deformations are generally built by the vanishing of a zero relation or the splitting of an idempotent. We will give one example of the first and two examples of the second.

The simplest way to construct a tilting module is to replace a simple projective by its inverse Auslander-Reiten translate. Two of our examples will be of this type, while the third will be more complicated. Example 2 is included for specialists in representations of quivers; other readers may skip it.

Example 1: The minimal example of a deformation arising from a vanishing zero relation is the deformation in Fig. 1 of $A$ to $\tilde{A}$. We take the tilting module obtained by replacing the simple projective $P_3$ in ${}_AA$ by $M_3 = \tau^{-1}P_3$, the cokernel of a homomorphism $u: P_3 \to P_1 \oplus P_2$. This module has a nontrivial endomorphism mapping the top of $P_2$ into the socle of $P_1$, so the tilting $B$ has a loop at vertex 3. Both $P_1$ and $P_2$ map into $M_3$, and $M_3$ maps into $P_1$. The deformed tilting module $\tilde{M}$ is a direct sum of $\tilde{M}_1 = \tilde{P}_1$, $\tilde{M}_2 = \tilde{P}_2$ and $\tilde{M}_3 \xrightarrow{\sim} \tilde{P}_1 \oplus S_2$, where $S_2$ is the simple at vertex 2, isomorphic to $\tilde{P}_2/(\tilde{P}_3)$.

The total commutative square is given by Fig. 1 below. Since both $\tilde{M}_1$ and $\tilde{M}_3$ contain copies of $\tilde{P}_1$, there are two isomorphic summands in $\tilde{M}$, which give rise to two isomorphic simples in $\tilde{B}$.

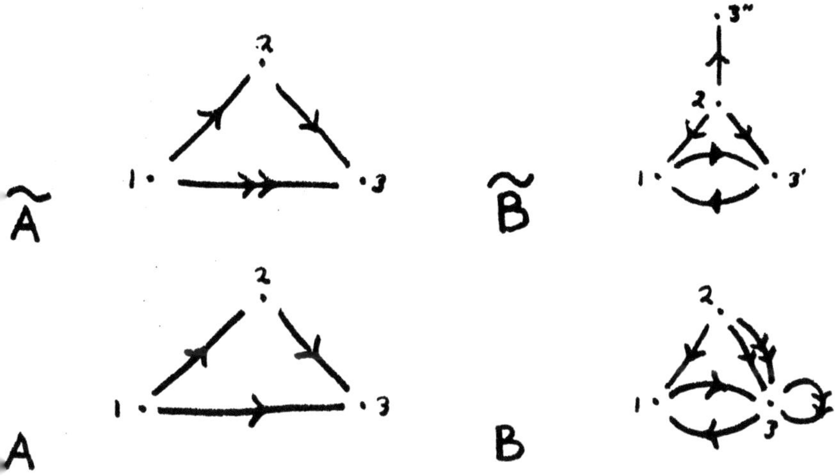


Fig. 1

Note that when 3' and 3'' coallesced into 3, a loop appeared. This illustrates the result in [S1], which says that if $\tilde{B}$ degenerates to B, then the basis graph $Q_B$ of B is obtained from the basis graph $Q_{\tilde{B}}$ of $\tilde{B}$ by coallescing idempotents, adding one loop for each idempotent which disappears.

<u>Example 2</u>: Let $\tilde{A}$ be the tame algebra $D_4$ with the orientation dual to the orientation of the 4-subspace problem, and let A be the degeneration of $\tilde{A}$ given by coallescing two of the simple projectives. Let M be the tilting module obtained by replacing another of the projectives by its inverse Auslander-Reiten translate. Then the tilting gives another $D_4$ with a different orientation, and B is a degeneration of it.

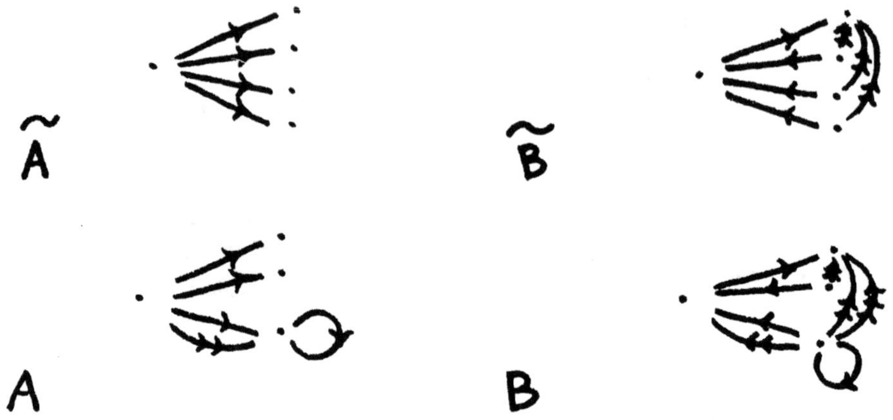

Fig. 2

Example 3: Let A be the algebra with the basis graph given in Fig 4 below. The Auslander-Reiten quiver is given in Fig. 3, where projectives are indicated by a line above the dimension at a given idempotent and injectives by a line below. We choose as tilting module the sum of the circled indecomposables, which include the two projective-injectives $P_1$ and $P_2$, and the simple module $S_2$ which is a quotient of $P_2$ by $P_3$.

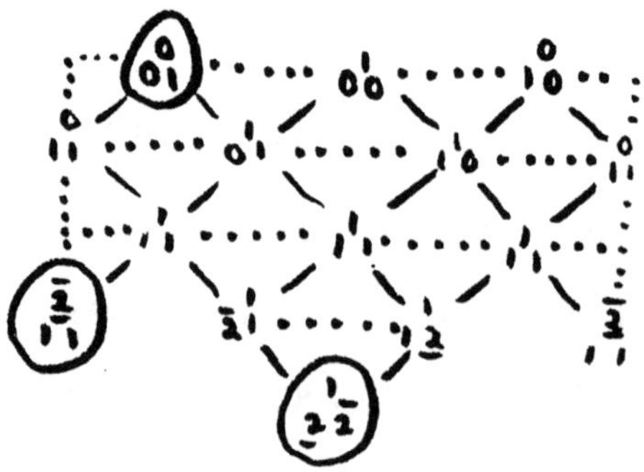

Fig. 3

Let us write $M_1 = P_1$, $M_2 = P_2$ and $M_3 = S_2$, with $M = M_1 \oplus M_2 \oplus M_3$. Then $B = \mathrm{End}_A(M)$ has the basis graph given in Fig. 4. $A$ has a unique deformation to $\tilde{A}$, which splits each idempotent into two, and converts some nonlooped arrows of $A$ into matrix units. In the tilting module both $\tilde{M}_1$ and $\tilde{M}_2$ decompose into $\tilde{M}_{1''} \oplus \tilde{M}_{1'}$ and $\tilde{M}_{2''} \oplus \tilde{M}_{2'}$, with $\tilde{M}_{1''} \rightleftarrows \tilde{M}_{2'}$ being the projective module associated with the matrix block in $\tilde{A}$. We also have a morphism from $\tilde{M}_{2''}$ into $\tilde{M}_3$. Thus $\tilde{B}$ has three components: (i) $\tilde{M}_{1''}$, (ii) a $2 \times 2$ matrix block whose idempotents represent projections onto $\tilde{M}_{1'}$ and $\tilde{M}_{2'}$, and (iii) a component whose idempotents represent projections onto $\tilde{M}_{2''}$ and $\tilde{M}_3$.

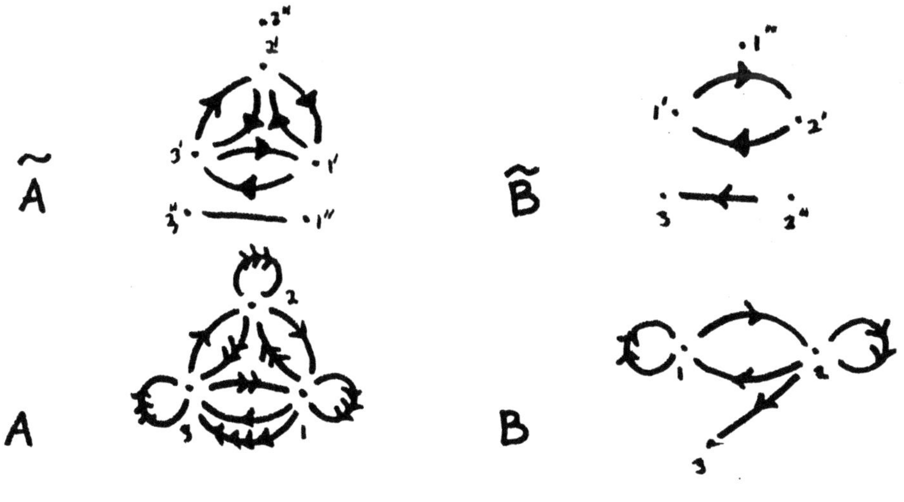


Fig. 4

References

[G] Gabriel P., Finite representation type is open, Representations of Algebras, Springer Lecture Notes 488, 132-155.

[HR] Happel D. and Ringel C.M., Tilted algebras, Trans. Am. Math. Soc. 274, No. 2 (1982), 399-443.

[S1] Schaps M., Deformations of finite dimensional algebras and their idempotents, to appear in Trans. Amer. Math. Soc.

[S2] Schofield A., Bounding the global dimension in terms of the dimension, Bull. London Math. Soc. 17 (1985), 393-394.

# RATIONAL INVARIANTS OF QUIVERS AND THE RING OF MATRIXINVARIANTS

Lieven Le Bruyn, University of Antwerp UIA-NFWO
Aidan Schofield, University College London

**Abstract :**

Let $V$ be a finite dimensional vectorspace over $\mathbb{C}$ and let $\alpha = (\alpha(1), ..., \alpha(m)) \in \mathbb{N}^m$. An action of $GL(\alpha) = \prod_{i=1}^{m} GL(\alpha(i))$ on $V$ is said to be Schurian if the stabilizer of a generic point is $\mathbb{C}^*$ embedded diagonally in $GL(\alpha)$. In this paper we show that the field of rational invariants for such an action is stably equivalent to the field of rational $n$ by $n$ matrixinvariants where $n = gcd(\alpha(i) : 1 \leq i \leq m)$.

## 1. The problem

Throughout this paper, we consider an algebraically closed field of characteristic zero and call it $\mathbb{C}$. Let $G$ be an affine linear reductive group acting allmost freely on a finite dimensional vectorspace $V$, that is, the stabilizer of a generic point is trivial. One of the main open problems in invariant theory is to determine for which groups $G$ the field of rational invariants $\mathbb{C}(V)^G$ is (stably) rational.

A first approach might be the following : if $G$ acts almost freely on $V$ then by Luna's results [Lu] we know that there exists an affine $G$-invariant open subvariety $U$ of $V$ consisting of points with trivial stabilizer and we can form the quotient variety $U/G$. The canonical map $\pi : U \to U/G$ is then a principal $G$-bundle in the étale



*F. van Oystaeyen and L. Le Bruyn (eds.), Perspectives in Ring Theory, 21–29.*

topology and such objects are classified by the cohomology group $H^1_{et}(U/G, G)$. Now, if the $G$-bundle is trivialisable in the Zariski topology then $U$ is birational to $U/G \times G$ and since reductive groups are rational we see that $U/G$ is stably rational. But of course, being Zariski trivialisable is a rather strong condition.

Another well known result concerning this problem is the following lemma which gives us some freedom on the particular choice of the vectorspace as long as we are interested in stable rationality :

**no-name lemma :** Let $G$ be a reductive group acting on two vectorspaces $V$ and $W$ such that the action on $V$ is almost freely. Then, the field of rational invariants of $G$ acting on $V \oplus W$ is rational over the field of rational invariants on $V$.

**Proof :** Since $\mathbb{C}[V]$ is a unique factorization domain having only trivial units we can apply [Lu,p 103,lemme 2] in order to obtain a non-empty affine $G$-invariant open subvariety $V'$ of $V$ such that generic orbits in $V'$ are closed. So, take a generic point $v \in V'$, then by the étale slice theorem [Lu,p.97] there exists an affine (!) $\pi_{V'}$-saturated subvariety $V$" of $V'$ containing $v$ such that the $G$- action on $V'$ induces an étale $G$-morphism

$$\psi : G \times V" \rightarrow V'$$

such that the image $U$ of $\psi$ is an open affine (!) $\pi_{V'}$- saturated subvariety of $V'$ and the canonical morphism

$$\psi/G : (G \times V")/G \cong V" \rightarrow U/G$$

is étale and gives rise to a $G$-isomorphism

$$G \times V" \cong U \times_{U/G} V"$$

Therefore, $\pi_U : U \rightarrow U/G$ is an affine (!) principal $G$-bundle in the étale topology.

But now we can apply an old result of J.P. Serre [Se] or [Lu,p.86] : if $U$ is a principal $G$-bundle in the étale topology and $W$ a variety with $G$-action. Define a $G$-action on $U \times W$ by $g.(u, w) = (ug^{-1}, gw)$ and denote by $U \times_G W = (U \times W)/G$, then $U \times_G W$ is the total space of a fibration of type $W$ with basis $U/G$. In the special case when $V$ is a vectorspace this implies that $\mathbb{C}(U \times_G W) = \mathbb{C}(V \oplus W)^G$ is a rational extension of $\mathbb{C}(U/G) = \mathbb{C}(V)^G$ done.

As an immediate consequence we get that if $G$ acts almost freely on both $V$ and $W$ then the rational invariants on $V$ are stably equivalent to the rational invariants on $W$.

## 2. Representations of quivers

Our main motivation comes from the study of representations of quivers. In this section we will briefly recall the setting. A quiver $Q$ is a quadruple $(Q_0, Q_1, t, h)$ consisting of a finite set $Q_0 = \{1, ..., m\}$ of vertices, a set $Q_1$ of arrows between these vertices and two maps $t, h : Q_1 \to Q_0$ assigning to an arrow $\phi$ its tail $t\phi$ and its head $h\phi$ respectively.

A representation $V$ of a quiver $Q$ is a family $\{V(i) : i \in Q_0\}$ of finite dimensional vectorspaces over $\mathbb{C}$ together with a family of linear maps $\{V(\phi) : V(t\phi) \to V(h\phi); \phi \in Q_1\}$. The $m$-tuple $dim(V) = (dim(V(i)))_i \in \mathbb{N}^m$ is called the dimension vector of $V$. A morphism $f : V \to W$ between two representations is a family of linear maps $\{f(i) : V(i) \to W(i); i \in Q_0\}$ such that for all arrows $\phi \in Q_1$ we have $W(\phi) \circ f(t\phi) = f(h\phi) \circ V(\phi)$.

Given a dimension vector $\alpha = (\alpha(1), ..., \alpha(m)) \in \mathbb{N}^m$ we define the representation space $R(Q, \alpha)$ to be the vectorspace of all representations $V$ of $Q$ such that $V(i) = \mathbb{C}^{\alpha(i)}$ for all $i \in Q_0$. Because $V \in R(Q, \alpha)$ is completely determined by the maps $V(\phi)$ we have that

$$R(Q, \alpha) = \oplus_{\phi \in Q_1} M_\phi(\mathbb{C})$$

where $M_\phi(\mathbb{C})$ denotes the vectorspace of all $\alpha(h\phi)$ by $\alpha(t\phi)$ matrices with entries in $\mathbb{C}$.

We will consider the representation space $R(Q, \alpha)$ as an affine variety with coordinate ring $\mathbb{C}[Q, \alpha]$ and functionfield $\mathbb{C}(Q, \alpha)$. We have a canonical action of the linear reductive group $GL(\alpha) = \prod_{i=1}^m GL(\alpha(i))$ on $R(Q, \alpha)$ by

$$(g.V)(\phi) = g_{h\phi} V(\phi) g_{t\phi}^{-1}$$

for all $g = (g_1, ..., g_m) \in GL(\alpha)$. The $GL(\alpha)$-orbits in $R(Q, \alpha)$ are precisely the isomorphism classes of representations.

Ultimately, one is interested in the description of this orbit structure. It suffices clearly to restrict attention to indecomposable representations. V. Kac [Ka] conjectured that the variety parametrizing isoclasses of indecomposable $\alpha$-dimensional representations admits a finite cellular decomposition into locally closed subvarieties each isomorphic to some affine space. Unfortunately there is, at this moment, not much evidence to support this conjecture. An immediate consequence of it would be that the field of rational invariants $\mathbb{C}(Q,\alpha)^{GL(\alpha)}$ is rational whenever $\alpha$ is a Schur root. Recall that $\alpha$ is said to be a Schur root if $\alpha$-representations in general position are indecomposable, or equivalently, if there exists an $\alpha$-dimensional representation with endomorphism ring reduced to $\mathbb{C}$.

Therefore, if we denote $PGL(\alpha) = GL(\alpha)/\mathbb{C}^*$ where $\mathbb{C}^*$ is embedded diagonally in $GL(\alpha)$ then Schur roots are precisely those dimensionvectors $\alpha$ such that $PGL(\alpha)$ acts almost freely on the representation space $R(Q,\alpha)$. In the special case of the two loop quiver (the classification of couples of $n$ by $n$ matrices under simultanous conjugation) such a rationality result would immediatly imply the Merkurjev-Suslin result, the lifting problem for crossed products over local algebras and the rationality of the moduli space of stable rank $n$ bundles over the projective plane with Chern-numbers $c_1 = 0$ and $c_2 = n$.

In this paper we will show that this special case is really the heart of the problem. More precisely we will prove that it $\alpha$ is a Schur root for the quiver $Q$ such that $gcd(\alpha(i); i \in Q_0) = n$, then the rational invariants are stably rational to the field of rational $n$ by $n$ matrixinvariants. The Schur root assumption is no real restriction since Kac [Ka] has indicated how the rational invariants of an arbitrary dimension vector can be computed in terms of the rational invariants for the Schur roots occuring in the generic decomposistion. Finally, we mention that C.M. Ringel [Ri] has proved rationality of the rational invariants in case $Q$ is a tame quiver.

## 3. Rational invariants and Azumaya algebras

Although we are primarely interested in the rational invariants of $GL(\alpha)$ acting on a representation space $R(Q,\alpha)$ where $\alpha$ is a Schur root for $Q$, our first result can be stated in a more general setting.

Let $V$ be an affine variety with a Schurian action of $GL(\alpha)$. Suppose there exist an open subvariety $V'$ of $V$ in which generic orbits are closed, then as in the proof of the no-name lemma we can find an affine open subvariety $U$ of $V$ such that the natural morphism $\pi : U \to X = U/GL(\alpha)$ is a principal $PGL(\alpha)$-bundle in the étale topology determining an element of the cohomology group $H^1_{et}(X, PGL(\alpha))$.

In this case,one can obtain this cohomology class in a more concrete way in terms of an Azumaya algebra whose triviality is equivalent to the existence of a Zariski cover splitting the cohomology class.

Recall that an $R$-ring $S$ is a ring with a specified homomorphism from $R$ to $S$. If $K = \times_{i\in Q_0}\mathbb{C}$ then the centre of $\times_{i\in Q_0}M_{\alpha(i)}(\mathbb{C})$is $K$ so we can regard it as a $K$-ring; in turn, if $m = \sum_{i\in Q_0}\alpha(i)$, there is an embedding of $\times_{i\in Q_0}M_{\alpha(i)}(\mathbb{C})$ in $M_m(\mathbb{C})$ along the diagonal and we regard $M_m(\mathbb{C})$ as a $K$-ring via this embedding.

The group of automorphisms of $M_m(\mathbb{C})$ that fix $K$ is isomorphic to $PGL(\alpha)$ since all automorphisms of $M_m(\mathbb{C})$ that fix the center are inner. Therefore, $H^1_{et}(X, PGL(\alpha))$ classifies twisted forms of $M_m(\mathbb{C})$ over $K$, that is, Azumaya algebras over $X$ with a distinguished embedding of $K$ that are split by an étale cover so that on the étale cover the embedding of $K$ in matrices is conjugate to the original embedding defined by $\alpha$.

So, let $\delta \in H^1_{et}(X, PGL(\alpha))$ and let $U(\delta) \to X$ be the corresponding principal $PGL(\alpha)$-bundle in the étale topology and let $A(\delta)$ be the sheaf of Azumaya algebras over $X$ determined by $\delta$. Then one can identify $A(\delta)$ with the ring of $GL(\alpha)$-concomitants from $U(\delta)$ to $M_m(\mathbb{C})$. This allows us at once to deduce the following result :

**Theorem 1 :** Let $GL(\alpha)$ act Schurian on an affine variety $V$ such that generic orbits are closed in an open affine subvariety. Let $U \to X = U/GL(\alpha)$ be the corresponding affine principal $PGL(\alpha)$- bundle. Then, if $gcd(\alpha(i) : 1 \leq i \leq m) = 1$ then $\mathbb{C}(V)$ is rational over $\mathbb{C}(X) = \mathbb{C}(V)^{GL(\alpha)}$.

**Proof :** In this case the corresponding Azumaya algebra must be split on a Zariski cover since the natural map from $K_0(K)$ to $K_0(M_m(\mathbb{C}))$ is surjective and this forces the same to be true for the Azumaya algebra. Since the Azumaya algebra is split on a Zariski cover the same thing is true for the principal $PGL(\alpha)$-bundle $U \to X$, which implies that $V$ is birational to $X \times PGL(\alpha)$ and $PGL(\alpha)$ being rational finishes the proof.

**Corollary 2 :**

(1) : Let $\alpha$ be a Schur root for the quiver $Q$ such that $gcd(\alpha(i); i \in Q_0) = 1$. Then, the corresponding field of rational invariants is stably rational.

(2) : Let $\alpha$ be a Schur root for the quiver $Q$ and let $Q'$ be a larger quiver on

the same vertexset. Then, the field of rational invariants for $\alpha$ on $Q'$ is rational over the field of rational invariants on $Q$.

(3) : Let $\alpha$ be a Schur root for the quivers $Q$ and $Q'$. Then, the field of rational invariants for $\alpha$ on $Q$ is stably equivalent to the field of rational invariants on $Q'$.

**Proof :** (1) : immediate from theorem 1 ; (2) : write $R(Q',\alpha) = R(Q,\alpha) \oplus W$ and apply the no-name lemma ; (3) : apply the no-name lemma twice.

One final construction is needed. Let $\alpha$ be a Schur root for the quiver $Q$ and let us denote $n = gcd(\alpha(i); i \in Q_0)$.We form a new quiver $Q^*$ by adjoining one vertex 0 and an arrow from 0 to some vertex $i \in Q_0$.Let $\alpha^*$ be the extended dimensionvector such that $\alpha^*(0) = n$ and $\alpha^*(i) = \alpha(i)$ for all $i \in Q_0$.Let $U$ be an affine open subvariety of $R(Q,\alpha)$ such that there is an orbit map $U \to X = U/GL(\alpha)$ which is a principal $PGL(\alpha)$- bundle in the étale topology (note that this is always possible by the argument in the proof of the no-name lemma). Let $A(\alpha)$ be the corresponding Azumaya algebra. Because $gcd(\alpha(i); i \in Q_0) = n$, we may assume by passing to a Zariski open subvariety that $A(\alpha)$ is isomorphic to $M_t(B(\alpha))$ where $m = tn$ and the embedding of $K$ in $A(\alpha)$ may be refined to a set of matrixunits.

Let $W$ be the open subvariety of $R(Q^*,\alpha^*)$ whose image in $R(Q,\alpha)$ is in $U$ and where the new arrow is injective and let

$$W/PGL(\alpha) \to U/PGL(\alpha)$$

be the induced map. Then, $W/PGL(\alpha)$ represents rank one $B(\alpha)$- submodules of free rank $s$ (where $sn = \alpha(i)$) $B(\alpha)$-modules, which is a rational projective variety. Then, by applying corrolary 2 we get :

**Theorem 2 :** Let $\alpha$ be a Schur root for the quiver $Q$ and let $n = gcd(\alpha(i); i \in Q_0)$. Let $Q'$ be a quiver with one extra vertex 0 and at least one more arrow. Let $\alpha'$ be the dimensionvector determined by $\alpha'(0) = n$ and $\alpha'(i) = \alpha(i)$ where defined. Then, the rational invariants for $\alpha'$ on $Q'$ are rational over the rational invariants for $\alpha$ on $Q$.

**Theorem 3 :** Let $\alpha$ be a Schur root for the quiver $Q$ and let $n = gcd(\alpha(i); i \in Q_0)$.

Then, the field of rational invariants for $\alpha$ on $Q$ is stably equivalent to the field of rational $n$ by $n$ matrixinvariants.

**Proof :** Consider the one point extension quiver $Q'$ where the number of arrows from the extra point 0 to the point $i$ is $\alpha(i)/n$. Consider the open subvariety of $R(Q',\alpha')$ where the $\alpha(i)/n$ maps from $V(0)$ to $V(i)$ define an isomorphism from $V(0)^{\alpha(i)/n}$ to $V(i)$. This reduces the classification problem of the quiver $Q'$ to representations of the original quiver $Q$ where each vertex space is in addition given a fixed representation as a vectorspace $V^{\alpha(i)/n}$ where $V$ is a vectorspace of dimension $n$. But this is the same as the classification of $\sum_{\phi\in Q_1} \alpha(t\phi)\alpha(h\phi)/n^2$ square $n$ by $n$ matrices upto simultanous conjugation. This shows that the rational invariants for $\alpha'$ on $Q'$ are stably equivalent to the rational invariants for $n$ by $n$ matrices and theorem 2 finshes the proof.

## 4. Rational invariants and reflection functors

In this section we will give a purely representation theoretic proof of theorem 3 using the Bernstein-Gelfand-Ponomarev theory of reflection functors. Let $i \in Q_0$ be a sink, that is for no $\phi \in Q_1$ we have $t\phi = i$, and let $\alpha$ be a dimensionvector. We form a new quiver $Q'$ by reversing the direction of all arrows connected to $i$ and define a new dimension vector $\alpha'$ by $\alpha'(j) = \alpha(j)$ whenever $i \neq j$ and $\alpha'(i) = \sum_{h\phi=i} \alpha(t\phi) - \alpha(i)$. Consider the open subvariety of $R(Q, \alpha)$

$$R^*(Q, \alpha) = \{V \in R(Q, \alpha) : \oplus V(\phi) : \oplus_{h\phi=i} V(t\phi) \to V(i) \text{ is surjective } \}$$

and similarly we consider the open subvariety of $R(Q', \alpha')$

$$R^*(Q', \alpha') = \{V \in R(Q', \alpha') : \oplus V(\phi) : V(i) \to \oplus_{t\phi=i} V(h\phi) \text{ is injective } \}$$

then there exists a homeomorphism between $R^*(Q, \alpha)/GL(\alpha)$ and $R^*(Q', \alpha')/GL(\alpha')$ such that corresponding representations have isomorphic endomorphism rings. In particular, if $\alpha$ is a Schur root for the quiver $Q$, then $\alpha'$ is a Schur root for the quiver $Q'$ and their corresponding fields of rational invariants are isomorphic.

**Proof of theorem 3 :** let $i \in Q_0$ be such that $\alpha(i) = kn$ is minimal and let $j \in Q_0$ be such that $\alpha(j) = ln$ with $k$ not dividing $l$ say $l = ak - b$ with $0 \neq b < k$. Form a new quiver $Q'$ on the same vertices with $a$ arrows pointing from $i$ to $j$ and all other arrows living on $Q_0 - \{j\}$ in such a way that $\alpha$ is a Schur root for $Q'$. Note that this can always be done by trowing in lots of loops. By corollary 2(3) the rational invariants for $\alpha$ on $Q$ are stably equivalent to the rational invariants for $\alpha$ on $Q'$ which are in turn isomorphic to the rational invariants for $\alpha'$ on the reflected quiver in $j$ where $\alpha'(j) = bn < kn$. So, by induction we may assume that the rational invariants for $\alpha$ on $Q$ are stably equivalent to the rational invariants for $\gamma$ on some quiver $Q^*$ with $gcd(\gamma(i) : i \in Q_0) = n$ and $\gamma(i) = n$ for some $i \in Q_0$.

Now, we can proceed by induction on the number of vertices. Take $j \neq i$ such that $\gamma(j) = kn$. Form a quiver $Q'$ with precisely $k$ arrows from $i$ to $j$ and the arrows in $Q'_) - \{j\}$ such that $\gamma$ is a Schur root for $Q'$. Then, after applying reflection with respect to the sink $j$ we get a quiver with one vertex less such that the rational invariants are still stably equivalent to the original.Continuing in this way we will end up with the two loop quiver in $i$ finishing the proof.

By appllying the no-name lemma we have therefore proved

**Theorem 4 :** Let $GL(\alpha)$ have a Schurian action on a finite dimensional vectorspace $V$. Then, the field of rational invariants for this action is stably equivalent to the field of rational $n$ by $n$ matrixinvariants where $n = gcd(\alpha(i) : 1 \leq i \leq m)$.

We recall that Ed Formanek proved rationality of the field of matrixinvariants for $n \leq 4$ and that David Saltman proved retract rationality for all squarefree $n$. Recall that the unramified Brauer group of a field $L$ is the Brauer group of a smooth projective model for $L$. D. Saltman proved that the unramified Brauer group of the field of rational $n$ by $n$ matrix- invariants is trivial. Hence we obtain :

**Corollary 5 :** Let $\alpha$ be a Schur root for the quiver $Q$ and let $n = gcd(\alpha(i) : i \in Q_0)$, then

(1) : if $n \leq 4$ the rational invariants for $\alpha$ on $Q$ are stably rational

(2) : if $n$ is squarefree , the rational invariants for $\alpha$ on $Q$ are retract rational

(3) : the Brauer group of a projective smooth model for the rational invariants for $\alpha$ on $Q$ is trivial

## References

[Ka] : V. Kac ; Root systems,representations of quivers and invariant theory,Springer LNM 996 74-108

[Lu] : D. Luna ; Slices étales , Bull.Soc.Math.France Mém 33 , 81-105 (1973)

[Ri] : C.M. Ringel ; The rational invariants of tame quivers, Invent Math 58 (1980) 217-239

[Se] : J.P. Serre ; Espaces fibrés algébriques , Sém C. Chevalley 21 avril 1958 E.N.S.

# The Hilbert Series of Matrix Concomitants and its Application

Yasuo Teranishi
Nagoya University
Nagoya, Japan

## 1. Invariant and concomitant.

Let $G$ be a semi-simple linear algebraic group over the complex number field $\mathbb{C}$. We denote by $g$ the Lie algebra of the group $G$ and by $W_m$ the direct sum of $m$ copies of the vector space.

We consider the simultanous adjoint action of $G$ on the vector space $W_m$ :

$$g(A_1, \ldots, A_m) = (Ad(g)A_1, \ldots, Ad(g)A_m)$$
$$\text{where } A_1, \ldots, A_m \in g, \text{ and } g \in G$$

This action of $G$ induces a rational action of $G$ on the polynomial ring $\mathbb{C}[W_m]$. We denote by $C(G, m)$ the fixed subring of $\mathbb{C}[W_m]$ under the action of $G$. By a natural grading, $C(G, m)$ is an $N^m$-graded ring.

A polynomial map $f : W_m \to g$ is called a **polynomial concomitant** if it satisfies :

$$f(g, v) = Ad(g)f(v), \text{ forany } v \in W_m \text{ and } g \in G$$

The set of polynomial concomitants will be denoted by $\widetilde{T}(G, m)$. For a polynomiant concomitant $f$, consider the polynomial $< f >$ on the vector space $W_{m+1}$ defined as follows :

$$< f > (A_1, \ldots, A_m, A_{m+1}) = (f(A_1, \ldots, A_m), A_{m+1})$$



*F. van Oystaeyen and L. Le Bruyn (eds.), Perspectives in Ring Theory, 31–36.*

where ( , ) denotes the Killing form on the Lie algebra $g$. Then $f$ is a polynomial invariant on $W_{m+1}$ of degree one with respect to te last argument vector on $W_{m+1}$. We denote by $T(G,m)$ the subspace of $C(G,m+1)$ consisting of invariants of degree one with respect to the last argument vector on $W_{m+1}$. Since the Killing form of $g$ is a nondegenerate bilinear form on $g \times g, f \to < f >$ induces an injective linear map $\widetilde{T}(G,m) \to T(G,m)$.

Inducing a gradation from that of $C(G,m), T(G,m)$ is a finitely generated $N^m$-graded module over $C(G,m)$. For each $d = (d_1, \ldots, d_m) \in N^m$ we denote by $C(G,m)_d$ and $T(G,m)_d$ the $\mathbb{C}$ vector space of $C(G,m)$ and $T(G,m)$ respectively spanned by all elements of degree $d$.

The Hilbert series of $C(G,m)$ and $T(G,m)$ are formal power series in $m$ variables $t_1, \ldots, t_m$ :

$$H(C(G,m),\underline{t}) = \sum \dim C(G,m)_d \underline{t}^d$$
$$H(T(G,m),\underline{t}) = \sum \dim T(G,m)_d \underline{t}^d$$

where $\underline{t}^d = t_1^{d_1}, \ldots, t_m^{d_m}$.

We shall use Hilbert series $H(C(G,m),t)$ and $H(T(G,m),t)$ in one variable $t$ :

$$H(C(G,m),t) = \sum \dim C(G,m)_d t^d$$
$$H(T(G,m),t) = \sum \dim T(G,m)_d t^d$$

where

$$C(G,m)_d = \sum_{k \in N^m} (C(G,m)_k, |k| = d)$$
$$T(G,m)_d = \sum_{k \in N^m} T(G,m)_k, |k| = d$$

**Theorem 1.** ([T2], [T3]).

(1) For $m \geq 2$, the Hilbert series of $C(G,m)$ satisfies the functional equation :

$$H(C(G,m),t^{-1}) = (-1)^d (t_1, \ldots, t_m)^{\dim G} H(C(G,m),t)$$

(2) For $m \geq 3$, the Hilbert series $H(T(G.m),t)$ satisfies the functional equation :

$$H(T(G,m),\underline{t}^{-1}) = (-1)^d (t_1, \ldots, t_m)^{\dim G} H(T(G,m),\underline{t})$$

where

$$d = K.\text{dim}.C(G,m) = (m-1)\text{dim } G, \text{ and } t^{-1} = (t_1^{-1}, \ldots, t_m^{-1})$$

As an application of the functional equation, we prove the following

**Theorem 2.**

(1) If $m \geq 3$, the rational expression of the Hilbert series $H(C(G,m),t)$ can never be an inverse of a polynomial with integer coefficients.

(2) Let $G$ be a simple group. Then if $m \geq 3$ and dim$G > 3$, the rational expression of the Hilbert series $H(T(G,m),t)$ can never be an inverse of a polynomial with integer coefficents.

**Proof.** Suppose that $H(C(G,m),t)$ (resp. $H(T(G,m),t)$) has the form $\frac{1}{f(t)}$ for some monic polynomial $f(t)$ with integer coefficients. The residue of the pole at $t=1$ equals to the Krull dimension of the graded ring $C(G,m)$. We set

$$K.\text{dim}.C(G,m) = (m-1)\text{dim } G = d$$

Then $f(t)$ is of the form

$$f(t) = (1-t)^d g(t)$$

for some polynomial; $g(t), (g(1) \neq 1)$, with integer coefficents.

It follows from the functional equation thet deg$g(t) = $ dim$G$. If the polynomial $g(t)$ has the form

$$g(t) = 1 + a_1 t + a_2 t^2 + \ldots,$$

one sees that the Hilbert series has the form

$$1 + (d - a_1)t + \text{heigher terms in } t$$

Since dim$C(G,m)_1 = 0$ (resp. dim $T(G,m)_1 = m$), one has

$$a_1 = d (\text{resp. } a_1 = d - m)$$

On the other hand, the Hilbert series has the form

$$\frac{p(t)}{(1-t^{\alpha_1})\ldots(1-t^{\alpha_k})}, p(t) \in N[t], \alpha_i \in N$$

Therefore $g(t)$ is a product of cyclotomic polynomials :

$$\Pi \text{ (cyclotomic polynomials } \Phi_r, \Phi_r(t) \neq 1 - t)$$

It is well known that the coefficient of $t$ in a cyclotomic polynomial $\Phi_r(t)$ is $1, -1$ of 0 and hence one has

$$a_1 \leq \deg g(t) = \dim G$$

Ihis yields an unequality :

$$(m-1)\dim G \leq \dim G \text{ (resp } > m(\dim G - 1) \leq 2 \dim G)$$

which contradicts the assumption in (1) (resp. (2)).

## 2. Trace Rings.

The set of matrix concomitants $\widetilde{T}(GL(n), m)$ is a noncommutative ring under pointwise sum and multiplication. The ring $T(GL(n), m)$ will be called the **trace ring** of $m$ generic $n$ by $n$ matrices and will be denoted by $T(n, m)$. Let $T^\circ(n, m)$ be the ring of polyniomial concomitants

$$f : \oplus sl(n) \longrightarrow M(n, \mathbb{C})$$

from $m$ copies of the $n$ by $n$ trace zero matrices $sl(n)$ to the space of $n$ by $n$ matrices $M(n, \mathbb{C})$. Then one has :

$$T(n, m) \simeq T^\circ(n, m)[Tr(X_1), \ldots, Tr(X_m)]$$

where $X_1, \ldots, X_m$ are $n$ by $n$ generic matrices, and

$$T^\circ(n, m) = T(SL(n), m) \oplus C(SL(n), m)$$

**Theorem 3.** If $n \geq 3, m \geq 2$ or $n = 2, m \geq 3$, the Hilbert series $H(T(SL(n), m), t)$ satisfies the functional equation

$$H(T(SL(n), m), \underline{t}^{-1}) = (-1)^d (t_1 \ldots t_m)^{n^2-1} H(T(SL(n), m), \underline{t})$$

where $d = (m-1)(n^2-1)$.

By Theorem 3, we obtain :

**Theorem 4.** ([L], [F], [T3])

If $n \geq 3, m \geq 2$ or $n = 2, m \geq 3$, the Hilbert series of $m$ generic $n$ by $n$ matrices satisfies the functional equation

$$H(T(n,m), \underline{t}^{-1}) = (-1)^d (t_1 \ldots t_m)^{n^2} H(T(n,m)\underline{t})$$

where $d = (m-1)n^2 + 1$.

As an application of the functional equation of trace rings, we give a proof of a theorem due to Le Bruyn and Procesi.

**Theorem 5.** (L. P., Theorem 3.1.). The trace ring $T(n,m), m \geq 2$, has finite global dimension if and only if

$$(n,m) = (2,2), (2,3) \text{ or } (3.2)$$

**Proof.** Le Bruyn and Procesi proved the theorem by using Luna's slice method. The following proof is a modification of their proof. Suppose that $T(n,m)$ has finite global dimension. Then the rational expression of the Hilbert series of $T(n,m)$ in one variable $t$ is a pure inverse of a monic polynomial $f(t)$ with integer coefficients. Then it follows from the functional equation of $H(T(n,m),t)$ that $\deg f(t) = m.n^2$. Since transcendence degree of the ring $C(GL(n),m)$ is $(m-1)n^2 + 1.f(t)$ has the form :

$$f(t) = (1-t)^d (1 + a_1 t + a_2 t^2 + \ldots)$$

where $d = (m-1)n^2 + 1$. If $m \geq 3$, the same argument as in the proof of Theorem 2 shows that the rational expression of $H(T(n,m),t)$ can never be a pure inverse of a monic polynomial with integer coefficents except $n = 2$. Assume now that $m = 2, n \geq 3$. Then the same argument as in the proof of theorem 2 shows that

$$a_1 = n^2 - 3$$

We set

$$g(t) = 1 + a_1 t + a_2 t^2 + \ldots = f(t)/(1-t)^d$$

Then $g(t)$ is a polynomial of degree $n^2 - 1$. Suppose that $g(t)$ has the form

$$g(t) = (1+t)^l \Phi_a(t) \Phi_b(t) \ldots \Phi_c(t)$$

Then comparing degree of both sides, one finds that

$$g(t) = (1+t)^{n^2-3}(1+t^2) \text{ or } (1+t)^{n^2-5}(1+t+t^2)^2$$

Therefore one has

$$a_2 = \frac{1}{2}(n^2-3)(n^2-4)+1 \text{ or } \frac{1}{2}(n^2-5)(n^2-6)+2n^2-7$$

On the other hand, $X^2, Y^2, XY, YX, \mathrm{Tr}(X^2), \mathrm{Tr}(Y^2), \mathrm{Tr}(XY)$ is a basis of the vector space $T(n,2)_2$, where $X$ and $Y$ are $n$ by $n$ generic matrices, and hence $\dim T^\circ(n.2)_2 = 7$. Therefore one has

$$H(T^\circ(n,2),t) = 1 + 2t + 7t^2 \ldots$$

Thus one has that $a_2 = \frac{1}{2}(n^2-1)(n^2-4) - 3$. Therefore

$$\begin{aligned}\frac{1}{2}(n^2-1)(n^2-4)-3 &= \frac{1}{2}(n^2-3)(n^2-4)+1 \text{ or} \\ &= \frac{1}{2}(n^2-5)(n^2-6)+2n^2-7\end{aligned}$$

This implies that $n = 3$, and hence if a trace ring $T(n,m)$ has a global dimension, $(n,m) = (2,2), (2,3)$ or $(3,2)$. For the converse see [S-S] and [L-V].

**References.**


[F] Formanek, *The functional equation for character series associated to n by n matrices*, Trans. Amer. Math. Soc. 295 (1986).

[L] Le Bruyn, *The functional equation for Poincaré series series of trace rings generic 2 by 2 matrices*, Israel J. Math. (1986).

[L-P] Le Bruyn and Procesi, *Etale local structure of matrix invariants and concomitants*, Springer LNM 1271 (1987).

[S-S] Small and Stafford, *Homological properties of generic matrices*, Israel J. Math. (1985).

[T1] Teranishi, *The ring of invariants of matrices*, Nagoya Math. J. 104 (1986).

[T2] Teranishi, *Linear diophantine equations and invariant theory of matrices*, Advanced Studies in Pure Math. 11(1987).

[T3] Teranishi, *The Hilbert series of rings of matrix concomitants*, to appear in Nagoya Math. J.

# A Theorem on Invariants of Semi-Simple Lie Algebras

Yasuo Teranishi
Nagoya University
Nagoya, Japan

Let $G$ be a semi-simple connected linear algebraic group over the complex number field $\mathbb{C}$. We denote by $g$ the Lie algebra of $G$. Let $g(m)$ denote the vector space of $m$ copies of $g$.

$$g(m) = g \oplus \ldots \oplus g$$

The group $G$ acts on $g(m)$ by the diagonal adjoint action. Let $\mathbb{C}\,[g(m)]$ be the polynomial ring on the vector space $g(m)$, and $\mathbb{C}\,[g(m)]^G$ the ring of invariants. We recall a well-known theorem of Chevalley (1).

**Chevalley's Theorem :** The ring of invariants $\mathbb{C}\,[g]^G$ is a polynomial ring generated by some algebraically independent polynomials.

In this note we shall prove the following

**Theorem.** The ring of polynomial invariants of $m \geq 2$ copies of the adjoint representation of $G$ is a polyunomial ring if and only if $G$ has the form

$$G = SL(2) \times \ldots \times SL(2)$$

**Acknowledgement.**

This work was done while the author was at the University of Mannheim. The author wishes to thank the department of mathematics and Prof. Popp for their hospitality.



*F. van Oystaeyen and L. Le Bruyn (eds.), Perspectives in Ring Theory, 37–40.*

Lemma 1. Let $G$ be a connected simple linear algebraic group. Then the ring of invariants $\mathbb{C}[g]^G$ is generated by invariants of degree 2 if and only if $G = SL(2)$.

**Proof.** According to Chevalley's theorem, $\mathbb{C}[g]^G$ is generated by some algebraically independent polynomials $f_1, \ldots, f_r$. We denote by $d_1, 1 \leq i \leq r$, the degree of the polynomial $f_i$ and by $W$ the order of the Weyl-group of $G$. Then, as is well-known,

$$d_1 \ldots d_r = W$$

If the order $W$ does not have the form $W = 2^m$, one of $d_i$ is an odd number and hence the ring of invariants is not generated by invariants of degree 2. The order of the Weyl group of a simple group $G$ satisfies $W = 2^m$ if and if $G = SL(2)$.

**Lemma 2.** The ring of invariants $\mathbb{C}[sl(2) \oplus sl(2)]^{SL(2)}$ is a polynomial ring.

**Proof.** See Procesi (2).

**Lemma 3.** Let $G_i, i = 1,2$ be a semi-simple group and $(G_i, V_i)$ a representation. Suppose that the group $G_1 \times G_2$ acts on $V_1 \oplus V_2$ as follows :

$$(g_1, g_2)(v_1 + v_2) = g_1 v_1 + g_2 v_2$$
$$\text{where } g_i \in G_i, v_i \in V_i \text{ for } i = 1,2$$

Then we have

$$\mathbb{C}[V_1 \oplus V_2]^{G_1 \times G_2} = \mathbb{C}[V_1]^{G_1} \otimes \mathbb{C}[V_2]^{G_2}$$

**Proof.** For $i = 1,2$, let $\natural_i : \mathbb{C}[V_i] \to \mathbb{C}[V_i]^{G_i}$ be a Reynolds operator. Let

$$f = \sum f_i g_i, f_i \in \mathbb{C}[V_1] \text{ and } g_i \in \mathbb{C}[V_2]$$

be an invariant. Then we have $f = \sum f_i^{\natural_1} g_i^{\natural_2}$ and the proof follows from this.

The ring $\mathbb{C}[g(m)]^G$ is a graded ring by the usual grading.

We consider the Hilbert series :

$$H(t) = \sum_{d \in N} \dim .\mathbb{C}[g(m)]_d^G t^d$$

where $\mathbb{C}[g(m)]_d^G$ denotes the space of degree $d$.

**Lemma 4.** (Theorem 2.1. in (3)). If $m \geq 2$, the Hilbert series $H(t)$ satisfies the following functional equation :

$$H(1/t) = \pm t^{m\ \mathrm{dim}.G} H(t)$$

**proof of the Theorem.**

Suppose that the ring $\mathbb{C}\,[g(m)]^G$ is a polynomial ring generated by some algebraically independent polynomial invariants $f_1, \ldots, f_N$. Since $K$ dim.$\mathbb{C}\,[g(m)]^G =$ $(m-1)$dim.$G$, we have $N = (m-1)$dim.$G$.

Therefore the Hilbert series $H(t)$ is given by

$$H(t) = \prod_{i=1}^{N} (1 - t^{\mathrm{deg(f)}} i)^{-1}$$

Then by Lemma 4, $m$dim.$G = \sum \deg(f_i)$.

Since $G$ is semi-simple, the center of $g$ is zero and hence there exists no invariant of degree 1. Therefore we have

$$(\#) \qquad m\ \mathrm{dim}.G = \sum \deg(f_i) \geq 2N = 2(m-1)\mathrm{dim}.G$$

In particular, we get $m \leq 2$. Thus we can assume $m = 2$. Suppose that the group $G$ has the following form :

$$G = G_1 \times \ldots \times G_r, G_i : \text{ simple group}$$

Let $g_i$ denote the Lie algebra of $G_i$. Then by Lemma 3,

$$\mathbb{C}\,[g(2)]^G = \mathbb{C}\,[g_1(2)]^{G_1} \otimes \ldots \otimes \quad \mathbb{C}\,[g_r(2)]^{G_r}$$

We have by (#) that $\deg(f_i) = 2$ for $i = 1, 2, \ldots, N$. The ring $\mathbb{C}\,[g_i(2)]^{G_i}$ contains no invariant of degree 1 and hence we see that, for $i = 1, 2, \ldots, N$

$$f_i \in \mathbb{C}\,[g_1(2)]_2^{G_1} \oplus \ldots \oplus \mathbb{C}\,[g_r(2)]_2^{Gr}$$

where $\mathbb{C}\,[g_i(2)]_2^{G_i}$ denotes the degree 2 part of $\mathbb{C}\,[g_i(2)]^{G_i}$.

Therefore $\mathbb{C}\,[g_i(2)]^{G_i}$ is generated by some polynomial invariants of degree 2. By Lemma 1, we see that $G_i = SL(2), 1 \leq i \leq r$. The "if" part follows from Lemma 2 and Lemma 3. This completes the proof.

**References.**


[1] Chevalley, *Invariants of finite groups generated by reflections*, Amer. J. Math. 77 (1955).

[2] Procesi, *Rings with polynomial identities*, Marcel Dekker (1973).

[3] Teranishi, *Linear diophantine equations and invariant theory of matrices*, Advanced Studies in Pure Math. 11 (1978).

# Deformations of Algebras and Hochschild Cohomology

Dr. Mary Schaps
Department of Mathematics
Bar-Ilan University
Ramat-Gan, Israel

Abstract

Since Hochschild cohomology is preserved under tilting, two algebras A and B related by tilting have infinitesimal deformation spaces of the same dimension. We show how, for each deformation of A, we can construct a deformation of the tilting module ${}_AT_B$ which will provide directly a deformation of B. We also show that the calculation of the vector space of infinitesimal deformations of a module is a feasible microcomputer calculation, and prove that this vector space is the tangent space to a versal deformation space.

## §1 Introduction

This talk was a continuation of a presentation by Dieter Happel of his work on preservation of Hochschild cohomology under tilting. The first sections report on joint work by him and the author which grew out of that work. The last section, work of the author on computation of dimensions of Hochschild cohomology and Ext groups, will not appear elsewhere, so proofs are included in an Appendix.



*F. van Oystaeyen and L. Le Bruyn (eds.), Perspectives in Ring Theory, 41–58.*

Let A be a finite dimensional unitary associative algebra over an algebraically closed field K. Let n be the number of non-isomorphic simple modules, and let d be the dimension of A. We consider two operations on A, the first belonging to representation theory and the second to algebraic geometry. Each has its own invariants, and its own "picture" of the algebra.

(a) Tilting: Let T be a left A-module with the following properties:

(i) proj dim $T \leq 1$.

(ii) $\mathrm{Ext}_A^1(T,T) = 0$.

(iii) The number of distinct direct summands of T is n, the number of non-isomorphic simples in A.

Then $B = \mathrm{End}_A(T)$ is a finite dimensional algebra with the same number n of non-isomorphic simples and by the Brenner-Butler theorem [H-R] $A \xrightarrow{\sim} \mathrm{End}_B({}_AT_B)$. T is called a tilting module for A. A module satisfying (i) and (ii) is called a partial tilting module.

The transition from A to B preserves important properties of the representation theory, and in discussions of tilting the algebras are generally represented by their quivers.

(b) Deformation: By continuously varying the structure constants of the multiplication in A, it can often be deformed into another algebra $\tilde{A}$ of the same dimension but a possibly different number of non-isomorphic simples. The study of such deformations belongs to algebraic geometry and involves an investigation of the decomposition of a scheme of structure constants into orbits under the action of base change. The appropriate "picture" of an algebra for the study of deformations is the basis graph, defined below in §2. See [S2] for more background.

The tilting-deformation theorem completes the following commutative diagram, in which the horizontal direction represents tilting and the vertical direction represents deformation.

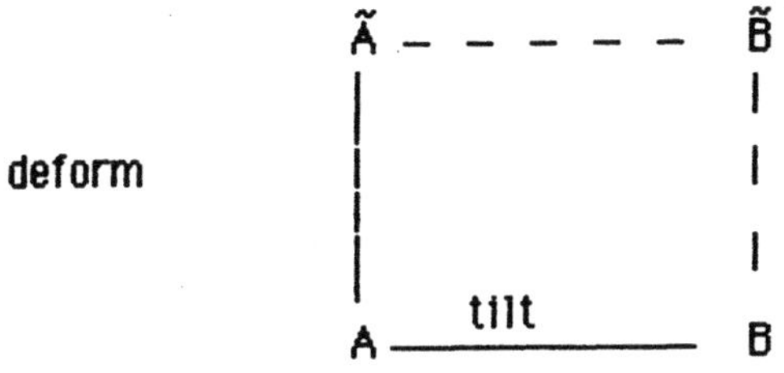


The proof will appear elsewhere. We give only the stages by which it was reached. All modules are finite dimensional.

§2 <u>Deformations of algebras</u>

We summarize briefly the straightening-out theorem and the definition of the basis graph from [S2], which are needed to deform the tilting modules and to give an example.

We may assume that $A$ is basic, i.e. that $A/\mathrm{Rad}\, A \xrightarrow{\sim} K^n$, since the deformation theory of an algebra is invariant under Morita equivalence [S2].

<u>Definition</u>: A deformation of $A$ over a nonsingular pointed curve $(C,t_0)$, $C = \mathrm{Spec}(R)$, is a flat $R$-algebra $\bar{A}$ together with an isomorphism $A \xrightarrow{\sim} \bar{A} \otimes R/m_0$, where $m_0$ is the maximal ideal corresponding to $t_0$.

<u>Theorem</u> (Straightening-out): If $\bar{A}$ is a deformation of $A$, and $e = \{e_1,\ldots,e_n\}$ is a complete primitive orthogonal set of idempotents for $A$, then there is an etale cover $p\colon (C',s_0) \to (C,t_0)$ such that $\bar{A} \times_C C'$ has sections $\bar{e}_1,\ldots,\bar{e}_n$ which induce a complete orthogonal (not necessarily primitive) set of idempotents in each fiber, and such that $\dim \bar{e}_i\bar{A}\bar{e}_j$ is

constant on closed fibers over $C'$.

<u>Definition</u>: The <u>basis graph</u> $Q_A$ of $A$ is the directed graph with $n$ vertices $v_i$ and $\dim e_iAe_j$ arrows from $v_i$ to $v_j$. In the <u>weighted</u> basis graph, each arrow is associated to a basis element in a basis filtered by powers of the radical $J$. An arrow corresponding to a basis element in $J^n - J^{n+1}$ is marked by $n$ barbs, and a matrix unit by a solid triangle.

Examples:

(a) $K[x]/x^3$ (b) $M_2(k)$

Fig. 1

From [S2] we know that if $\bar{A}$ is a deformation of $A$, and we let $Q'$ be the basis graph of a general fiber of $\bar{A}$, then $Q_A$ is obtained from $A'$ by coallescing some of the vertices, adding $m-1$ new loops wherever $m$ vertices coallesce to one. This result is useful in determining whether a given algebra has deformations at all, and if so what they might be. Two examples of deformations and the corresponding basis graphs appear in Fig. 2 below.

<u>Remark</u>: The weights are generally not preserved under deformation.

§3 Relevance to representation theory

In the course of the proof of the tilting deformation theorem, we show that every partial tilting module deforms in an essentially unique way to a partial tilting module over the deformed algebra. This provides a connection between the representation theories of the two algebras. The deformations $\bar{e}_1,\ldots,\bar{e}_n$ of the idempotents determine deformations $\bar{P}_i = \bar{e}_i\bar{A}$ of the projective modules. A different choice of idempotent sections would give isomorphic projective modules which differ from these by conjugation by a unit section. If $\bar{e}_i$ is not primitive over a general point, then $\bar{P}_i$ will decompose over a general point.

§4 The tilting theorem

The motivation for trying to prove the tilting theorem was Dieter Happel's result that Hochschild cohomology is invariant under tilting. Since $H^2(A,A)$ measures the infinitesimal deformations of the algebra $A$, the equality $\dim H^2(A,A) = \dim H^2(B,B)$ seemed to mean that $A$ and $B$ should have related deformations.

Stage 1: During a visit to Israel the conjecture was formulated and we proved:

Theorem: Let $T$ be a partial tilting module for $A$, and let $\bar{A}$ be a deformation for which the idempotents can be deformed. Then one can define a module $\bar{T}$ over $\bar{A}$ such that $\bar{B} = \mathrm{End}_{\bar{A}}(\bar{T})$ is a flat deformation of $B$, and the general fiber of $\bar{T}$ is a partial tilting module for the corresponding fiber of $\bar{A}$.

Construction of $\bar{T}$: Take a short projective resolution

$0 \to Q_1 \to Q_0 \to T \to 0$, lift the projective modules as in §3, lift the map arbitrarily and define $\bar{T}$ to be the cokernel

$$0 \to \bar{Q}_1 \to \bar{Q}_1 \to \bar{T} \to 0.$$

One uses Schofield's result in [S4] on the upper semicontinuity of $\mathrm{Ext}_A^{1}(M,M)$ to show that $\bar{T}$ is a partial tilting module.

<u>Stage 2</u>: The difficult remaining part of the theorem was to show that if $T$ had the proper number of direct summands to be a tilting module, so did $\bar{T}$. This part of the proof was considerably more expensive, involving several long distance phone calls. The result required the semicontinuity of $\mathrm{Ext}_A^{i}(M,N)$ in order to show that proj dim $\leq i$ is an open condition, and thus that $\bar{T}$ is essentially the <u>unique</u> deformation of $T$. For simplicity we proved the theorem for deformations in which almost all fibers of $\bar{A}$ are isomorphic to some fixed algebra $A'$.

<u>Example</u>:

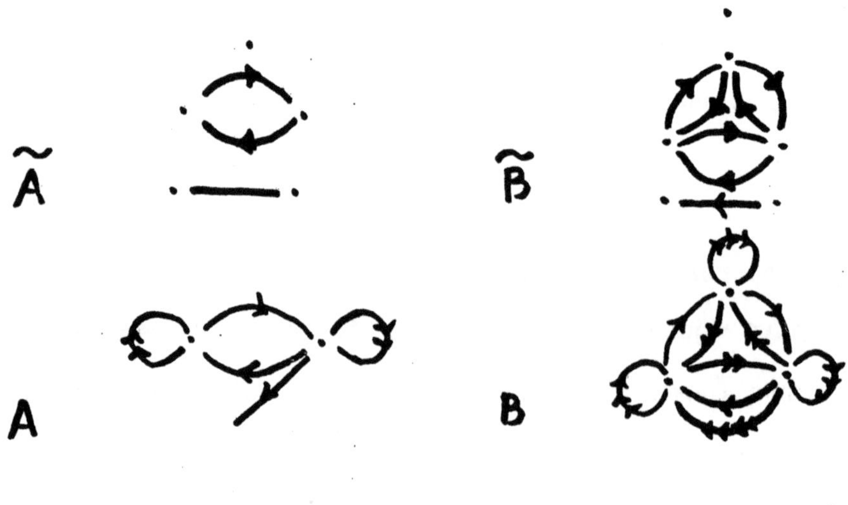


Fig. 2

In this example, $A$ is an algebra of dimension 8 with $n = 3$ non-isomorphic simples. $A$ deforms into the 8 dimensional algebra $A'$ with 5 simples: one isolated idempotent, one matrix block $M_2(k)$ with two isomorphic simples, and one copy of $A_2$, giving 4 classes of non-isomorphic simples.

Now we consider a tilting module which is the direct sum of three indecomposables: two projectives corresponding to the vertices with loops, and one indecomposable of proj. dim. = 1 which is a quotient of the projective on the rightmost idempotent by the simple projective. The algebra B which is the endomorphism algebra of this module has dimension 13 and the weighted basis graph shown in Fig. 2. B has three non-isomorphic simples corresponding to the three indecomposable components of the tilting module. $A'$ can be tilted to the 13 dimensional deformation $B'$ of B, with one isolated idempotent, one matrix block $M_3(K)$, and one copy of $A_2$, again giving 4 classes of non-isomorphic simples. The appearance of non-isomorphic simples in $\tilde{B}$ reflects the appearance of multiple copies of the indecomposable summands of the tilting module ${}_{\tilde{A}}\tilde{T}_{\tilde{B}}$ as a module over $\tilde{A}$.

## §5 Computation of Hochschild cohomology modules and Ext groups

This section is independent of the others and describes a different application of algebraic geometry to representation theory. In the Appendix we describe how Schlessinger's theory of functions on Artin rings can be applied to construct a parameter space for the infinitesimal deformations of a pair consisting of an algebra and a module over that algebra. Here we describe how the calculation of the first order deformations of an algebra or representation over $k[\varepsilon]$, $\varepsilon^2 = 0$ can be used to calculate various cohomology groups. The calculations have been implemented on a microcomputer.

Deformations of algebras:

Let an algebra of dimension $d$ be given by structure constants $(a_{ij}{}^k)$

$$x_i \cdot x_j = \sum a_{ij} x_k.$$

A deformation over $k[\varepsilon]$ is given by considering $d^3$ variables $b_{ij}{}^k$ and requiring the multiplication

$$x_i \cdot x_j = \sum (a_{ij}{}^k + \varepsilon b_{ij}{}^k) x_k$$

to satisfy the associativity equations

$$(x_i \cdot x_j) x_k = x_i (x_j \cdot x_k),$$

a total of $d^4$ equations.

Definition: A basis $x_1, \ldots, x_d$ respects a set of idempotents $e_1, \ldots, e_n$ if $x_i = e_i$ for $i = 1, \ldots, n$ and each $x_k$ lies in a Peirce component $x_i A x_j$.

Applying the material in §2 above, one can assume that $x_1, \ldots, x_n$ are idempotents and that the basis respects them even after deformation, with a considerable reduction in the number of variables and equations which must be considered. In spite of its enormous size, this system can usually be solved quite quickly (less that a minute for $d \leq 10$) because most equations are of length 1 or 2 and can be solved as soon as they are produced, progressively lowering the number of parameters. Let $f_1$ be the dimension of the solution space of this system of equations, the total number of free parameters remaining.

Some of these deformations are trivial, being produced by a sequence of base changes of the form

$$x_j \to x_j + \varepsilon x_k.$$

Since our deformations leave the Peirce decomposition and idempotents fixed, we need only consider such automorphisms when $x_k$ lies in the same Peirce block as $x_j$, and where $x_j$ is not an idempotent.

Letting $d_{ij} = \dim x_i A x_j$, and letting $\delta = (d_{ij})$, we have a subgroup $GL_\delta$ of $GL_d(k)$ of dimension Aut $\dim(\delta) = \sum d_{ii}(d_{ii} - 1) + \sum d_{ij}{}^2$, for $j \neq i$, which contains all the automorphisms of $\langle x_1,\ldots,x_d \rangle$ preserving this Peirce decomposition and idempotent set.

Since our calculations are all made over the rational numbers $\mathbb{Q}$, we divide out by the action of this automorphism group by taking the perpendicular subspace to this automorphism group. This generates Aut $\dim(\delta)$ new equations (many trivial) to be solved, and the total number of free parameters is reduced to some new value $f_2$. From $\delta$, $f_1$, and $f_2$ we can now calculate the following quantities:

1) $\dim H^2(A,A) = f_2$, the total dimension of the space of nontrivial deformations module deformations obtained by base change. (In [D-S] it was proven that it suffices to consider only the deformations which respect

$(x_1,\ldots,x_n)$). This is the codimension to the orbit of A in the structure constant variety.

2) $\dim \mathrm{Stab}_{GL(d)}(A) = (\mathrm{Aut}\dim \delta - (f_1 - f_2)) + \sum d_{ij}$, for $j \neq i$. The component $(\mathrm{Aut}\dim \delta - (f_1 - f_2))$ in this sum is the dimension of the stabilizer of A in $GL_\delta$, since Aut dim $\delta$ is the dimension of $GL_\delta$, and $(f_1 - f_2)$ is the dimension of the orbit of A under the action of $GL_\delta$, being the total number of nontrivial associative deformations produced by applying elements of $GL_\delta$. The second part of this is the dimension of all the infinitesimal automorphisms which are obtained by conjugating by a unit, and which do not leave the idempotents $x_1,\ldots,x_n$ fixed, so that they do not lie in $GL_\delta$.

3) The dimension of the orbit $O(A) = \dim GL(d)/\mathrm{Stab}_{GL(d)}(A) = d^2 - \dim \mathrm{Stab}_{GL(d)}(A)$.

4) The dimension of the Zariski tangent space to the structure constant variety is $\dim H^2(A,A) + \dim O(A)$.

The computer program, having been supplied with $\delta$ and with the structure constants $a_{ij}{}^k$, computes $f_1$ and $f_2$, and all the dimensions listed above.

Deformations of representations:

The calculations of the infinitesimal deformations of a representation

$R$ of a fixed algebra $A$ is very similar to the above. Let

$$y_i \cdot y_j = \sum c_{ij}{}^k y_k$$

be the structure constants for the algebra, and let

$$y_i \cdot x_j = \sum a_{ij}{}^k x_k$$

be the structure constants for the representation, where now $i = 1,\ldots,d$ but $j,k = 1,\ldots,m$, where $m$ is the total dimension of the representation.

Let $y_1,\ldots,y_n$ be idempotents, and define the dimension vector $\alpha = (\alpha_1,\ldots,\alpha_n)$, where $\alpha_i = \dim y_i R$. Let $GL(\alpha) = GL(\alpha_1) \times \ldots \times GL(\alpha_n)$ be the group of dimension $\sum \alpha_i^2$ of possible automorphisms of this algebra. As before, we consider all variables $(b_{ij}{}^k)$ and all equations resulting from the requirement that

$$y_i \cdot x_j = \sum (a_{ij}{}^k + \varepsilon b_{ij}{}^k) x_k$$

satisfy the $d^2m^2$ equations

$$(y_i \cdot y_j) x_k = y_i (y_j x_k).$$

As before, denote the total number of such deformations by $f_1$. Then generate the deformations obtained from automorphisms in $GL(\alpha)$, calculate the dimension of the perpendicular space and denote the result by $f_2$.

1) $\dim O(R) = f_1 - f_2$.

2) $\dim \mathrm{Stab}_{GL(\alpha)}(R) = \dim GL(\alpha) - \dim O(R)$

$$= \sum \alpha_i^2 - (f_1 - f_2)$$

3) $\dim \mathrm{End}_A(R) = \dim \mathrm{Aut}_A(R)$

$= \dim \mathrm{Stab}_{GL(\alpha)}(R)$

$= \sum \alpha_i^2 - (f_1 - f_2)$

4) codim $O(R)$ in the Zariski tangent space to the structure constant variety is $f_2$.

5) If the structure constant variety is smooth, then $\mathrm{Ext}_A^1(R,R) = f_2$. (This property holds whenever proj dim $R \leq 1$, and in particular if $A$ is the path algebra of a quiver.) In this case the Euler characteristic is just $\sum \alpha_i^2 - f_1$.

<u>Remark</u>: By considering deformations of direct sums of representations and setting some of the $b_{ij}^k$ to zero, one can also calculate $\dim \mathrm{Hom}_A(R,S)$ and $\dim \mathrm{Ext}^1(R,S)$.

## <u>Appendix</u>:

## Local parameters for representation of algebras

We apply Schlessinger's theory of functors of Artin rings to give a local step-by-step method for calculating the variety $\text{alg-mod}_{m,n}$ of $n$ dimensional representations of $m$ dimensional algebras. The method is being implemented on the computer and calculations of the deformations of a given representation can be requested from the author.

Let $R$ be a commutative ring, $U = Rx_1 \oplus \tilde{U}$, a free $R$ module of rank $m$ and $V$ a free $R$-module of rank $n$. A normalized <u>alg-mod</u> structure on the pair $(U,V)$ consists of an associative algebra structure on $U$ for which $x_1$ is the identity element, together with an action of this algebra on $V$ which makes it a module. An alg-mod structure is determined by a pair of tensors $\alpha \in U^* \otimes U^* \otimes U$ and $\beta \in U^* \otimes V^* \otimes V$, where $\alpha$ defines the multiplication

$$u \cdot u' = u''$$

and $\beta$ determines the action

$$u \cdot v = v'$$

If we let $x_1, \ldots, x_m$ be a basis for $U$ and $v_1, \ldots, v_n$ a basis for $V$, then the coordinates of $\alpha$ and $\beta$ can be given by two contravariant indices and one covariant index.

$$x_i \cdot x_j = \sum_{p=1}^{m} a_{ij}{}^{p} x_p$$

$$x_j \cdot v_k = \sum_{p'=1}^{n} b_{jk}{}^{p'} v_{p'}$$

These tensors must satisfy the identity conditon $a_{i\ell}{}^{p} = a_{\ell 1}{}^{p} = b_{1\ell}{}^{p} = \delta_{\ell p}$.

The associativity conditions are

$$0 = (x_i x_j)x_k - x_i(x_j x_k) = \sum_{q=1}^{n} [\sum_{p=1}^{m} (a_{ij}{}^{p} a_{pk}{}^{q} - a_{ip}{}^{q} a_{jk}{}^{p})]x^q$$

$$0 = (x_i x_j)v_k - x_i(x_j v_k) = \sum_{q=1}^{n} [\sum_{p=1}^{m} (a_{ij}{}^{p} b_{pk}{}^{q} - \sum_{p'=1}^{n} b_{ip'}{}^{q} b_{j'k}{}^{p'})]v^q$$

For brevity we denote these by $\alpha_{..}{}^{*} \alpha_{*.}{}^{\cdot} - \alpha_{.*}{}^{\cdot} \alpha_{..}{}^{*} = \langle\alpha,\alpha\rangle$.

$$\alpha_{..}{}^{*} \beta_{*.}{}^{\cdot} - \beta_{.*}{}^{\cdot} \beta_{..}{}^{*} = \{\alpha,\beta\} - [\beta,\beta]$$

and all pairings are bilinear.

<u>Definition</u>: Let $A(A) \subset GL_m(A) \times GL_n(A)$ be the algebraic subgroup stabilizing $x_1$. These are all the base changes in $U,V$ fixing the identity. Let $A_1(A) \subset A(A)$ be the subgroup reducing to the identity over $K$.

If $R' \to R$ is a surjective ring homomorphism, then a <u>deformation</u> of a structure $(\alpha,\beta)$ over $R'$ is a structure $(\alpha',\beta')$ for $(U \otimes_R R', V \otimes_R R')$ such that $\alpha' \otimes_{R'} R = \alpha$ and $\beta' \otimes_{R'} R = \beta$. An isomorphism of deformations $(\Theta,\varphi)$ is a pair of isomorphisms of $(U \otimes_R R', V \otimes_R R')$ which reduce to the identity when tensored with R.

Definition: Given an alg-mod structure $(\alpha,\beta)$ over a field K, and an Artin ring R' over K, let

$AM_{\alpha,\beta}(R')$

be the set of isomorphism classes of deformations of $(\alpha,\beta)$ over R'. $AM_{\alpha,\beta}$ is a covariant functor from the category of Artin rings over K to sets, and $AM_{\alpha,\beta}(K)$ consists of a single point.

Proposition: $AM_{\alpha,\beta}$ has a prorepresentable hull, represented by a complete local ring $R_\infty$ over K.

Proof: We will write AM for brevity. Since AM satisfied the conditions of Schlessinger's theory of functors on Artin rings ,[S3], it suffices to check conditions (H1) - (H3). Let $u': R' \to R$ be a homomorphism of Artin rings and $u'': R'' \to R$ a surjective homomorphism. There is a natural map

$h: AM(R' \times_R R'') \to AM(R')\times_{AM(R)}AM(R'')$

(H1) h is surjective: Let $(\gamma',\delta')$, $(\gamma,\delta)$, $(\gamma'',\delta'')$ be representatives of isomorphism classes $\eta',\eta,\eta''$ in AM(R'), AM(R), AM(R''). There is a base change $Q \in A(A)$ carrying $(\gamma',\delta')\otimes_{A'}A$ into $(\gamma,\delta)$. Let $Q' \in A(A')$ be an arbitrary lifting of this base change and replace $(\gamma',\delta')$ by $Q'\circ(\gamma',\delta')$, so that $(\gamma',\delta')\otimes_{A'}A = (\gamma,\delta)$. Similarly, we may assume that $(\gamma'',\delta'')\otimes_{A''}A = (\gamma,\delta)$. Let $(\tilde{\gamma},\tilde{\delta})$ be the fiber product of $(\gamma',\delta')$ and $(\gamma'',\delta'')$. Consider the bilinear forms

$$\langle\tilde{\gamma},\tilde{\gamma}\rangle$$

and

$$\{\tilde{\gamma},\tilde{\delta}\} - [\tilde{\delta},\tilde{\delta}]$$

with components in $R' \times_R R"$. The images of the components in $R'$ and $R"$ are

$$\langle\gamma',\gamma'\rangle,\ \langle\gamma",\gamma"\rangle$$

and

$$\{\gamma',\delta'\} - [\delta',\delta'],\ \{\gamma",\delta"\} - [\delta",\delta"]$$

Since these are zero, so are the first two forms. Thus $(\tilde{\gamma},\tilde{\delta})$ is an algebra module structure. If we let $\tilde{\eta}$ be its isomorphism class then

$$h(\tilde{\eta}) = \eta' \times_{AM(R)} \eta"$$

<u>(H2) If $R = K$, $R" = K[\varepsilon]$, $\varepsilon^2 = 0$, then $h$ is bijective</u>: Let $K[\varepsilon]$, $\varepsilon^2 = 0$ be the algebra of the two-fold point. We need to show that if $A = K$ and $A" = K[\varepsilon]$, then

$$h\colon AM(A' \times_A A") \to AM(A') \times_{AM(A)} AM(A")$$

is a bijection. We will in fact show this when $A = K$ and $A" \to K$ is an arbitrary homomorphism of $K$-algebras, necessarily surjective. Let $\bar{A} = A' \times_K A"$. Since we have already shown that $h$ is a surjection, it remains to show that it is one-to-one. More specifically, if $\bar{\eta} \in AM(\bar{A})$ is a representative of the corresponding orbit isomorphism class of tensor pairs, then we need to show that the orbit of $(\bar{\gamma},\bar{\delta})$ is completely determined by the orbits of its images $(\gamma',\delta') = (\bar{\gamma},\bar{\delta}) \bullet_{\bar{A}} A'$ and $(\gamma",\delta") = (\bar{\gamma},\bar{\delta}) \bullet_{\bar{A}} A"$.

Suppose $Q' \in A_1(A')$ and $Q'' \in A_1(A'')$. Since both $Q'$ and $Q''$ reduce to the identity $I$ in $A_1(K)$, we can form the fiber product $\bar{Q} = Q' \times_I Q''$ which will carry $(\bar{\gamma},\bar{\delta})$ to $\bar{Q}\circ(\bar{\gamma},\bar{\delta})$, a structure constant tensor which reduces to $Q'\circ(\gamma',\delta')$ over $A'$ and to $Q''\circ(\gamma'',\delta'')$ over $A''$.

<u>(H3) AM(K[ε]) is finite dimensional</u>: $U^*\otimes U^*\otimes U$ and $U^*\otimes V^*\otimes V$ are finite dimensional, and the representatives of the classes in AM(K[ε]) are chosen from these sets, tensored with the two dimensional ring K[ε].

Although the simultaneous deformation of algebra and module structure has considerable theoretical importance, the module deformations place no restriction on the algebra deformations, and thus from a practical point of view, one may as well deform the algebras first and then look at the module deformations along a selected family of algebras, usually one parametric. We hope that the computer system will eventually calculate deformations as a function of continuous parameters in the algebraic structure constants. In the meantime we check the behavior of the module deformations along a continuous family of algebras by substituting "general" values (usually "2" is sufficiently general) for the continuous algebra parameters.

## Bibliography

[D-S] Th. Dana-Picard and M. Schaps, Counting Generic Algebras, preprint, Bar-Ilan Univerity, 1987.

[H-R] D. Happel and C. Ringel, Tilted Algebras, Trans. Am. Malth. Soc. 274 (1982), no.2, 399-443.

[S1] M. Schaps, Moduli of commutative and non commutative covers, Israel Journal of Mathematics, Vol. 58, No.1, 1987.

[S2] ————, Deformations of finite dimensional algebras and their idempotents, to appear in Trans. Am. Math. Soc.

[S3] M. Schlessinger, Functions on Artin rings, Trans. Am. Math. Soc. 130 (1968), 208-222.

[S4] A.H. Schofield, Bounding the global dimension in terms of the dimension, Bull. London Math. Soc. 17 (1985), 393-394.

# REPORT AND OPEN QUESTIONS FROM A WORKSHOP ON GEOMETRY AND INVARIANT THEORY OF QUIVERS

compiled by M. Schaps,Bar Ilan University,Israel

## 0. Introduction.

The general theory of representations of classical 'matrixproblems' and questions about root systems of Lie algebras. The theory of representations of quivers forms an important special case, frequently used as a test case, in the larger study of representations of algebras.

In the recent years, as the theory of quivers with a finite number of indecomposable representations has been better understood, new attention has been turned to quivers with infinite families of representations, both to the geometry and stratification of the representation space and also to the search for invariants or semi-invariants which will allow the representations lying over a particular stratum of this scheme to be parametrized upto isomorphism.

A notable breakthrough in this direction was the work of V. Kac , who established a general correspondence between the dimension vectors of indecomposable representations of quivers and the positive roots in a Kac-Moody Lie algebra. He established the dimension of the parameter space and made a number of conjectures about the rationality of this variety and about the decomposition of a generic representation.

The Coxeter reflection functors were used successfully to obtain an equivalence of derived categories between representations of quivers with different orientations. They have been generalized by the theory of tilting modules, which establish such



*F. van Oystaeyen and L. Le Bruyn (eds.), Perspectives in Ring Theory, 59–68.*

equivalences with representations of algebras which are not quiver algebras.

The theory of invariants of the ring of $m$ generic $n$ by $n$ matrices provides another test case which has been well-researched in pi-theory and general ring theory,since such questions can be considered questions about the $n$-dimensional representations of a quiver with a single point and $m$ loops.

The main themes of the conference were continuation of Kac's work, tilting modules and invariant theory of matrices. After a section of notation and definitions, the suggested research directions and open questions from the problem section at the end of the conference will be organized according to these themes. The problem session was intented,not as an agenda, but as an aid to reserchers and students.

The contributors to the problem session were E. Formanek (Penn.State), D. Happel (Bielefeld),L. Le Bruyn (Antwerp),J. Rickard (London), C.M. Ringel (Bielefeld),A. Schofield (London) and Y. Teranishi (Nagoya). They will be identified by initials.

## 1.Notations and definitions.

(1.0) : Quivers and representations : A quiver $Q$ is a graph $G$ which we will take to be finite, together with an orientation $\Omega$ of the edges. Over a field $k$,usually algebraically closed, the set of all finite oriented paths can be taken as a basis of an hereditary algebra called the quiver algebra or path algebra and denoted by $kQ$. If $Q$ has no oriented cycles and in particular no loops, then this algebra has finite dimension over $k$. A quiver algebra always has global dimension $\leq 1$ so it is hereditary and all modules have proj. dim. $\leq 1$.

The representations of the quiver $Q$ which will be defined in the next paragraph, correspond to modules over this algebra. Every finite dimensional algebra can be represented as the quotient of a quiver algebra by an ideal contained in the radical squared.

Let $v_1, ..., v_n$ be the vertices of the quiver $Q$, and for each arrow $a$, let $i(a)$ be its initial vertex and $t(a)$ its terminal vertex. A representation $R$ of $Q$ over $k$ is an $n$-tuple $(R(1), ..., R(n))$ of $k$-vectorspaces and a set of $k$-linear transformations $R(a) : R(i(a)) \rightarrow R(t(a))$ for each arrow $a$ in the quiver. A morphism of representations $F : R \rightarrow S$ is a set of $n$ linear transformations $F(i) : R(i) \rightarrow S(i)$ such that the obvious diagrams commute.

Given a representation $R$, let $\alpha(i) = dim_k R(i)$. The vector $\alpha = (\alpha(1), ..., \alpha(n))$ $\mathbb{N}^n$ is called the dimension vector of the representation. For a given dimension $\alpha$, the algebraic group $GL(\alpha) = \prod GL(\alpha(i))$ acts on the representation space by basis change : $R(a) \rightarrow g(t(a))^{-1} R(a) g(i(a))$. The stabilizer of $R$ is called the isotropy subgroup $H$ of $R$, it consists of all $h \in GL(\alpha)$ such that $hR = R$. Since this always contains the multiplicative subgroup $k^*$ of $k$, embedded as an $n$- tuple of scalar matrices, the group $GL(\alpha)$ is sometimes replaced by the quotient group $PGL(\alpha) = GL(\alpha)/k^*$.

The representation space is $R(Q, \alpha) = \prod_a Hom(R(i(a)), R(t(a)))$ and is a vectorspace with $GL(\alpha)$- action. The $GL(\alpha)$ orbit $O(R)$ is isomorphic to $GL(\alpha)/Stab(R)$.

For some dimension vectors there is a single representation whose orbit is dense in $R(Q, \alpha)$ and such a representation is called 'generic'. For other $\alpha$ one must seek a family of 'generic' representations, such that the union of their orbits is dense in $R(Q, \alpha)$. Members of such a generic family all decompose in the same way, and the dimension vectors of the components of a generic representation give the generic decomposition of $\alpha$.

For any two dimension vectors we define a form $< \alpha, \beta >= \sum \alpha(n)\beta(n) - \sum_a \alpha(ia)\beta(ta)$, so that $< \alpha, \alpha >= dimGL(\alpha) - dimR(Q, \alpha)$. The dimension vector of an indecomposable representation is called a root and always satisfies $< \alpha, \alpha > \leq 1$. It is called real if $< \alpha, \alpha >= 1$, imaginary if $< \alpha, \alpha > \leq 0$ and isotropic if $< \alpha, \alpha >= 0$. A root is called a Schur root if $End(R) = k$ for some representation $R$ and Kac proved that in the generic decomposition $\alpha = \sum \beta_i$ each $\beta_i$ is a Schur root.

A variety (i.e. a reduced irreducible scheme) is rational if it is birationally equivalent to a projective space. Two varieties are stably equivalent if their functionfields have a common rational extensionfield. A variety which is stably equivalent to a rational variety is called stably rational.

The Auslander-Reiten quiver of an algebra $A$ consists of all indecomposable modules over $A$, connected by maps representing homomorphisms which are irreducible, ie. cannot be factored.

The AR-quiver is equipped with a translation functor $\tau$ and its inverse translation $\tau^{-1}$ defined by AR-split exact sequences.

$Mod(A)$ is the category of all finitely generated modules and $\underline{Mod}(A)$ is obtained from $Mod(A)$ by factoring out all homomorphisms which factor through projectives. The derived category $D^b(A)$ is obtained from the homotopy category of all bounded

copmplexes by formally inverting all morphisms which induce isomorphisms on homology. For a fixed module $M$, $Add(M)$ is the full subcategory of $Mod(A)$ generated by sums of direct summands of $M$.
For a selfinjective algebra $A$, $\underline{Mod}(A)$ has the structure of a triangulated category,where the distinguished triangles have the form $X \to Y \to Z \to \tau X$. For selfinjectives, we say that $A$ and $A'$ are stably equivalent if we have an equivalence of $\underline{Mod}(A)$ with $\underline{Mod}(A')$ as triangulated categories.

## 2. Continuations of the Kac theorems

(geometry and invariant theory of 'generic' representations, connections with Lie algebras)
The original conjectures of Kac have required considerable modification. Those interested in seeing various counterexamples might try to acquire the preprint of A. Schofield. What follows is a modified set of questions.

2.1 : Generic decomposition questions

Question 1 (A.S.) : If for a root $\alpha$ we can find a decomposition $\alpha = \sum \beta_i$ into Schur roots for which $< \beta_i, \beta_j > \geq 0$ for all $i \neq j$, does this mean that $\alpha$ is not a Schur root. (In general, for $\alpha$ not a root, there may be several ways of decomposing $\alpha$ such that $< \beta_i, \beta_j > \geq 0$ for all $i \neq j$).

Question 2 (A.S.) : Let $\gamma$ be a real Schur root such that $< \alpha, \gamma >$ and $< \gamma, \alpha >$ are strictly positive. Does this mean that $\alpha$ is not a Schur root. (This is true for the $n$-subspace problem and if true in general it would give a simple picture of the Schur roots).

The implication Quest 2 $\to$ Quest 1 is as follows : Kac has shown that every generic decomposition contains at least one real Schur root $\beta_j$. Suppose $\alpha$ satisfies the hypotheses of question 1. Then take $\gamma = \beta_i$ in question 2 and one has the desired conclusion that $\alpha$ is not a Schur root

2.2 Rationality questions

Let $\alpha$ be a root. Given a finite field of $q$ elements, let $p(q)$ be the number of isomorphism classes of indecomposable representations of dimension vector $\alpha$. This is a polynomial

$$p(q) = q^n + c_{n-1}q^{n-1} + \dots + c_0$$

(Kac is reworked by Kraft and Riedtmann). Kac conjectured 1. all $c_i \geq 0$ and 2. $c_0$ is the multiplicity of $\alpha$ as a root in the Kac-Moody Lie algebra build from the symmetrization of $< -, - >$

Question 1 (A.S.) : Can any relatively simple calculation be found to chech these conjectures ( in 2. the multiplicity of $\alpha$ can be checked but the computation of $c_0$ is very difficult except in simple cases)

Question 2 (A.S.) : For any Schur root $\alpha$ , is the variety of isomorphism classes of indecomposable representations with this dimension vector a rational variety ?

(By joint work of Le Bruyn and Schofield presented at the conference, if $t$ is the greatest common divisor of the components $\alpha(i)$ then this variety is stably equivalent to the variety of matrixinvariants for the $t$ by $t$ generic matrices and thus either both or neither are stably rational. Thus stable rationality problems for general quivers and Schur roots are equivalent to the classical matrixinvariant case).
Given a quiver with lots of oriented cycles,it suffices to check rationality for a simple representation vector. If $\alpha$ is the dimensionvector of a simple representation, the functionfield of the quotientvariety $R(Q, \alpha)/GL(\alpha)$ coincides with the field of rational invariants for $\alpha$. The coordinatering of $R(Q, \alpha)/GL(\alpha)$ is the ring of invariants under the action of the groupo $GL(\alpha)$. For any $\alpha$ not necessarely a root, the variety $S(\alpha) = R(Q, \alpha)/GL(\alpha)$ parametrizes isomorphism classes of semisimple representations. Luna showed that this variety has a canonical stratification with strata corresponding to stabilizer subgroups $H$ of semi-simple representations $R$ where $S(\alpha)_H = \{\xi \in S(\alpha) : \text{the stabilizer subgroup of } \xi \text{ is conjugated to } H\}$. There is a rather strong conjecture of Kac that $R(Q, \alpha)$ has a cellular decomposition into cells which are affine spaces

Question 3 (L.L) : If the variety of all isomorphism classes of $\alpha$-representations has a cellular decomposition, can one find such a decomposition and an embedding

$$S(\alpha) \to IsoR(Q, \alpha)$$

such that different strata in the Luna stratification go to different cells.

In addition, we have the following information which gives the local structure of $S(\alpha)$ near a point $\xi$. THere exists a quiver $Q'$ , a dimension vector $\alpha'$ and an etale extension

$$S_{Q'}(\alpha') \to S_Q(\alpha)$$

such that the origin in $S_{Q'}(\alpha')$ goes to $\xi$. This essentially acts as a transversal to the stratum $S(\alpha)_H$ for $\xi$ . The dimension vector $\alpha'$ has strictly smaller greatest common divisor and this might form the basis for an inductive rationality argument

Question 4 (L.L.) : If all these transversals varieties for strata larger than the origin are stably rational, does this imply that $S(\alpha)$ is itself stably rational

## 3. Tilting modules

The Coxeter reflection functor replaces a quiver $Q$ with a sink $v_i$ by a new quiver $Q'$ with a source at $v_i$.Whereas

$$A = kQ \simeq End_A(P_1 \oplus ... \oplus P_n)$$

the new algebra is

$$B = kQ' \simeq End_A(P_1 \oplus ... \oplus \tau^{-1} P_i \oplus ... \oplus P_n)$$

where $\tau^{-1}$ is the inverse of the Auslander-Reiten translation.
Letting $T = P \oplus ... \oplus \tau^{-1} P_i \oplus ... \oplus P_n)$ be the corresponding $A - B$-bimodule, $T$ determines a functor

$$M \to Hom_A(T, M)$$

which gives an equivalence between the corresponding derived categories $D^b(A)$ and $D^b(B)$. The Coxeter functors correspond to reflections in the corresponding Lie algebra.
Define a partial tilting module or a partial 1-tilting module to be a module $T$ such that
(i) projdim $T \leq 1$
(ii) $Ext^1(T, T) = 0$
If, in addition, $T$ satisfies
(iii) there exists an exact sequence of left $A$-modules

$$0 \to A \to T' \to T" \to 0$$

then $T$ is called a tilting module or a 1-tilting module. Then we have the Brenner-Butler theorem, that if $B = End_A({}_AT)$, then $A = End_B(T_B)$ and an equivalence of

derived categories. Algebras obtained from quiver algebras of tame type are called tilted algebras and played an important role in the classification of algebras of finite representation type.
As a further generalization we can define a partial $r$-tilting module to be a module $T$ satisfying
(i) pr-dim $T \leq r$
(ii) $Ext^i(T,T) = 0$ for $i = 1, ..., r$
If, in addition,
(iii) there is an exact sequence of $A$-modules

$$0 \to A \to T^{(0)} \to ... \to T^{(r)} \to 0$$

then $T$ is called an $r$-tilting module and we have

$$D^b(A) \simeq D^b(End_A(T))$$

The number $t$ of direct summands in an $r$-tilting module is always $\leq rkG_0(A)$, the Grothendieck group. Condition (iii) always implies equality.
If $r = 1$ then condition (iii) is equivalent to $t = rkG_0(A)$ a condition which is very easy to check.

Question 1 (D.H.) : For a partial $r$-tilting module $T$, does $t = rkG_0(A)$ imply that $T$ is an $r$-tilting module

This would follow if one could show the following stronger result

Question 2 : If $T$ is a partial tilting module, is there a module ${}_AX$ such that $T \oplus X$ is an $r$-tilting module ?
(the implication from Quest 2 to Quest 1 comes from an embedding of derived categories). Quest 2 also hold for $r = 1$. As a preliminary to tackling these two questions, one might want to test the following consequence of Quest 1 :

Question 3 (D.H.) : If $prdim_A(D(A_A) < \infty$ then $ind_A A < \infty$
(This holds if the finitistic dimension of $A$ is finite. This is the supremum over all modules with finite projective dimension and there is a conjecture that this is always finite)

An invariant of the derived category : a partial tilting complex is a complex of projective modules over $A$ such that
(i) it is bounded

(ii) satisfies $Hom(T, T[n]) = 0$ whenever $n \neq 0$
A partial tilting complex is called a tilting complex if it generates as a triangulated category the subcategory of the derived category generated by projectives.

Question 1' (J.R.) : Does the analog of question 1 above hold for tilting complexes ?

There are counterexamples to an analoque of question 2.
Define the tilting simplicial complex of an algebra to be the simplicial complex whose simplices are partial 1-tilting modules and $T \leq T'$ if $T$ is a direct summand of $T'$.

Question 4 (C.R.) : What kind of information does this complex carry ?

Example : For the Dynkin diagram $A_n$ with the linear orientation , this complex is a cone with vertex the unique projective injective,whose boundary consists of all non faithful p.t. modules

Question 4 (C.R.) : Do there exist algebras which are stably equivalent but not derived eqyuivalent ?

A positive solution would solve the Auslander-Alperin conjecture that if two algebras are stably equivalent the number of non projective simples are the same.

## 5. Matrixinvariants

Let $Gr(2, m)$ be rthe Grassmannian of 2-dimensional subspaces in an $m$-dimensional vectorspace $V$ and let $p_{ij}$ for $1 \leq i < j \leq m$ be independent variables. Let $\mathbb{C}[Gr(2, m)] = \mathbb{C}[p_{ij}]/I$ where $I$ is the ideal of the Plucker conditions. Define a standard Young tableau (resp. semistandard) of degree $d$ by giving

$$\begin{matrix} i_1 & i_2 & \dots & i_d \\ j_1 & j_2 & \dots & j_d \end{matrix}$$

where the numbers horizontally are nondecreasing and vertically strictly (resp. nondecreasing) increasing.
With $Y_d^s$ one denotes the set of all standard (resp. semistandard) Young tableaux of degree $d$. The monomial $m(Y_\sigma)$ associated to a Young tableau $Y_\sigma$ is

$$p_{i_1 j_1} p_{i_2 j_2} \cdots p_{i_d j_d}$$

We also define

$$Tr(Y_\sigma) = Tr(X_{i_1}X_{j_1})...Tr(X_{i_d}Y_{j_d})$$

Then Procesi has shown that if $\mathbb{T}^0_m$ is the trace ring of generic 2 by 2 trace zero matrices and if $H$ is its Hilbert series, then

$$H(t) = \frac{1}{(1-t)^m} H(\mathbb{C}\,[Gr(2,m)], t)$$

and the set $m(Y_\sigma)$ for $Y_\sigma$ in $Y_d$ is a basis for the vectorspace $\mathbb{C}\,[Gr(2,m)]_{2d}$

Question 1 (Y.T.) : Is the following true? Let $X_1$,...,$X_m$ be 2 by 2 trace zero matrices, then $\{X_{i_1}...X_{i_j}Tr(Y_\sigma) : 0 \le j \le m, degY_\sigma + j = d\}$ is a basis for $\mathbb{T}^0(2m)_d$.

Question 2 (Y.T.) : What is $c(n)$, the minimal degree for which the traces generate the entire trace ring of $n$ by $n$ matrices.

It is known that $c(1) = 1$,$c(2) = 3$,$c(3) = 6$. We have two conjectures
(E.F.) :$c(n) = \frac{n(n+1)}{2}$
(Y.T.) : $c(n) = n(n-1)$

Question 3 (Y.T.) : $tr(XYX^2Y^2...X^{n-1}Y^{n-1})$ can never be written as a combination of lower degree traces

Question 4 (E.F. - L.L.) : Is the trace ring of $m$ generic $n$ by $n$ matrices a Cohen-Macaulay module over its center ? (true for $n = 2$)

Question 5 (A.S.) : Let $C(m,n)$ be the ring of invariants of $m$ generic $n$ by $n$ matrices. Let $C_t$ be the subring generated by all Cayley Hamilton coefficients of degree $\le t$. For which $t$ is $C_t$ the whole ring (Maybe $t = n + 1$ ? )

**Bibliography**

This brief list of articles is included only as a starting point for a literature search


Formanek E. : Invariants and polynomial invariants of $n \times n$ matrices , in Invariant theory,West Chester Univ (1985)

Happel D.,Ringel C.M. : Tilted Algebras, Trans AMS 274 (1982) 399-443

Kac V. : Root systems,representations of quivers and invariant theory, Springer LNM 996 (1984) 74-108

Kraft H. : Geometric methods in reporesentation theory, Springer LNM

Le Bruyn L. : Counterexamples to the Kac conjecture on Schur roots,Bull Sc Math 110 (1986) 437-448

Schofield A. : Generic representations of quivers , preprint London (1986)

# AUTOMORPHISMS OF GENERIC 2 BY 2 MATRICES

Jak Alev [(1)] and Lieven Le Bruyn [(2)]
(1) : Université Paris VI, (2) : Universiteit Antwerpen UIA-NFWO

## 1. Introduction.

The purpose of this paper is to investigate the automorphisms of the ring of 2 generic 2 by 2 matrices $G$. Along the way, we will have to analyze the automorphisms of related algebras as well : the trace ring $T$, its center $C$, the ring of 2 generic 2 by 2 trace zero matrices $T^0$ and some Clifford algebras.

In [Be], G. Bergman has constructed wild automorphisms of rings of generic matrices in connection with the commutative tame automorphism problem that we recall in section 1. Concentrating on the ring of 2 generic 2 by 2 matrices, one can get more specific information on the automorphisms. According to [FHL] and [Pro], the center of the corresponding trace ring is a polynomial algebra in 5 variables and one can then investigate the induced automorphisms. The main result is that all automorphisms of $G$ that we can construct induce tame automorphisms of $C$.

On the other hand, the study of $T^0$ leads to automorphisms that can reasonably be considered as wild as is defined in section 4 and such that the induced automorphisms on the 3-dimensional polynomial center is Nagata's automorphism up to a linear change of coordinates. This last automorphism is commonly considered as a good candidate for a wild automorphism on 3-dimensional affine space; and it appears as the restriction to the center of a wild automorphism of a 3-dimensional, yet 2-generator non-commutative algebra $T^0$. Perhaps this adds evidence to its conjectural wildness.



*F. van Oystaeyen and L. Le Bruyn (eds.), Perspectives in Ring Theory, 69–83.*

## 2. Tameness and Nagata's automorphism.

(2.1) : Let $k$ be a reduced commutative ring and let $GA_n(k)$ be the group of $k$-algebra automorphisms of $k[X_1, ..., X_n]$. $GA_n(k)$ has two natural subgroups ; $Af_n(k)$ the affine automorphisms and $BA_n(k)$, the triangular (or Jonquière) automorphisms. They are defined by

$$Af_n(k) : \begin{pmatrix} x_1 \\ x_2 \\ \vdots \\ x_n \end{pmatrix} \to \sigma . \begin{pmatrix} x_1 \\ x_2 \\ \vdots \\ x_n \end{pmatrix} + \begin{pmatrix} a_1 \\ a_2 \\ \vdots \\ a_n \end{pmatrix}$$

$$BA_n(k) : \begin{pmatrix} x_1 \\ x_2 \\ \vdots \\ x_n \end{pmatrix} \to \begin{pmatrix} \alpha_1 x_1 + P_1(x_2, ..., x_n) \\ \alpha_2 x_2 + P_2(x_3, ..., x_n) \\ \vdots \\ \alpha_n x_n \end{pmatrix}$$

where $\sigma \in GL_n(k)$, $a_i \in k$ , $\alpha_i \in k^* = G_m(k)$ and $P_i(x_{i+1}, ..., x_n) \in k[x_{i+1}, ..., x_n]$ for all $1 \leq i \leq n-1$.

The subgroup of $GA_n(k)$ generated by $Af_n(k)$ and $BA_n(k)$ is called the group of tame automorphisms. The main open problem in this context is

**Problem 1 :** If $k$ is a field, are all automorphisms in $GA_n(k)$ tame ?

For $n = 2$, this is a classical result which describes $GA_2(k)$ as an amalgamated product of $Af_2(k)$ and $BA_2(k)$ along their intersection ([Ju],[Na],[Re],[Va]). If $k$ is not a field, Nagata considers in [Na] the following automorphism of $K[x, y, z]$ where $K$ is a field :

$$\sigma : \begin{array}{ccc} x & \to & x - 2y(y^2 + xz) - z(y^2 + xz)^2 \\ y & \to & y + z(y^2 + xz) \\ z & \to & z \end{array}$$

**Theorem (Nagata) :** $\sigma$ considered as an element in $GA_2(K[z])$ is not tame, but it is tame considered in $GA_2(K[z, z^{-1}])$

To the best of our knowledge, the following problem is still open

**Problem 2 :** Is $\sigma$ tame considered as an element in $GA_3(K)$ ?

(2.2) : Non commutative analoques of this problem have been studied extensively. The tameness notion is adapted to the considered non commutativity, the central idea being to be generated by the most natural automorphisms. In the sequel we take $k = \mathbb{C}$ in order to simplify the notation

(a) : The free algebra of rank 2 : In this case, tameness is defined as in the commutative case and one has the following theorem ([Co],[Cz],[Di],[Ma1])

**Theorem :** There is a natural isomorphism between $Aut_{\mathbb{C}}\, \mathbb{C} < X, Y >$ and $Aut_{\mathbb{C}}\, \mathbb{C}\,[x, y]$

(b) : The Weyl algebra $A_1(\mathbb{C})$ : $A_1(\mathbb{C}) = \mathbb{C}\,[p, q]$ where $[p, q] = 1$. In this case, tameness is defined by considering the subgroups $S$ and $B$ defined as follows

$$S : \begin{array}{ccc} p & \rightarrow & \alpha p + \beta q + \gamma \\ q & \rightarrow & \alpha' p + \beta' q + \gamma' \end{array}$$

where $\alpha\beta' - \beta\alpha' = 1$ and $\gamma, \gamma' \in \mathbb{C}$ and

$$B : \begin{array}{ccc} p & \rightarrow & \alpha p + P(q) \\ q & \rightarrow & \alpha^{-1} q \end{array}$$

where $P(q) \in \mathbb{C}\,[q]$ and $\alpha \in \mathbb{C}^*$.Then, one has the following theorem ([A],[Dix],[Ma2])

**Theorem :** $Aut_{\mathbb{C}}(A_1(\mathbb{C}))$ is the amalgamated product of $S$ and $B$ along their intersection

(c) : Low dimensional enveloping algebras :

(i) : $g$ soluble , $dim(g) = 2$ or $3$ and $g$ not nilpotent. In this case tameness is defined replacing $Af_n(\mathbb{C})$ by the group generated by $Aut_{\mathbb{C}}\, g$ and the translations, and $BA_n(\mathbb{C})$ by automorphisms triangular with respect to a basis of $g$ adapted to the derived series of $g$. In [Sm1], the automorphism groups of $U(g)$ are determined for different $g$'s and simple inspection reveals tameness in all cases.

(ii) : $g = sl(2, \mathbb{C})$. In this case, tame automorphisms are defined to be the ones which are generated by automorphisms of $U(sl(2, \mathbb{C}))$ fixing an element $X \in sl(2, \mathbb{C})$. In [Jo], it is shown that there exists wild automorphisms of $U(sl(2, \mathbb{C}))$

(iii) : $g = g_3$ , the nilpotent 3-dimensional Heisenberg algebra. In this case, tameness is defined as in (i) , the derived series being replaced by the central descending series. In [A], it is shown that a modified version of Nagata's automorphism which accounts for the non commutativity gives an example of a wild automorphism of $U(g_3)$

The case of generic matrix rings was considered by Bergman [Be] and we will briefly recall some of his results in the next section.

## 3. Bergman's wild automorphisms of generic matrices.

In [Be], G. Bergman constructs different types of wild automorphisms of generic matrix rings.In the special case of 2 generic 2 by 2 matrices, we will compute the extension of certain automorphisms to the trace ring and their restrictions to the 5-dimensional polynomial center. The wild automorphisms at the non commutative level induce then Nagata like automorphisms which become tame as we will see in the last section.

(3.1) : Let $G$ be the ring of 2 generic 2 by 2 matrices , that is , the subring of $R = M_2(\mathbb{C}[x_1, x_2, x_3, x_4; y_1, y_2, y_3, y_4])$ generated by the generic matrices

$$X = \begin{pmatrix} x_1 & x_2 \\ x_3 & x_4 \end{pmatrix}; Y = \begin{pmatrix} y_1 & y_2 \\ y_3 & y_4 \end{pmatrix}$$

With $Z$ we will denote the center of $G$ and $T$ will be the trace ring of $G$ , that is , the subalgebra of $R$ generated by $G$ and the traces of its elements. With $C$ we denote the center of $T$. Finally we introduce the ring $T^0$ of 2 generic 2 by 2 trace zero matrices, that is, the subalgebra of $R$ generated by $X^0 = X - \frac{1}{2}T(X)$ and $Y^0 = Y - \frac{1}{2}T(Y)$ where $T(-)$ is the trace map. The following result describes the relationship between these algebras [FHL]

**Theorem (Formanek,Halpin,Li) :**

(i) : The commutator ideal of $G$ is equal to $T.[X, Y]$

(ii) : $C = \mathbb{C}[T(X), T(Y), D(X), D(Y), T(XY)]$ and $T$ is a free $C$-module with basis $1, X, Y, XY$

(iii) : $G/[G, G] = \mathbb{C}[x, y]$ a polynomial ring in 2 variables

(3.2) : For our computations, it is more convenient to consider $T^0$ and its center $C^0 = \mathbb{C}[D(X^0), D(Y^0), T(X^0Y^0)]$. One has the following relations

(i) : ${X^0}^2 = -D(X^0)$, ${Y^0}^2 = -D(Y^0)$, $X^0Y^0 + Y^0X^0 = T(X^0Y^0)$

(ii) : $D(X) = D(X^0) + \frac{1}{4}T(X)^2$ , $D(Y) = D(Y^0) + \frac{1}{4}T(Y)^2$, $T(XY) = T(X^0Y^0) + \frac{1}{2}T(X)T(Y)$

(iii) : $[X,Y] = [X^0,Y^0] = T(X^0Y^0) - 2Y^0X^0 = 2X^0Y^0 - T(X^0Y^0)$, $[X^0,Y^0]^2 = T(X^0Y^0)^2 - 4D(X^0Y^0)$

Hence in particular, $C = \mathbb{C}\,[T(X),T(Y),D(X^0),D(Y^0),T(X^0Y^0)]$ and $T$ is a free $C$-module with basis $1, X^0, Y^0, X^0Y^0$

(3.3) : Following Bergman, consider the diagram

$$\begin{array}{ccc} \mathbb{C} < X,Y > & \rightarrow & G \\ & & \downarrow \\ & & \mathbb{C}\,[x,y] \end{array}$$

which induces the diagram

$$\begin{array}{ccc} Aut_{\mathbb{C}}\,\mathbb{C} < X,Y > & \rightarrow & Aut_{\mathbb{C}}\,G \\ & & \downarrow \pi \\ \psi & & Aut_{\mathbb{C}}\,\mathbb{C}\,[x,y] \end{array}$$

According to 2.2 (a) , $\psi$ is an isomorphism so $\phi$ is mono and $\pi$ is epi. Now, let us define the tame automorphisms of $G$ to be the ones which are induced by automorphisms of $\mathbb{C} < X,Y >$ (which are all tame); they are generated by the following types

$$\eta_f : \begin{array}{ccc} X & \rightarrow & X + P(Y) \\ Y & \rightarrow & Y \end{array}$$

where $P(Y) \in \mathbb{C}\,[Y]$

$$\epsilon_c : \begin{array}{ccc} X & \rightarrow & cX \\ Y & \rightarrow & Y \end{array}$$

where $c \in \mathbb{C}^*$

$$\theta : \begin{array}{ccc} X & \rightarrow & Y \\ Y & \rightarrow & X \end{array}$$

Therefore, any non-trivial automorphism of $G$ in the kernel of $\pi$ will be wild. Let us give a few examples :

(1) : Bergman gives the following wild automorphism of $G$

$$\sigma_1 : \begin{array}{ccc} X & \rightarrow & X + [X,Y]^2 \\ Y & \rightarrow & Y \end{array}$$

One easily checks that $\sigma_1$ fixes $T^0$ and that the induced automorphism on $C$ is the triangular one fixing all variables except $T(X)$ which is mapped to $T(X) + 2[X,Y]^2 = T(X) + 2T(X^0Y^0)^2 - 8D(X^0)D(Y^0)$.

(2) : A slightly more complicated example which gives rise to Nagata like automorphisms is

$$\sigma_2 : \begin{array}{ccc} X & \to & X + Y[X,Y]^2 \\ Y & \to & Y \end{array}$$

Then, one verifies that $\sigma_2$ induces an automorphism on $T^0$. In the next section we will give a more consistent procedure to produce automorphisms of the generic trace zero matrices. Moreover, the induced automorphism by $\sigma_2$ on the center of the trace algebra $C$ is given by

$$\begin{array}{ccc} T(X) & \to & T(X) + T(Y)[X,Y]^2 \\ T(Y) & \to & T(Y) \\ D(X^0) & \to & D(X^0) - T(X^0Y^0)[X,Y]^2 + D(Y^0)[X,Y]^2 \\ D(Y^0) & \to & D(Y^0) \\ T(X^0Y^0) & \to & T(X^0Y^0) - 2D(Y^0)[X,Y]^2 \end{array}$$

The restriction of $\sigma_2$ to $C^0$ is then a Nagata like automorphism of $C^0$. In the last section we will see that $\sigma_2 \mid C$ is a tame automorphism.

## 4. Constructing weird automorphisms.

In this section we will present a method to construct automorphisms of $T^0$, the generic trace zero ring for 2 $2 \times 2$ matrices. This method rests on the description of $T^0$ as a generic Clifford algebra, see [LB]. Our construction can be used also to construct weird automorphisms on commutative polynomial ring and on the ring of $m$ generic 2 by 2 trace zero matrices $T^0_m$.

(4.1) : Let us first recall some basic facts on quadratic forms and their Clifford algebras, see for example [Ba] for more details.Let $R$ be a commutative $\mathbb{C}$ -algebra. Then any quadratic form

$$q = \sum_{i,j=1}^{m} \alpha_{ij} X_i X_j$$

with $\alpha_{ij} = \alpha_{ji} \in R$ induces a symmetric bilinear form on a free $R$-module of rank $m$

$$P = Re_1 \oplus ... \oplus Re_m$$

by defining $B(e_i, e_j) = \alpha_{ij}$. The Clifford algebra of $P$ associated to the quadratic form $q$ is defined to be the quotient of the tensor $R$-algebra $T(P)$ of $P$ by the ideal generated by the elements of the form

$$p \otimes p - B(p,p)$$

for all $p \in P$. If we give the tensor-algebra the usual gradation, it follows that the Clifford algebra $Cl(P,q)$ has an induced $\mathbb{Z}/2\mathbb{Z}$-gradation, i.e. $Cl(P,q) = C_0 \oplus C_1$. There is a canonical $R$-algebra automorphism on $Cl(P,q)$ sending $c_0 \oplus c_1$ to $c_0 \oplus (-c_1)$ which is called the main automorphism.

In [LB] the so called generic Clifford algebras $Cl_m$ were introduced. $Cl_m$ is the Clifford algebra over the polynomial algebra

$$S_m = \mathbb{C}\,[a_{ij} : 1 \leq i \leq j \leq m]$$

that is, the homogeneous coordinate ring of the variety of all symmetric $m$ by $m$ matrices, corresponding to the generic quadratic form

$$q_m = \sum_{i,j=1}^{m} a_{ij} X_i X_j$$

Genericity here means that any Clifford algebra of an $m$-ary quadratic form over $\mathbb{C}$ can be obtained as a specialization of $Cl_m$. One can show, [LB] , that $Cl_m$ is isomorphic to the iterated Öre extension

$$\mathbb{C}\,[a_{ij} : 1 \leq i < j \leq m][a_1][a_2, \sigma_2, \delta_2]...[a_m, \sigma_m, \delta_m]$$

where $\sigma_j(a_i) = -a_i$ and $\delta_j(a_i) = 2a_{ij}$ for all $i < j$ and trivial actions on the other variables and $a_{ii} = a_i^2$. In particular, $Cl_m$ has finite global dimension equal to $\frac{m(m+1)}{2}$, is a maximal order ans has p.i.-degree equal to $2^\alpha$ where $\alpha$ is the largest natural number smaller or equal to $\frac{m}{2}$.

Restriction of automorphisms of $Cl_m$ to $S_m$ gives an exact sequence

$$1 \to \mathbb{Z}/2\mathbb{Z} \to Aut_{\mathbb{C}}\, Cl_m \to Aut_{\mathbb{C}}\, S_m = GA_{\frac{m(m+1)}{2}}(\mathbb{C}\,)$$

the kernel being generated by the main automorphism. Therefore, in order to describe the automorphism group of the generic Clifford algebra we aim to compute $Aut(S_m : Cl_m)$ which is the subgroup of $Aut_{\mathbb{C}}\, S_m$ consisting of those automorphisms which can be extended to $Cl_m$. A large subgroup of them can be described in the following way :

Let $\sigma$ be an automorphism of $S_m$ , then $\sigma$ extends to $Cl_m$ if and only if $Cl_m \cong_\sigma Cl_m$ as $S_m$-algebras where ${}_\sigma Cl_m$ is equal to $Cl_m$ as an abelian group and with $S_m$-action given via $s * c = \sigma^{-1}(s).c$ for all $s \in S_m$ and $c \in Cl_m$. Now, it follows from the definition that ${}_\sigma Cl_m$ is the Clifford algebra over $S_m$ associated to the $m$-ary quadratic form

$$\sigma^{-1}(q) = \sum_{i,j=1}^{m} \sigma^{-1}(a_{ij})X_iX_j$$

Now, it is well known that two Clifford algebras are isomorphic if their corresponding matrices are congruent.That is, if there exists an invertible matrix $A \in GL_m(S_m)$ such that

$$A^\tau.(a_{ij})_{i,j}.A = (\sigma^{-1}(a_{ij}))_{i,j}$$

These observations make it possible to determine lots of elements of $Aut(S_m : Cl_m)$ by determining which of the endomorphisms of $S_m$ determined by sending $a_{ij}$ to the $(i,j)$-entry of the matrix $A^\tau(a_{ij})_{i,j}A$ for $A \in GL_m(S_m)$ are automorphisms (which can be tested by calculating the Jacobian).

(4.2) : We will now see what the above general procedure gives us in the special case that $m = 2$. For notational simplicity we will let $x = a_{11}, y = a_{12}$ and $z = a_{22}$.Let us consider the easiest case of an elementary matrix

$$A = \begin{pmatrix} 1 & 0 \\ f & 1 \end{pmatrix}$$

for some $f \in \mathbb{C}[x,y,z] = S_2$. As we have seen above, this matrix induces an endomorphism on $\mathbb{C}[x,y,z]$ which is given by

$$\begin{array}{ccc} x & \to & x + 2fy + f^2z \\ y & \to & y + fz \\ z & \to & z \end{array}$$

To check when this is an automorphism we can calculate its Jacobian

$$\begin{pmatrix} 1 + 2y\partial_x f + 2fz\partial_x f & 2f + 2y\partial_y f + 2fz\partial_y f & * \\ z\partial_x f & 1 + z\partial_y f & * \\ 0 & 0 & 1 \end{pmatrix}$$

If we set the determinant equal to 1 we get the equation

$$2y\partial_x f + z\partial_y f = 0$$

which implies that $f \in \mathbb{C}[z, y^2 - xz]$ as can be readily verified. Moreover, any such $f$ clearly induces an automorphism on $\mathbb{C}[x, y, z]$ .

(4.3) : Note that for any $m$ there is a natural epimorphism

$$\pi : \mathbb{C} < X_1, ..., X_m > \rightarrow Cl_m$$

which is determined by $\pi(X_i) = a_i$. Therefore, the natural notion of tame automorphisms of $Cl_m$ is that they are the automorphisms which can be lifted to tame automorphisms of the free algebra $\mathbb{C} < X_1, ..., X_m >$. Using this convention we can now prove

**Theorem :** If $f \in \mathbb{C}[y^2 - xz]_+$ then the induced automorphism on $Cl_2$ is wild

**Proof :** In general,if an endomorphism of $S_m$ determined by an element $A \in GL_m(S_m)$ is an automorphism, the extension of it to $Cl_m$ is given by sending $a_i$ to the $i$-th entry of the vector

$$A^\tau . \begin{pmatrix} a_1 \\ \vdots \\ a_m \end{pmatrix}$$

Now, if $f \in \mathbb{C}[y^2 - xz]_+$ the induced automorphism on $Cl_2$ is given by

$$\begin{array}{ccc} a_1 & \rightarrow & a_1 + f a_2 \\ a_2 & \rightarrow & a_2 \end{array}$$

It is east to verify that this automorphism fixes the normalizing element $[a_1, a_2]$ and hence induces an automorphism on the quotient $Cl_2 / Cl_2[a_1, a_2]$ which is a polynomial ring in the images of the $a_i$ say $\mathbb{C}[u, v]$. Since $[a_1, a_2]^2 = y^2 - xz$ it is clear that this induced automorphism on $\mathbb{C}[u, v]$ is the identity. But we have seen before that the natural morphism

$$Aut_{\mathbb{C}}\, \mathbb{C} < X_1, X_2 > \rightarrow Aut_{\mathbb{C}}\, \mathbb{C}[u, v]$$

is an isomorphism, so the automorphism on $Cl_2$ being non-trivial cannot be lifted to an automorphism of $\mathbb{C} < X_1, X_2 >$ and hence is not tame.

Remark that the special case when $f = y^2 - xz$ gives us the Nagata automorphism so the foregoing result may add some evidence to the conjectural wildness of this automorphism.

Of course, one can repeat the same argumentation for more general elements of $GL_2(\mathbb{C}[x,y,z])$. Note that there is a fairly precise description of this group as an amalgamated product with $GL_2(\mathbb{C})$ as one of the components. This prompts the following question

**Problem 3 :** Is the subgroup $Aut(S_2 : Cl_2)$ of $GA_3(\mathbb{C})$ an amalgamated product with $GL_2(\mathbb{C})$ as one of the components ?

At a time, we had the following fairly optimistic procedure to find a counterexample to the Jacobian conjecture in three variables : consider the elements $A \in GL_2(\mathbb{C}[x,y,z])$ such that the Jacobian of the associated endomorphism of $S_2$ is invertible. Then (modulo the Jacobian conjecture) the endomorphism of $Cl_2$ given by sending $a_i$ to the $i$-th entry of $A.(a_i)_i$ should be an automorphism fixing the normalizing element $[a_1, a_2]$ and hence should induce an automorphism on the quotient $Cl_2/Cl_2[a_1,a_2] = \mathbb{C}[u,v]$ for which there exists a test by computing the Jacobian. At first sight there is not much relation between these two Jacobians but , due to lacking technical support , we were not able to construct interesting $A$'s to verify whether this aproach has any chance.

(4.4) : Of course, the general method can be used to provide weird automorphisms on arbitrary polynomial algebras. Let us consider, as an example, the case $m = 3$ and denote $a_{12} = u, a_{13} = v, a_{23} = w, a_{11} = x, a_{22} = y, a_{33} = z$ and consider an elementary matrix

$$A = \begin{pmatrix} 1 & 0 & 0 \\ 0 & 1 & 0 \\ f & 0 & 1 \end{pmatrix}$$

where $f \in \mathbb{C}[x,y,z,u,v,w]$. Then the induced endomorphism is given by

$$\begin{array}{ccc} x & \longrightarrow & x + 2fv + f^2 z \\ y & \longrightarrow & y \\ z & \longrightarrow & z \\ u & \longrightarrow & u + fw \\ v & \longrightarrow & v + fz \\ w & \longrightarrow & w \end{array}$$

Setting the Jacobian equal to one gives this time the condition

$$w\partial_u f + z\partial_v f + 2v\partial_x f = 0$$

which gives as possible solutions $f \in \mathbb{C}[uz - vw, v^2 - xz, y, z, w]$.For example, is the special case when $f = uz - vw$ a tame automorphism ?

(4.5) : It is about time to clarify what the above has to do with our original topic. If $T_m^0$ denotes the ring of $m$ generic 2 by 2 trace zero matrices, then there is a natural epimorphism

$$\pi_m : Cl_m \to T_m^0$$

sending $a_i$ to $X_i^0$ and $a_{ij}$ to $\frac{1}{2}T(X_i^0 X_j^0)$.This epimorphism is an isomorphism in case $m = 2$ or 3 and in general the kernel is the unique graded ideal of $Cl_m$ lying over the ideal of $S_m$ generated by the four by four minors of the generic symmetric matrix $(a_{ij})_{i,j}$. Note that this ideal is invariant for all automorphisms of the form described above, therefore any automorphism of $Cl_m$ corresponding to a matrix $A \in GL_m(S_m)$ induces an automorphism on $T_m^0$, thus providing a large class of interesting examples. In particular, the automorphism on $Cl_2 = T^0$ induced by the triangular matrix with $f = y^2 - xz$ is the restriction of $\sigma_2$ determined in the previous section to $T^0$.

## 5. Stable tameness and central tameness.

(5.1) : The following stable tameness result is due to M. Smith.

**Definition :** (i) Let $D$ be a derivation of $\mathbb{C}[x_1, ..., x_n]$. $D$ is said to be triangular if $D(x_i) = \phi_i(x_{i+1}, ..., x_n)$ for all $1 \leq i \leq n-1$ and $D(x_n) = 0$

(ii) $D$ is said to be locally nilpotent if for every $P \in \mathbb{C}[x_1, ..., x_n]$ we can find $n(P) \in \mathbb{N}^*$ such that $D^n(P) = 0$

**Theorem (M. Smith,[Sm2]) :** Let $D$ be a triangular,locally nilpotent derivation of $\mathbb{C}[x_1, ..., x_n]$ and $u \in Ker(D)$. Then, $uD$ is locally nilpotent and the automorphism $exp(uD)$ becomes tame when extended to $\mathbb{C}[x_1, ..., x_n, t]$ by fixing $t$.

(5.2) : Let $\sigma$ be an automorphism of $G$. Then, theorem 3.1 implies that $\sigma$ induces an automorphism of $G/[G,G]$ which has to be tame by the Jung-Van der Kulk result (2.1). Up to a tame automorphism of $G$, $\sigma$ can then be brought in the following form

$$\begin{array}{lcl} X & \to & X + t_1[X,Y] \\ Y & \to & Y + t_2[X,Y] \end{array}$$

where $t_1, t_2 \in T$. Unfortunately, we cannot continue the analysis in this generality. Restricting to automorphisms of the form

$$\begin{array}{lcl} X & \to & X + t[X,Y] \\ Y & \to & Y \end{array}$$

for $t \in T$, one can show the following lemma

**Lemma :** $t$ has to be of the form $(P_1 + P_2Y)[X,Y]$ with $P_i \in C$

**Proof:** First, remark that $\sigma([X,Y]) = \lambda[X,Y]$ for some $\lambda \in \mathbb{C}^*$.This implies the following equivalences

$[X,Y] + [t[X,Y],Y] = \lambda[X,Y]$

$:: [X^0,Y^0] + [t[X^0,Y^0],Y^0] = \lambda[X^0,Y^0]$

$:: [t[X^0,Y^0],Y^0] = (\lambda - 1)[X^0,Y^0]$

$:: [t,Y^0][X^0,Y^0] + t[[X^0,Y^0],Y^0] = (\lambda - 1)[X^0,Y^0]$

$:: [t,Y^0][X^0,Y^0] - 2tY^0[X^0,Y^0] = (\lambda - 1)[X^0,Y^0]$ (since $[[X^0,Y^0]Y^0] = -2Y^0[X^0,Y^0]$)

$:: [t,Y^0] - 2tY^0 = \lambda - 1$ (since $T$ is a domain)

Now, if we let $t = a + bX^0 + cY^0 + dX^0Y^0$ with $a,b,c,d \in C$ then this is equivalent to

$:: [bX^0,Y^0] + d[X^0Y^0,Y^0] - 2tY^0 = \lambda - 1$

$:: b(2X^0Y^0 - T(X^0Y^0)) + d(2X^0Y^0 - T(X^0Y^0))Y^0 - 2aY^0 - 2bX^0Y^0 + 2cD(Y^0) + 2dD(Y^0)X^0 = \lambda - 1$

which entails then that

$2cD(Y^0) = bT(X^0Y^0)$

$2a = -dT(X^0Y^0)$

By factoriality of $C$, there exists an element $\mu \in C$ such that

$b = 2\mu D(Y^0)$ , $c = \mu T(X^0Y^0)$

We then have that $t$ is equal to

$a + dX^0Y^0 + 2\mu D(Y^0)X^0 + \mu T(X^0Y^0)Y^0$

$= \frac{d}{2}(-T(X^0Y^0) + 2X^0Y^0) + \mu(-2{Y^0}^2X^0 + T(X^0Y^0)Y^0)$

$= \frac{d}{2}[X^0,Y^0] + \mu Y^0[X^0,Y^0] = (\frac{d}{2} + \mu Y^0)[X^0,Y^0]$

**Remark :** Unfortunately, not every $t$ of this form will give an automorphism as the following example shows : consider the endomorphism $\sigma$ of $G$ defined by

$$\begin{array}{ccc} X & \rightarrow & X + T(X)[X,Y]^2 \\ Y & \rightarrow & Y \end{array}$$

Then, $T(X)$ is send under the extension of $\sigma$ to $T$ to $T(X)(1 + 2[X,Y]^2)$ whereas $T(Y), D(X^0), D(Y^0)$ and $T(X^0Y^0)$ are fixed. But by computing the Jacobian on $C$ we see that this is not an automorphism. So, additional restrictions on $t$ are necessary. The following result summarizes the most general automorphisms we can construct and their behaviour on $C$

**Theorem :** (i) : All tame automorphisms of $G$ induce tame automorphisms on $C$

(ii) : All wild automorphisms $\sigma$ of $G$ of the form

$$\begin{array}{ccc} X & \rightarrow & X + (P_1 + P_2Y)[X,Y]^2 \\ Y & \rightarrow & Y \end{array}$$

where $P_1 \in \mathbb{C}\,[T(Y), D(X^0), D(Y^0), T(X^0Y^0)]$ and $P_2 \in \mathbb{C}\,[T(Y), D(Y^0)]$ induce tame automorphisms on $C$

**Proof :** (i) : This is an easy verification on the generators $\eta_f, \epsilon_c, \theta$ of the tame automorphisms given in 3.3

(ii) : Remark that $\sigma = \sigma_1 \circ \sigma_2$ where $\sigma_i$ fixes $Y$ and $\sigma_1(X) = X + P_1[X,Y]^2$ whereas $\sigma_2(X) = X + P_2Y[X,Y]^2$ and both $\sigma_1$ and $\sigma_2$ are automorphisms of $G$. We will first consider $\sigma_1$ :

$\sigma_1$ fixes $T(Y), D(X^0), D(Y^0)$ and $T(X^0Y^0)$ and it sends $T(X)$ to $T(X) + 2P_1[X,Y]^2$ , hence $\sigma_1 \mid C$ is a triangular automorphism.

On the other hand, $\sigma_2$ has the following action on $C$

$$\begin{array}{ccc} T(X) & \rightarrow & T(X) + T(Y)P_2[X^0,Y^0]^2 \\ T(Y) & \rightarrow & T(Y) \\ D(X^0) & \rightarrow & D(X^0) - T(X^0Y^0)P_2[X^0,Y^0]^2 + D(Y^0)(P_2[X^0,Y^0]^2)^2 \\ D(Y^0) & \rightarrow & D(Y^0) \\ T(X^0Y^0) & \rightarrow & T(X^0Y^0) - 2D(Y^0)P_2[X^0,Y^0]^2 \end{array}$$

Now, we define

$$\Delta = -2D(Y^0)\frac{\partial}{\partial T(X^0Y^0)} - T(X^0Y^0)\frac{\partial}{\partial D(X^0)}$$

Then $\Delta$ is a triangular, locally nilpotent derivation of $C$. Moreover, $\Delta(P_2[X^0,Y^0]^2) = P_2\Delta(T(X^0Y^0)^2 - 4D(X^0)D(Y^0)) = 0$. Now, put $C_1 = \mathbb{C}[T(Y), D(X^0), D(Y^0), T(X^0Y^0)]$ then $\sigma_2 \mid C_1 = exp(P_2[X^0,Y^0]^2\Delta)$. By M. Smith's stable tameness result in 5.1 , $\sigma_2 \mid C_1$ becomes tame when extended to $C$ by fixing $T(X)$. Composing this last automorphism of $C$ with the triangular one which sends $T(X)$ to $T(X) + T(Y)P_2[X^0,Y^0]^2$ and fixes the other variables one gets $\sigma \mid C$, finishing the proof of the tameness of $\sigma$

**References .**


[A] : J. Alev ; Un automorphisme non modéré de $U(g_3)$ , Comm.Alg. 14 (1986) no 8

[Ba] : H. Bass ; Clifford algebras and spinor norms over a commutative ring; Amer.J.Math.(1974)

[Be] : G. Bergman ; Wild automorphisms of free p.i. algebras and some new identities , preprint

[Co] : P.M. Cohn ; Free rings and their relations, 2nd edition , LMS monograph 2 Academic Press London (1985)

[Cz] : A.J. Czerniakiewitz ; Automorphisms of a free associative algebra of rank 2 II , Trans AMS 171 (1972) 309-315

[Di] : W. Dicks ; Contemp Math 43 (1985) 63-68

[Dix] : J. Dixmier ; Sur les algèbres de Weyl , Bull Soc Math France 96 (1968) 209-242

[FHL] : E, Formanek,P. Halpin,W.W. Li ; The Poincaré series of the ring of 2 by 2 generic matrices, J.Alg. 69 (1981) No 1

[Jo] : A. Joseph ; A wild automorphism of $U(sl(2,\mathbb{C}))$, Math Proc Cambr Phil Soc 80 (1976) 61-65

[Ju] : H.W.E. Jung, Einfuhrung in der theorie der algebraishen functionen zweier veränderlicher, Akadem Verlag Berlin (1951)

[LB] : L. Le Bruyn ; Trace rings of generic 2 by 2 matrices ; Mem AMS 363 (1987)

[Ma1] : L.G. Makar-Limanov ; On automorphisms of free algebra with two generators, Func Anal i ego prilizeniya 4 (1970) 107-8

[Ma2] : L.G. Makar-Limanov ; On automorphisms of Weyl algebras,Bull Soc Math France 112 (1984) 359-363

[Na] : M. Nagata ; On the automorphism group of $k[x,y]$,Lect Math Kyoto University (1972)

[Pr] : C. Procesi ; Rings with polynomial identities , Marcel Dekker (1973)

[Re] : R. Rentschler ; Opérations du groupe additif sur le plan affine , CRASc Paris 267 (1968) série A

[Sml] : M.K. Smith ; Automorphisms of enveloping algebras, Comm. Alg. 11 (1983) No 16

[Sm2] : M.K. Smith ; Stable tame automorphisms,preprint

[Va] : W. Van der Kulk ; On polynomial rings in two variables , Nieuw Archief voor Wiskunde 3 (1953) 33-41

# An Example in Central Division Algebras.

S. A. Amitsur

Hebrew University, Jerusalem, Israel

1. *Introduction:* Given r, n two relatively prime integers, and m be the minimal integer such that $r^m \equiv 1$ (mod n). We construct, for each $s = r^t$, $(t,m) = 1$ a central division algebra $A_s$ of dimension $m^2$ over a field of rational function $K(X)$ in a commutative indeterminate $X$, and $K$ a finite extension of the rationals $Q$. These algebras have peculiar properties: The ring of rational functions $A_s(\xi)$ over a commutative indeterminate $\xi$-are isomorphic as algebras, but belong to different Brauer classes over their center $K(X,\xi)$ (theorem 4.1). (b) The division algebras $A_s$ are non-isomorphic (as rings) except that $A_s \cong A_v$ if $sv \equiv 1$(mod m), yet for $m \neq 2$-they belong to different Brauer classes. (c) Nevertheless, there exist a chain of ring isomorphism maps into: (but not onto!) $A_r \to A_{r^{v_1}} \to \ldots \to A_{r^{v_t}} \to A_r$ where $\{v_i\}$ is a set of integers relatively prime to $m$. (Corollary 4.4). (d) By Tsen's theorem, the algebras $A_s$ have splitting fields of the form $L(X)$ where $L/K$ is an algebraic extension. (Theorem 5.1) For all these algebras there exists an algebraic extension $F \supset K$ such that $F(X)$ is a common maximal subfield of all $A_s$, and $L(X)$ splits $A_s$ if and only if $L \supseteq F$. Further relations between these divisions algebras are discussed (§5).

It seems that properties of this type are common to all generic abelian division algebras ([1]), and we hope to present it in a future paper.

2. *The algebras* $A_r = A(\sigma)$. $\mathbf{Q}$ be the field of rationals, and $\omega$ a fixed primitive n-th root of unity. The Galois group of the cyclotomic extension $\mathbf{Q}(\omega)/\mathbf{Q}$ is abelian and its automorphisms $\sigma$ are determined by maps $\sigma\colon \omega \to \omega^r$, where $(r,n) = 1$. The automorphism $\sigma$ is of order $m$ where $m$ is the minimal integer such that $r^m \equiv 1$(mod n). Let $K = \mathbf{Q}(\omega)^\sigma$ be the invariant field of $\sigma$. $\mathbf{Q}(\omega)/K$ is a cyclic extension of order $m$, whose group is generated by $\sigma$.



*F. van Oystaeyen and L. Le Bruyn (eds.), Perspectives in Ring Theory, 85–92.*

2.1 Let $R(\sigma) = \mathbf{Q}(\omega)[x;\sigma]$ be the ring on non-commutative polynomials in $x$ with coefficients in $\mathbf{Q}(\omega)$, defined by the relation: $xa = \sigma(a)x$ for every $a \in \mathbf{Q}(\omega)$. Denote by $A(\sigma)$ the ring of quotients of $R(\sigma)$, whose existence is well known. The following is a summary of known results:

**Theorem** : $A(\sigma)$ is a division algebra, and a cyclic crossed product $A(\sigma) = (\mathbf{Q}(\omega,X)/K(X),\sigma,X)$ where $X = x^m$. $A(\sigma)$ is of degree $m$ and exponent $m$ over the center $K(X)$.

A consequence of the theorem is that every element $q \in A(\sigma)$ can be written uniquely in the forms:

$$q = \sum_{j=0}^{m-1} p_j x^j, p_j \in \mathbf{Q}(\omega,X).$$

3. A *basic property of* $A(\sigma)$: Let $\tau$: $\omega \to \omega^s$ be another automorphism of $\mathbf{Q}(\omega)$; construct similarly the division algebra $A(\tau) = \mathbf{Q}(\omega)(y;\tau)$. We distinguish the non commutative indeterminate of $A(\tau)$ by using the letter $y$ i.e. $ya = \tau(a)y$ for $a \in \mathbf{Q}(\omega)$.

3.1 Let $\tau = \sigma^t$, with $(t,m) = 1$ and we then choose $s = r^t$. Then $\sigma,\tau$ generates the same group of automorphisms of $\mathbf{Q}(\omega)$, and in this case, $\mathbf{Q}(\omega)^\sigma = \mathbf{Q}(\omega)^\tau = K$, and $A(\tau)$ is also a central division algebra of dimension $m^2$ over a center $K(Y)$ with $Y = y^m$. We can identify $X$ with $Y$ and consider both algebras $A(\sigma),A(\tau)$ to be central division algebras over a common center $K(X)$, and so also $A(\tau) = (\mathbf{Q}(\omega,X)/K(X),\tau,X)$.

For these algebras, we prove:

**Theorem** : Let $s = r^t,(t,m) = 1, 1 \le t < m$. The algebras $A(\sigma),A(\tau)$ are isomorphic as rings (algebras over $\mathbf{Q}$), if and only if $t \equiv \pm 1 (\text{mod m})$. For $t = 1, A(\sigma) = A(\tau)$ and for $t \equiv -1 (\text{mod m})$, $A(\sigma)$ and $A(\tau)$ are isomorphic over $\mathbf{Q}$ but not over the center $K(X)$, more over they are anti-isomorphic over the center.

We begin with a lemma:

**Lemma** : Any automorphism of a subfield $L \subseteq \mathbf{Q}(\omega)$ can be extended to an automorphism of $A(\sigma)$ (over $\mathbf{Q}$).

Indeed, the Galois group of $\mathbf{Q}(\omega)/\mathbf{Q}$ is abelian and hence $L/\mathbf{Q}(\omega)$ is Galois and its automorphisms can be extended to an automorphism of $\mathbf{Q}(\omega)$. Thus, we can assume that we can start with $L = \mathbf{Q}(\omega)$, and with an automorphism $\rho$ of the whole cyclotomic field.

First we extend $\rho$ to the polynomial ring $\mathbf{Q}(\omega)[x;\sigma]$ by setting $\rho(x) = x$; that is, $\rho(\Sigma a_j x^j) = \Sigma\rho(a_j)x^j$. To prove that $\rho$ is an automorphism it suffices to show that $\rho(xa) = \rho(x)\rho(a)$, and in fact:

$$\rho(xa) = \rho(\sigma(a)x) = \rho\sigma(a)\cdot x = x\sigma^{-1}\rho\sigma(a) = x\rho(a) = \rho(x)\rho(a)$$

since $\rho\sigma = \sigma\rho$ as the Galois group is abelian. We then extend $\rho$ to the ring of quotient $A(\sigma)$ in the obvious way: $\rho(fg^{-1}) = \rho(f)\rho(g)^{-1}$.

3.2 Let $A(\sigma) = \mathbf{Q}(\omega)(x), A(\tau) = \mathbf{Q}(\omega)(y); x^m = y^m = X$ be isomorphic rings, by the isomorphism $\varphi: A(\sigma) \rightarrow A(\tau)$.

The fields $\mathbf{Q}(\omega)$ and $\varphi(\mathbf{Q}(\omega))$ are isomorphic subfields in $A(\sigma)$, and $\varphi$ induces also an automorphism of the center $K(X)$. The field $K$ is algebraically closed in $K(X)$, hence $\varphi$ induces an automorphism $\rho$ of $K$ (over $\mathbf{Q}$). Apply the lemma to extend $\rho$ to $A(\sigma)$, and replace $\varphi$ by $\rho^{-1}\varphi$. This allows us to assume that $\varphi$ leaves the elements of $K$ invariant. But then $\omega$ and $\varphi(\omega)$, as elements of $A(\sigma)$ satisfy the same irreducible equation $g[\lambda]$ over $K$, but as such they are conjugate in $A(\sigma)$ (e.g. by the Skolem- Noether theorem). That is, there exists $d \in A(\sigma)$ satisfying $\omega = d\varphi(\omega)d^{-1}$. Let $I_d$ be the inner automorphism of $A(\sigma)$ determined by d. Next we replace $\varphi$ by $I_d\varphi$, which enable us to assume that $\varphi$ leaves the elements of $\mathbf{Q}(\omega)$ invariant.

3.3 Next we determine the form of the element $\varphi(y)$: By (2.1)

$$\varphi(y) = \sum_{j=0}^{m=1} p_j x^j, \; p_j \in \mathbf{Q}(\omega)(X)$$

From the relation in $A(\tau)$: $ya = \tau(a)y$ for $a \in Q(\omega)$, we obtain

$$\varphi(ya) = \varphi(y)a = \sum_{j=0}^{m-1} p_j x^j \cdot a = \sum_{j=0}^{m-1} p_j \sigma^j(a) x^j$$

which must be equal to:

$$\varphi(\tau(a)y) = \tau(a)\cdot\varphi(y) = \tau(a)\Sigma p_j x^j$$

The uniqueness of representation of (2.1), shows that for $j = 0,1, \ldots, m-1$, $p_j(\tau(a)-\sigma^i(a)) = 0$. Now, if $\tau \neq \sigma^j$ we can find $a \in \mathbf{Q}(\omega)$, such that $\sigma^j(a)-\tau(a) \neq 0$, and therefore $p_j = 0$. Thus, if $\tau = \sigma^t$, we obtain $\varphi(y) = px^t$, $p \in \mathbf{Q}(\omega)(X)$.

3.4 Finally, $\varphi$ induces an automorphism of $K(X)$, which is the center of both $A(\sigma)$ and $A(\tau)$; hence it follows by Luroth theorem that $\varphi(X) = (aX+b)/(cX+d)$ with $ad-bc \neq 0$. But then:

$$\varphi(X) = \varphi(y^m) = (px^t)^m = p\tau(p)\tau^2(p)\cdots\tau^{m-1}(p)x^{tm},$$

where $\tau = \sigma^t$ denotes here the extension of $\tau$ to the rational function field $\mathbf{Q}(\omega)(X)$ which induce the original $\tau$ on $\mathbf{Q}(\omega)$ and $\tau(X) = X$. Let $p = P[X]/Q[X]$ be the representation of p as a quotient of two relative prime polynomials in $X$ with coefficients in $\mathbf{Q}(\omega)$.

We obtain from the previous equation:

$$(aX+b)Q[X](\tau Q[X])\cdots(\tau^{m-1}Q[X]) = (cX+d)P[X](\tau P[X])\cdots(\tau^{m-1}P[X])X^t$$

All polynomials $\tau^j Q[X]$ have the same degree, and similarly the polynomials $\tau^j P[X]$. Thus,

$$\deg(aX+G)+m\deg Q = t+\deg(cX+d)+m\deg p$$

which leads to:

$$t+\deg(cX+d) \equiv \deg(aX+\ell)(mod\ m)$$

Now $0 < t < m$, and the linear factors are of deg 0 or 1, hence one of the linear factors is of degree 0 and the other of degree 1, so either $t+1 \equiv 0$(mod m) i.e. $t = m-1$, or $t \equiv 1$(mod m) i.e. $t = 1$.

Conversely, if $t = 1$ then $r = s$ and so $A(\sigma) = A(\tau)$. If $t = m-1$, then $\tau = \sigma^{-1}$, and $A(\sigma) = \mathbf{Q}(\omega)(x) \cong A(\tau) = \mathbf{Q}(\omega)(y)$ by setting $\varphi(y) = x^{-1}$. The proof is a straightforward computation.

Finally, if $\sigma = \tau^{-1}$, then the map $\psi: A(\sigma) \to A(\tau)$ which is the identity on $\mathbf{Q}(\omega)$, and $\psi(y)=x$, will be an anti-isomorphism (over the center), and generally define: $\psi(\Sigma a_i y^i) = \Sigma x^i a_i$. The proof follows easily from the basic relation:

$$\psi(ya)=\psi(\sigma^{-1}(a)y) = x\sigma^{-1}(a) = ax=\psi(a)\psi(y).$$

4. *The algebras* $A(\sigma)(\xi)$. Let $\xi$ be a commutative indeterminate over all $A(\tau), \tau = \sigma^s$, $s = r^t$ and let $A(\tau)(\xi)$ be the ring of all rational functions in $\xi$ with coefficients in $A(\tau)$. In contrast to theorem 3.1, we have the following result:

4.1 **Theorem** : All the division algebras, $A(\tau)(\xi)$ - are isomorphic rings; but they belong to different Brauer classes as central division algebra over the common center $K(X,\xi)$.

**Proof** : If $s = r^t, (t,m) = 1$, then choose integer $j,k$ such that $jt-km = 1$. The integral matrix $P = \begin{pmatrix} t & k \\ m & j \end{pmatrix}$ is then invertible in $GL_2(\mathbf{Z})$, with $P^{-1} = \begin{pmatrix} j & -k \\ -m & t \end{pmatrix}$.

With the aid of P we define an automorphism $\varphi: A(\sigma)(\xi) \to A(\tau)(\xi)$. First we define an isomorphism $\varphi: \mathbf{Q}(\omega)[x^{\pm 1}, \xi^{\pm 1}] \to \mathbf{Q}(\omega)[y^{\pm 1}, \xi^{\pm 1}]$, by using the rows of $P^{-1}$ and setting

$$\varphi(x) = y^j \xi^{-k} \quad ; \varphi(\xi) = y^{-m} \xi^t \quad ; \varphi(a) = a \text{ for } a \in \mathbf{Q}(\omega)$$

and generally $\varphi(\Sigma a_{\lambda\mu} x^\lambda \xi^\mu) = \Sigma a_{\lambda\mu} \varphi(x)^\lambda \varphi(\xi)^\mu$.

One easily verifies, by using the rows of P to obtain an inverse map $\Psi$:

$$\Psi(y) = x^t \xi^k \quad , \Psi(\xi) = x^m \xi^j, \quad \Psi(a) = a \quad \text{for } a \in \mathbf{Q}(\omega)$$

Indeed,

$$\varphi\Psi(y) = \varphi(x)^t \varphi(\xi)^k = (y^t \xi^k)^j (y^m \varphi^j)^{-k} = y^{jt-mk} \xi^0 = y$$

$$\varphi\Psi(\xi) = \varphi(x)^m \varphi(\xi)^j = (y^j \xi^{-k})^m (y^{-m} \xi^t)^j = \xi^{jt-km} = \xi$$

and the rest is immediate.

To prove that $\varphi$ is an isomorphism, it is sufficient to show first that $\varphi$ satisfies $\varphi(xa) = \varphi(x)\varphi(a)$ for $a \in \mathbf{Q}(\omega)$. First note that $\sigma^t = \tau$, implies that $\tau = \sigma^j$ and, hence:

$$\varphi(xa) = \varphi(\sigma(a)x) = \sigma(a)\varphi(x) = \sigma(a) y^j \xi^{-k} =$$
$$= \tau^j(a) y^j \xi^{-k} = y^j a \xi^{-k} = y^j \xi^{-k} a = \varphi(x)\varphi(a)$$

since $ya = \tau(a)y\xi$ belongs to the center, and $\varphi(a) = a$.

And also,

$$\varphi(\xi a) = \varphi(a\xi) = a\varphi(\xi) = ay^{-m}\xi^t = y^{-m}\xi^t a = \varphi(\xi)\varphi(a)$$

since $y^m, \xi$ belong to the center, and finally it is clear that $\varphi(\xi y) = \varphi(\xi)\varphi(y)$, etc.

4.2 The algebras $A(\tau)(\xi), \tau = \sigma^s, s = r^t$ are in difference Brauer classes. Indeed we proved in section (5.2) that $A(\tau)(\xi)$ is in the Brauer class of $A(\sigma)(\xi)^j$, where $jt \equiv 1$ (mod m); on the other hand, $A(\sigma)(\xi)$ generates in the Brauer group a element of order m by theorem 2.1, and hence $A(\sigma)(\xi)^j \sim A(\tau)(\xi)^{j'}$ if and only if $j \equiv j'$ (mod m), which yields $j = j'$ for integers $<m$, and this shows that all $A(\tau)(\xi)$ are in different Brauer classes

Clearly

4.3 The isomorphism $\varphi: A(\sigma)(\xi) \to A(\tau)(\xi)$ induce an automorphism $\varphi_0$ on the center $K(X,\xi)$, hence we have an interesting consequence of our theorem:

**Corollary** : $\varphi_0$ cannot be extended to an automorphism of $A(\sigma)$. Indeed, if this is the case and $\overline{\varphi}$ is an extension to $A(\sigma)$ then $\varphi\overline{\varphi}^{-1}: A(\sigma)(\xi) \to A(\tau)(\xi)$ will be an isomorphism over the center, which will imply that $A(\sigma)(\xi)$ and $A(\tau)(\xi)$ belong to the same Brauer class. Contradiction!

5 *Properties of* $A(\tau)$. There are various relations between the different rings $A(\tau)$, two of them are given in the following theorem:

5.1 **Theorem** : Let $\tau = \sigma^t$, $jt \equiv 1$(mod m), then:

1) $A(\tau)$ is equivalent to $A(\sigma)^j$ in the Brauer group of $K(X)$.

2) $A(\tau) \cong A(\sigma) \otimes K(X^{\frac{1}{j}})$ as algebras over $\mathbf{Q}$.

**Proof** : Note that $\tau = \sigma^t$, yields $\sigma = \tau^j$, where $\sigma, \tau$ are the automorphism of $\mathbf{Q}(\omega)$, used to define $A(\sigma)$ and $A(\tau)$, respectively. We point also that in $A(\tau)$ we have the relation: $y^j a = \tau^j(a) y^j = \sigma(a) y^j$. Hence,

$$A(\tau) = \mathbf{Q}(\omega, y) = \mathbf{Q}(\omega, y^j, X) = \mathbf{Q}(\omega, y^j) \underset{K(X^j)}{\otimes} K(X)$$

since $X = y^m$, and so $y = (y^j)^t X^{-k} \in \mathbf{Q}(\omega, y^j, X)$, which proves the second equality. The third equality is a consequence of the fact that $(y^j)^m = X^j$, and $K(X^j)$ is the center of $\mathbf{Q}(\omega, y^i)$. Finally we note that $K(X) \cong K(X^{\frac{1}{j}})$ and $\mathbf{Q}(\omega, y^j) \cong \mathbf{Q}(\omega, x) = A(\sigma)$, and thus the map: $y^j \to x$ and

$X \to X^{\frac{1}{j}}$ yield the isomorphism $A(\tau) \cong A(\sigma) \underset{K(X)}{\otimes} K(X^{\frac{1}{j}})$ given in (2).

To prove (1) we use the same representation: $A(\tau) = \mathbf{Q}(\omega, y^j, X)$ and we note that since $(y^j)^m = X^j$, the algebra can be written as a crossed product $(\mathbf{Q}(\omega,X))/K(\omega,X), \sigma, X^j)$ and so it is equivalent to the power $(\mathbf{Q}(\omega,X)/K(\omega,X),\sigma,X)^j = A(\sigma)^j$, which proves (2).

5.2 **Remark** : The preceding proof will hold also if we replace $\mathbf{Q}(\omega)$ by any composite L $\mathbf{Q}(\omega)$ where L and $\mathbf{Q}(\omega)$ are linearly disjoint over K, and $\tau, \sigma$ are extended to act as the identity on L. In particular if $L = K(\xi)$, we get the completion of the proof of 3.3.

5.3 Although the algebras $A(\tau)$ are non isomorphic, there exist isomorphic maps of each into the other:

**Theorem** : Let $\tau = \sigma^{\nu}, \rho = \sigma^{\mu}$, and $\nu, \mu$ relatively prime to m, then there exists an isomorphism $\varphi = \varphi_{\rho,\tau}: A(\tau) \to A(\rho)$ of $A(\tau)$ into (!) $A(\rho)$.

**Proof** : Let $A(\tau) = \mathbf{Q}(\omega, y)$ with the relation $ya = \tau(a)y$, $a \in \mathbf{Q}(\omega)$ and $y^m = X$, and let $A(\rho) = \mathbf{Q}(y,z)$, with $za = \rho(a)z$. Since $(\nu, m) = 1$, the automorphism $\tau$ generates the same group as $\sigma$, hence there exists and integer $\lambda$ with $\rho = \tau^{\lambda}$. We then define $\varphi$ to be the identity on $\mathbf{Q}(\omega)$ and $\varphi(z) = y^{\lambda}$, and $\varphi$ extended to the ring of polynomial $\mathbf{Q}(\omega)[z^{\pm 1}]$ will be an isomorphism, since

$$\varphi(za) = \varphi(\rho(a)z) = \rho(a)\varphi(z) = \rho(a)y^{\lambda} = \tau^{\lambda}(a)y^{\lambda} = y^{\lambda}a = \varphi(z)\varphi(a).$$

Its extension to the ring of quotients $\mathbf{Q}(y,z)$ yields the required map.

4.4 **Corollary** . Let $m-1 = \nu_{\ell} > \nu_{\ell-1} > \cdots > \nu_1 = 1$ be the set of all relatively prime integers to m arranged in a decreasing order, then we have a circle of automorphisms (into !):

$$A(\sigma) \to A(\sigma^{\nu_2}) \to \ldots \to A(\sigma^{\nu_{\ell-1}}) \to A(\sigma^{\nu_\ell}) \to A(\sigma)$$

Indeed, by the previous result choose $\varphi: A(\sigma^{\nu_i}) \to A(\sigma^{\nu_{i+1}})$, and the last isomorphism by theorem 3.1 since $\nu_\ell = m-1 \equiv -1 \pmod{m}$, or we can take the map $\varphi: A(\sigma^{\nu_{\ell-1}}) \to A(\sigma)$.

6. *Algebraic splitting fields.* All the algebras $A(\tau)$, have a center $K(X)$ of trancendence degree 1 over K, hence by Tsen's theorem ([2] p.375), $A(\tau)$ has splitting fields of the form $L(x)$ where $L/K$ is an

algebraic extension. Actually, one of them - is $\mathbf{Q}(\omega)(X)$ which is a maximal subfield, and hence every $L \supseteq \mathbf{Q}(\omega)$, will be a splitting field $L(X)$.

Surprising is the converse:

6.1 **Theorem** $L(X)$ splits $A(\tau)$ if and only if $L \supseteq \mathbf{Q}(\omega)$. Indeed, we use the representation of $A(\tau)$ as a crossed product, and the tensor product formula (e.g. [2] p. 273):

$A(\tau) \otimes L(X) = (\mathbf{Q}(\omega, X)/K(X), \tau, X) \sim (\frac{L\mathbf{Q}(\omega)(X)}{[L \cap \mathbf{Q}(\omega)](X)}, \tau^t, X)$ where $(\tau^t)$ generates the Galois group of the composite $L(X)\mathbf{Q}(\omega)(X) = L(\omega)(X)$ over the intersection $L(X) \cap \mathbf{Q}(\omega)(X) = (L \cap \mathbf{Q}(\omega))(X)$. The right side is again a ring of quotients of polynomial ring, and hence a division algebra. Now, $L(X)$ splits $A(\tau)$ if and only if the latter is commutative, and this happens only if $\tau^t =$ identity; namely if and only if $(L \cap \mathbf{Q}(\omega))(X) = L\mathbf{Q}(\omega)(X)$ which is equivalent to $L \cap \mathbf{Q}(\omega) = L\mathbf{Q}(\omega)$ i.e. $L \supseteq \mathbf{Q}(\omega)$.

7. **Remark** : (1) One can replace $\mathbf{Q}(\omega)K$ by any cyclic extension $F/E$ of a field E, but then one has to assume in theorem 3.1, that the isomorphism $A(\sigma) \cong A(\tau)$ are algebra isomorphism over the field E.

(2) The algebras $A_r = A(\sigma)$ are a special case of a generic abelian division algebras, where the defining group is cyclic ([1]). It seems that properties of this type is common to all generic abelian division algebra, which we hope to present in another paper.

## References


[1] S. A. Amitsur, D. Saltman, *"Generic abelian crossed products and p-algebras"*. Journal of Algebra 51 (1978) pp. 76-87.

[2] Richard S. Pierce, *Associative Algebras*. Springer-Verlag 1982

COHOMOLOGIE LOCALE ET COHOMOLOGIE D'ALGEBRES DE LIE

Geneviève Barou
Université de Caen
14000 Caen

Marie-Paule Malliavin
Université Pierre et Marie Curie
4, Place Jussieu
75005 Paris

1. Introduction.

Bien que de nombreux résultats rappelés ici sont valables sous des hypothèses plus faibles, nous supposerons toujours que les anneaux considérés sont Noéthériens à droite et à gauche.

Si R est un anneau premier, un R-module à gauche M est dit sans-torsion si la multiplication de M dans lui même définie par un élément régulier de R est injective. On note, si P est un idéal premier de R, par $\mathscr{C}(P)$ l'ensemble des éléments de R réguliers modulo P. Si P et Q sont deux idéaux premiers d'un anneau R, on dit qu'il existe un lien de P vers Q (notation : $P \rightsquigarrow Q$), s'il existe un idéal bilatère A de R tel que $PQ \subsetneqq A \subseteq P \cap Q$ et tel que $(P \cap Q)/A$ est sans torsion comme $R/_P$-module à gauche et $R/_Q$-module à droite On définit ainsi un graphe orienté $\mathfrak{L}$ dont les sommets sont les idéaux premiers de R et dont les arêtes sont les liens [11], [14]. Une clique de R est une composante connexe de ce graphe.

En ce qui concerne la structure du graphe orienté $\mathfrak{L}$, J.T. Stafford [15] établit que si P est idéal premier d'un anneau R, si n est un entier $\geqslant 1$, l'ensemble des $Q \in \operatorname{Spec}(R)$ tels que $P \rightsquigarrow Q$ et pour lesquels le rang de Goldie de $R/_Q$ est majoré par n, est fini.



*F. van Oystaeyen and L. Le Bruyn (eds.), Perspectives in Ring Theory, 93–104.*

En particulier, si les rangs de Goldie des idéaux premiers de R sont uniformément bornés, chaque clique de R est un graphe localement fini. Les anneaux premiers héréditaires n'admettent que des cliques finies et une description complète en est donné en (Th. A-1.15 [11]). D'autre part, [8], K. Brown et F du Cloux se sont attaqué à la description de la clique d'un idéal primitif de l'algèbre enveloppante d'une algèbre de Lie $\mathscr{G}$ résoluble de dimension finie sur $\mathbb{C}$, description faisant intervenir les poids de $\mathscr{G}$ et de certaines de ses sous-algèbres.

La notion de clique permet de développer une théorie raisonnable de la localisation pour certains anneaux. Dans ce but, A.V. Jategaonkar [11] introduit la notion importante suivante; si un anneau R satisfait <u>la condition de second niveau à gauche forte</u> alors pour tout R-module à gauche M, de type fini et uniforme, dont l'annulateur est un idéal premier P, l'annulateur de chaque sous-module non nul de M coïncide avec P. Cette condition est la traduction en non commutatif du fait suivant : si A est un anneau (commutatif) et M un A-module de type fini, le support de M coïncide avec l'ensemble des idéaux premiers de A qui contiennent l'annulateur de M.

Des exemples d'anneaux qui satisfont à la fois la condition de second niveau forte et la finitude locale du graphe $\mathfrak{L}$ sont les anneaux de groupes presque-polycycliques et les algèbres de Lie résolubles de dimension finie sur un corps algébriquement clos, les anneaux noethériens complètement bornés et certains anneaux d'opérateurs différentiels plus ou moins tordus ([2] et [3]).

Nous montrons ici que si $\mathscr{G}$ est une algèbre de Lie résoluble sur un corps algébriquement clos de caractéristique nulle, si P est un idéal premier cofini, i.e. si $U(\mathscr{G})/_P$ est une représentation de dimension 1 de $\mathscr{G}$, si $\mathcal{E}(P)$ désigne la famille des idéaux bilatères de $U(\mathscr{G})$ qui contiennent un produit fini d'éléments de

$C\ell(P)$, on peut définir [1] de foncteur de cohomologie locale à support dans $\mathcal{E}(P)$ et nous montrons que cette cohomologie s'interprête comme une cohomologie $\mathcal{G}$-finie définie en [10] (cf. §.2).

Soit $\mathcal{G}$ une algèbre de Lie résoluble de dimension finie sur le corps $\mathbb{C}$ (ou n'importe quel corps algébriquement clos de caractéristique zéro), $U = U(\mathcal{G})$ l'algèbre enveloppante de $\mathcal{G}$, c'est un anneau noethérien (à droite et à gauche) dont tout idéal premier est complètement premier. On note Spec U le spectre premier de U. Il existe un lien entre P et Q, ($P \rightsquigarrow Q$) ($P,Q \in$ Spec U) si et seulement si $\ell.\mathrm{Ann}_U (P\cap Q/_{PQ}) = P$ et $r.\mathrm{Ann}_U (P\cap Q/_{PQ}) = Q$ avec la convention que $(0) \rightsquigarrow (0)$ est la seule arête arrivant ou partant de (0) et en notant $\ell\,\mathrm{Ann}_U$ (resp. $r.\mathrm{Ann}_U$) l'annulateur à gauche (resp. à droite) d'un module. On a toujours $P \rightsquigarrow P$, pour tout $P \in$ Spec $U(\mathcal{G})$ comme le montre un raisonnement par récurrence sur la dimension de $\mathcal{G}$ et le fait que les cliques de $U(\mathcal{G})$ sont localisables ([6]). Cela signifie en particulier que, si $C\ell(P)$ désigne la clique de P, l'ensemble $\mathcal{S} = \cup\{U\setminus Q \mid Q \in C\ell(P)\}$ est une partie de Ore à droite et à gauche dans $U(\mathcal{G})$.

Un exemple bien connu de clique non réduit à un point est le suivant ; soit $\mathcal{G} = \mathbb{C}x \oplus \mathbb{C}y$ l'algèbre de Lie résoluble de dimension 2, $[x,y] = y$ alors le graphe des idéaux premiers de $U(\mathcal{G})$ est formé de $(0) \rightsquigarrow (0)$, $(y) \rightsquigarrow (y)$ et, pour chaque $\lambda \in \mathbb{C}$, la clique de $P_\lambda = (y, x-\lambda)$ est :

$$\cdots \rightsquigarrow \underset{\circlearrowleft}{P_{\lambda-1}} \rightsquigarrow \underset{\circlearrowleft}{P_{\lambda}} \rightsquigarrow \underset{\circlearrowleft}{P_{\lambda+1}} \rightsquigarrow \cdots$$

On remarque que, dans cet exemple, si l'on fait abstraction des boucles $P_\lambda \circlearrowleft P_\lambda$ , la clique de $P_\lambda$ est un arbre. Cependant une clique n'est pas toujours un arbre; en effet si $\mathcal{G}$ est l'algèbre de Lie de dimension 3 engendrée par x, y, t avec les relations $[x,y] = y$ $[x,t] = -t$ $[t,y] = 0$, et si $\mathcal{M} = \mathcal{G}U(\mathcal{G})$ est l'idéal d'augmentation on a: $\mathcal{M} \rightsquigarrow (x+1,y,t) \rightsquigarrow \mathcal{M}$ . ([6]).

Nous suivons les notations de [4] et [8].

1.0 On pose $\Lambda = (\mathcal{G}/_{[\mathcal{G},\mathcal{G}]})^\star = \mathrm{Hom}_{\mathbb{C}}(\mathcal{G}/_{[\mathcal{G},\mathcal{G}]}, \mathbb{C})$, c'est l'espace vectoriel des caractères de $\mathcal{G}$.
Si $\lambda \in \Lambda$, on note $\tau_\lambda : \mathcal{G} \to U(\mathcal{G})$ l'application définie par $\tau_\lambda(X) = X - \lambda(X)$ , $X \in \mathcal{G}$ ; $\tau_\lambda$ se prolonge en un automorphisme d'algèbre de U. Si M un $\mathcal{G}$-module de dimension finie, un <u>poids</u> de M est un caractère $\lambda \in (\mathcal{G}/_{[\mathcal{G},\mathcal{G}]})^\star$ donnant l'action de $\mathcal{G}$ dans un sous-quotient simple N de M , i.e. $N = N^\lambda = \{x \in N$ $X.x = \lambda(X)x$ pour tout $X \in \mathcal{G}\}$. En particulier les <u>racines de</u> $\mathcal{G}$ sont les poids de $\mathcal{G}$ qui apparaissent dans les sous-quotients simples de $\mathcal{G}$ pour la représentation adjointe.
On note $\Gamma_{\mathcal{G}}(=\Gamma)$ le sous-groupe de $\Lambda$ engendré par les racines de $\mathcal{G}$ et par $\Sigma_{\mathcal{G}}(=\Sigma)$ l'ensemble des combinaisons linéaires à coefficients entiers positifs des racines de $\mathcal{G}$ . On a donc $\Sigma \subset \Gamma \subset \Lambda$.
Rappelons que ([6]) : <u>Si</u> P <u>et</u> $Q \in \mathrm{Spec}\, U$ <u>et si</u> $P \rightsquigarrow Q$, <u>il existe</u> $\lambda \in -\Sigma$ <u>tel que</u> $Q = \tau_\lambda(P)$. Il en résulte alors que $U/_P \simeq U/_Q$ dans l'isomorphisme d'algèbre induit par $\tau_\lambda$.

§.2 <u>Cohomologie locale</u>.

2.0 On notera $\mathcal{E}(P)$ la famille des idéaux bilatères de $U(\mathcal{G})$ qui contiennent un produit fini d'éléments de $C\ell(P)$ et pour tout U-module (à gauche) $M$ on posera $H^{o}_{C\ell(P)}(M) = \varinjlim_{I\in\mathcal{E}(P)} \mathrm{Hom}_U(U/_I, M)$ ; $H^{o}_{C\ell(P)}(-)$ est un foncteur covariant exact à gauche dont les dérivées sont notés $H^{i}_{C\ell(P)}(M)$.

Nous supposerons que $P$ est cofini ; alors il existe ([9]) $f\in\Lambda = (\mathcal{G}/_{[\mathcal{G},\mathcal{G}]})^{\star}$ tel que $P = \tau_f(\mathcal{M})$ où $\mathcal{M} = \mathcal{G}U(\mathcal{G})$ est l'idéal d'augmentation et les cliques de $P$ et de $\mathcal{M}$ sont isomorphes (en tant que graphe) par l'automorphisme $\tau_f$ .

2.1 Rappelons le résultat suivant de K.A. Brown [6] .

Lemme. La clique de $\mathcal{M}$ est en correspondance bijective avec $\Gamma = \Gamma_{\mathcal{G}}$ Plus précisément si $\lambda\in\Gamma$, l'idéal primitif associe à $-\lambda$ par [9] et noté $I(-\lambda)$, appartient à $C\ell(\mathcal{M})$ et si $I(-\lambda) = I(-\mu)$ $\lambda$ et $\lambda\in\Gamma$, on a $\lambda = \mu$.

2.2 Une famille $\mathcal{F}$ d'idéaux bilatères de $U(\mathcal{G})$ est dite d'Artin Rees [1] si elle vérifie les conditions suivantes

(i) Si $I\in\mathcal{F}$ et si $I\subseteq J$ où $J$ est un idéal bilatère de $U$ alors $J\in\mathcal{F}$

(ii) Le produit de deux idéaux de $\mathcal{F}$ appartient à $\mathcal{F}$.

(iii) Pour tout U-module de type fini $N$, tout sous-module $M$ et $N$ et tout $I\in\mathcal{F}$, il existe $J$ dans $\mathcal{F}$ tel que $JN\cap M\subseteq IM$. Cette dernière condition étant équivalente à la même condition où l'on remplace $N$ par $U$ et $M$ par un idéal à gauche de $U$.

2.3 On notera $\underline{\mathrm{Mod}}_{\mathcal{G}}$ la catégorie des $\mathcal{G}$-modules à gauche. Un $\mathcal{G}$-module $M$ est localement fini si pour tout $x\in M$, $\dim_{\mathbb{C}} U(\mathcal{G})x$ est fini. On note [10] $\mathrm{Mod}_{(\mathcal{G},\Gamma)}$ la sous-catégorie pleine de $\underline{\mathrm{Mod}}_{\mathcal{G}}$ formée des $\mathcal{G}$-modules localement finis dont tous les poids appartiennent à $\Gamma$.

Lemme. 1) Si $M\in\underline{\mathrm{Mod}}_{(\mathcal{G},\Gamma)}$ et si $0\to M\to E$ est une extension essentielle de $M$ dans $\underline{\mathrm{Mod}}_{\mathcal{G}}$ alors $E\in\underline{\mathrm{Mod}}_{(\mathcal{G},\Gamma)}$

2) Si $M\in\underline{\mathrm{Mod}}_{(\mathcal{G},\Gamma)}$ et si $E$ est l'enveloppe invective de $M$ dans $\underline{\mathrm{Mod}}_{\mathcal{G}}$ alors $E\in\underline{\mathrm{Mod}}_{(\mathcal{G},\Gamma)}$ .

Preuve. 1) Soit $x \in E$, $x \neq 0$ ; il existe $a \in U(\mathscr{G})$, $0 \neq ax \in M$. Donc $U(\mathscr{G})ax$ est de dimension finie et les poids qui apparaissent dans $U(\mathscr{G})ax$ sont dans $\Gamma$. On est ramené au cas où $M = U(\mathscr{G})ax$ est essentiel dans $U(\mathscr{G})x = E$. Soit $P$ un idéal premier cofini tel que $Py = 0$, $y \in E, y \neq 0$ ; alors $Pby = 0$, avec $0 \neq by \in M$, $b \in U$. Donc $P$ correspond à un $\lambda \in \Gamma$. La partie 2°) résulte facilement de 1°).

2.4 Fokko du Cloux [10] définit un foncteur covariant exact à gauche $M \Rightarrow M^{(\mathscr{G},\Gamma)}$ de la catégorie $\mathrm{Mod}_{\mathscr{G}}$ dans $\mathrm{Mod}_{(\mathscr{G},\Gamma)}$. En posant $M^{(\mathscr{G},\Gamma)} = \{x \in M,\ U(\mathscr{G})x \in \mathrm{Mod}_{(\mathscr{G},\Gamma)}\}$. On note $H^i_{(\mathscr{G},\Gamma)}$ les dérivées de $(-)^{(\mathscr{G},\Gamma)}$. Si $E$ est un $\mathscr{G}$-module injectif alors $E^{(\mathscr{G},\Gamma)}$ est injectif [10].
On note $R(\mathscr{G},\Gamma)$ le $\mathscr{G}$-module $\mathrm{Hom}_{\mathbb{C}}(U,\mathbb{C})^{(\mathscr{G},\Gamma)}$.

2.5 Proposition. 1°) La famille $\mathscr{E}(\mathscr{G})$ des idéaux de $U(\mathscr{G})$ qui contiennent un produit fini d'éléments de $C\ell(\mathscr{M})$ est une famille d'Artin Rees.
2°) Pour tout $U$-module $M$ et tout $i \geq 0$ on a $H^i_{C\ell(\mathscr{M})}(M) = H^i_{(\mathscr{G},\Gamma)}(M) = \varinjlim_{I \in \mathscr{E}(\mathscr{M})} \mathrm{Ext}^i_U(U/_I, M)$.
3°) Le $\mathscr{G}$-module $R(\mathscr{G},\Gamma)$ est isomorphe à la somme directe

$\bigoplus_{\lambda \in \Gamma} E(U/_{\tau_\lambda(\mathscr{M})})$ est le dernier terme de la résolution injective minimale du $\mathscr{S}^{-1}U$ où $\mathscr{S} = \cap\{\mathscr{C}(Q), Q \in C\ell(\mathscr{M})\}$.

Preuve. 2°) Montrons que $H^o_{C\ell(\mathscr{M})}(M)$ est contenu dans $H^o_{(\mathscr{G},\Gamma)} = M^{(\mathscr{G},\Gamma)}$. Soit $x \in M$ tel que $\tau_{\lambda_1}(\mathscr{M})\ldots\tau_{\lambda_k}(\mathscr{M})x = 0$, pour $\lambda_i \in \Gamma$, $i=1,\ldots,k$, c'est-à-dire $\tau_{\lambda_1}(\mathscr{M})\ldots\tau_{\lambda_k}(\mathscr{M})$ est contenu dans l'annulateur à gauche de $x$. Comme tous les $\tau_{\lambda_i}(\mathscr{M})$ sont cofinis, il en est de même de leur produit, donc $\dim U(\mathscr{G})\,x < \infty$. Pour montrer que $U(\mathscr{G})x \in M^{(\mathscr{G},\Gamma)}$ il suffit de vérifier que les poids de $\mathscr{G}$ dans $U/_I$, où $I$ est un idéal à gauche de $U$ contenant $\tau_{\lambda_1}(\mathscr{M})\ldots\tau_{\lambda_k}(\mathscr{M})$, sont contenus dans l'ensemble $\{\lambda_1,\ldots,\lambda_k\}$ et pour cela, il suffit de vérifier que les poids de

$N = (\tau_{\lambda_{i-1}}(\mathcal{M})\ldots\tau_{\lambda_k}(\mathcal{M})/\tau_{\lambda_i}(\mathcal{M})\ldots\tau_{\lambda_k}(\mathcal{M})$ sont $\lambda_i$ ou $0$.
Supposons que $N = J/\tau_{\lambda_i}(\mathcal{M})J \neq (0)$ et posons $N^{\lambda_i} =$
$= \{y \in N \quad X.y = \lambda_i(X)y$ pour tout $X \in \mathcal{G}\}$ ; il est clair que $N = N^{\lambda_i}$. Réciproquement soit $x \in M$ telque $\dim U(\mathcal{G})x < \infty$ et $N = U(\mathcal{G})x \in \underline{\mathrm{Mod}}_{(\mathcal{G},\Gamma)}$. Alors $\ell.\mathrm{Ann}_U(N)$ est cofini et il suffit de vérifier qu'il contient un produit $\tau_{\lambda_i}(\mathcal{M})\ldots\tau_{\lambda_k}(\mathcal{M})$. On considère une suite de Jordan-Hölder de $N, (0) \subset N_1 \subset \ldots \subset N = N_s$. Il est clair qu'il existe $\lambda_1 \in \Gamma$ tel que $\tau_{\lambda_1}(\mathcal{M})N \subseteq N_{s-1}$, et ensuite on procède par récurrence. De l'égalité $H^o_{C\ell(\mathcal{M})}(M) = H^o_{(\mathcal{G},\Gamma)}(M)$ résulte la partie 2°).

1°) Soit $K$ un idéal à gauche de $U(\mathcal{G})$ et $I \in \&(\mathcal{M})$. Il s'agit de vérifier qu'il existe $J \in \&(\mathcal{M})$ tel que $J \cap K \subseteq IK$. Notons $F$ l'enveloppe injective du $\mathcal{G}$-module $K/_{IK}$. Alors on a
$H^o_{C\ell(\mathcal{M})}(K/_{IK}) = K/_{IK} \subset H^o_{C\ell(\mathcal{M})}(F) = H^o_{(\mathcal{G},\Gamma)}(F)$ ; d'après les remarques 2.4, $F' = H^o_{(\mathcal{G},\Gamma)}(F)$ est injectif et par définition $F' = \bigcup_{J \in \&(\mathcal{M})} \mathrm{Ann}_{F'}(J)$. Soit $\dot{f} : K \to F'$ la composée de la projection $\pi : K \to K/_{IK}$ et de l'injection $\underset{\sim}{i} : K/_{IK} \to F'$. Comme $F'$ est un $\mathcal{G}$-module injectif, il existe $\underset{\sim}{f} : U \to F'$ qui prolonge $\dot{f}$ est donc il existe $J \in \&(\mathcal{M})$ tel que $J\,\tilde{f}(U) = 0$, c'est-à-dire $\tilde{f}(J) = 0$. Donc $0 = \tilde{f}(J \cap K)$. Donc $i(\frac{J \cap K + IK}{IK}) = 0$ et $J \cap K \subseteq IK$.

3°) Le $\mathcal{G}$-module $R(\mathcal{G},\Gamma)$ étant injectif est facteur direct du $\mathcal{G}$-modules des fonctions représentatives $V(\mathcal{G})$, c'est-à-dire du sous-module formé des éléments $\mathcal{G}$-finis de $\mathrm{Hom}_{\mathbb{C}}(U,\mathbb{C})$. On a $V(\mathcal{G}) = \oplus\, E(U/_P)$, où $P$ parcourt une fois et une seule l'ensemble des idéaux premiers cofinis de $U(\mathcal{G})$ [1]. Donc $R(\mathcal{G},\Gamma) = \oplus\, E(U/_P)$ où $U/_P \in \underline{\mathrm{Mod}}_{(\mathcal{G},\Gamma)}$ et un tel premier $P$ est de la forme $\tau_\lambda(\mathcal{M})$ pour un $\lambda \in \Gamma$. La fin de la démonstration résulte de 1°) et de [13].

2.6 Si $f \in (\mathcal{G}/_{[\mathcal{G},\mathcal{G}]})^\star$ et $M$ est un $\mathcal{G}$-module, on notera $\mathbb{C}_{-f} \otimes M$ le $\mathcal{G}$-module qui a même espace sous-jacent que $M$ et dont la structure de $U$-module est définie par $a.m = \tau_f(a)m$ si $a \in U$, $m \in M$, où $\tau_f$ est l'isomorphisme de $U(\mathcal{G})$ défini par $f : \tau_f(X) - X - f(X)$ si $X \in \mathcal{G}$.

Si $P$ est un idéal premier cofini de $U(\mathcal{G})$, on fixera un $f \in (\mathcal{G}/_{[\mathcal{G},\mathcal{G}]})^{\star}$ tel que $P = I(f)$ ([9]).

<u>Corollaire</u>. <u>Soit</u> $P = I(f)$, $f \in (\mathcal{G}/_{[\mathcal{G},\mathcal{G}]})^{\star}$, <u>un idéal premier cofini de</u> $U(\mathcal{G})$. <u>La clique de</u> $P$ <u>est en correspondance bijective avec</u> $\Gamma$ ; $\mathcal{E}(P)$ <u>est une famille d'Artin-Rees et l'on a</u>
$H^{i}_{C\ell(I(f))}(M) \cong H^{i}_{(\mathcal{G},\Gamma)}(\mathbb{C}_{-f} \otimes M)$.
<u>Preuve</u>. Ceci résulte du fait que la clique de $P$ est isomorphe en tant que graphe à la clique de $\mathcal{M}$, par l'isomorphisme $\tau_{+f}$. L'espace $H^{o}_{(\mathcal{G},\Gamma)}(\mathbb{C}_{-f} \otimes M) = H^{o}_{C\ell(\mathcal{M})}(\mathbb{C}_{-f} \otimes M)$ est l'ensemble des $x \in M$ tels qu'il existe $\lambda_1,\ldots,\lambda_k \in \Gamma$ avec $\tau_{\lambda_1}(\mathcal{M})\ldots\tau_{\lambda_k}(\mathcal{M}).x = 0$ c'est-à-dire $\tau_{-f}(\tau_{\lambda_1}(\mathcal{M})\ldots\tau_{\lambda_k}(\mathcal{M}))x = 0$ ou encore
$\tau_{\lambda_1}(I(f))\ldots\tau_{\lambda_k}(I(f))x = 0$ d'où l'égalité.

2.7 Il est facile d'interpréter la cohomologie locale en la clique d'un idéal premier cofini de $P$ en terme de la cohomologie ordinaire de $\mathcal{G}$, soit en utilisant l'argument de Fokko du Cloux [10] soit celui légèrement modifié de D. Wigner [17]. Les résultats de [17], concernent une algèbre de Lie $\mathcal{G}$, deux sous-algèbres de Lie $\underline{k}$ et $\mathfrak{h}$ de $\mathcal{G}$, $\mathfrak{h} \subset \underline{k}$ et étudient le foncteur qui à un $\mathcal{G}$-module $M$ fait correspondre le sous $\mathcal{G}$-module des éléments $\underline{k}$-finis de $M$ et ses dérivées relatifs par rapport à $\mathfrak{h}$. Supposons $\mathfrak{h} = (0)$, $\underline{k} = \mathcal{G}$ résoluble et soit $M$ un $\mathcal{G}$-module, $\lambda$ la structure de $\mathcal{G}$-module sur $\mathrm{Hom}_{\mathbb{C}}(U(\mathcal{G}),M)$ définie par $(\lambda(X)\varphi)(u) = \varphi(-Xu)$ si $x \in \mathcal{G}$ $u \in U(\mathcal{G})$, $\varphi \in \mathrm{Hom}_{\mathbb{C}}(U(\mathcal{G}),M)$.
Il existe [17] un isomorphisme de $V(\mathcal{G}) \otimes_{\mathbb{C}} M$ sur le sous-module module des éléments $\lambda(\mathcal{G})$-finis de $\mathrm{Hom}_{\mathbb{C}}(U(\mathcal{G}),M)$ défini par $(\mathcal{Y} \otimes m)(u) = \mathcal{Y}(u)m$, $m \in M$, $u \in U(\mathcal{G})$, $\mathcal{Y} \in V(\mathcal{G})$ et il est facile de voir que la restriction de cet isomorphisme à $R(\mathcal{G},\Gamma) \otimes_{\mathbb{C}} M$ a pour image $\mathrm{Hom}_{\mathbb{C}}(U(\mathcal{G}),M)^{(\mathcal{G},\Gamma)}$ où l'action de $\mathcal{G}$ sur $\mathrm{Hom}_{\mathbb{C}}(U(\mathcal{G}),M)$ est donnée par $\lambda$.
On notera $(R(\mathcal{G},\Gamma) \otimes_{\mathbb{C}} M,\mu)$ le $\mathcal{G}$-module d'espace sous-jacent $R(\mathcal{G},\Gamma) \otimes_{\mathbb{C}} M$ et sur lequel $\mathcal{G}$ opère, compte tenu de l'identification

précédente, par :

$$(\mu(X)\varphi)(u) = X\varphi(u) + \varphi(uX)$$

si $\varphi \in \mathrm{Hom}_{\mathbb{C}}(U(\mathcal{G}),M)$, $u \in U(\mathcal{G})$ et $X \in \mathcal{G}$.
On a alors :

<u>Proposition</u>. <u>Pour tout</u> $\mathcal{G}$-<u>module</u> M, <u>on a un isomorphisme</u> $\theta$ <u>de</u> $\mathcal{G}$ <u>modules</u> :

$$H^{\star}_{C\ell(\mathcal{M})}(M) \simeq H^{\star}(\mathcal{G},(R(\mathcal{G},\Gamma) \otimes_{\mathbb{C}} M,\mu))$$

<u>où la cohomologie ordinaire de</u> $\mathcal{G}$ <u>à coefficients dans</u> $R(\mathcal{G},\Gamma) \otimes_{\mathbb{C}} M$ <u>est calculée pour l'action</u> $\mu$ <u>et où l'isomorphisme</u> $\theta$ <u>utilise la structure de</u> $\mathcal{G}$-<u>module définie par l'action à gauche de</u> $\mathcal{G}$ <u>sur</u> $R(\mathcal{G},\Gamma)$.

<u>Preuve</u>. Ceci n'est autre que la proposition de [17] ou encore la proposition 3.5 de [10].

2.8 <u>Corollaire</u>. <u>Si</u> $P = I(f)$, $f \in (\mathcal{G}/_{[\mathcal{G},\mathcal{G}]})^{\star}$, <u>on a, avec les notations de 2.7, un isomorphisme</u> :

$$H^{\star}_{C\ell(P)}(M) \simeq H^{\star}(\mathcal{G},(R(\mathcal{G},\Gamma) \otimes_{\mathbb{C}} \mathbb{C}_{-f} \otimes M,\mu))$$

2.9 Supposons l'idéal premier P non plus cofini mais primitif. Notons $\mathcal{A} = \mathrm{Ad}(\mathcal{G})$ le groupe algébrique adjoint de $\mathcal{G}$ et $\tilde{\mathcal{G}}$ l'algèbre de Lie de $\mathcal{A}$ ; alors $P = I(f)$ où $f \in \mathcal{G}^{\star} = \mathrm{Hom}_{\mathbb{C}}(\mathcal{G},\mathbb{C})$, est l'idéal primitif de U associé par l'application de Dixmier : $\mathcal{G}^{\star}/_{\mathcal{A}} \to \mathrm{Prim}\, U$. On note $\mathcal{G}(f)$ le stabilisateur de f, i.e.

$$\mathcal{G}(f) = \{X \in \mathcal{G}\ , f([Y,X]) = 0 \text{ pour tout } y \in \mathcal{G}\}$$

et $\tilde{\mathcal{G}}(f) = \{X \in \mathcal{G}\ , f([Y,X]) = 0$ pour tout $Y \in \tilde{\mathcal{G}}\}$. On a toujours $\tilde{\mathcal{G}}(f) \subset \mathcal{G}(f)$ mais $\tilde{\mathcal{G}}(f)$ peut être strictement plus petite que $\mathcal{G}(f)$. Evidemment si P est cofini, $\tilde{\mathcal{G}}(f) = \mathcal{G}(f) = \mathcal{G}$. L'algèbre $\tilde{\mathcal{G}}(f)$ (notée aussi $\mathcal{G}[f]$ dans [16]) joue un rôle important dans l'étude des idéaux primitifs de $U(\mathcal{G})$ ([16],[8]) en partie en raison du lemme [4] :

Lemme. Si $P = I(f) \in \mathrm{Prim}\, U$. On a $\tau_\lambda(P) = P$, $\lambda \in \Lambda$ si et seulement si $\lambda_{|\tilde{\mathcal{G}}(f)} = 0$.

D'autre part la dimension de $\tilde{\mathcal{G}}(f)$ est égale à la hauteur de $P = I(f)$ [16]. En général nous n'avons pas pu rattacher la cohomologie locale en $C\ell(P)$ à la cohomologie $\tilde{\mathcal{G}}(f)$-finie. Cependant lorsque $\mathcal{G}$ est nilpotente, (alors $C\ell(P) = \{P\}$, $P$ est localisable, $\tilde{\mathcal{G}}(f) = \mathcal{G}(f)$, et $\Gamma = \{0\}$), on peut vérifier que $H^{\star}_{\{P\}}(U(\mathcal{G}))$ s'interprête comme espace de cohomologie finie d'une algèbre de Lie de dimension inférieure à $\dim \mathcal{G}$, éventuellement tensorisé par une algèbre de Weyl.

2.10 Remarque. Soit $P$ un idéal maximal de $U(\mathcal{G})$ (non nécessairement cofini) et telle que $\mathcal{E}(P)$ soit une famille d'Artin Rees. Alors $\mathcal{E}(P)$ est une famille d'Artin Rees minimale, c'est-à-dire que toute famille engendré comme l'est $\mathcal{E}(P)$ par un ensemble d'idéaux premiers $\subsetneq C\ell(P)$ n'est pas d'Artin Rees.

Preuve. Il suffit de prouver que la famille $\mathcal{F}$ des idéaux premiers de $U(\mathcal{G})$ qui contiennent un produit fini d'éléments de $C\ell(P)$ distincts de $P$ ne vérifie pas la condition d'Artin Rees et ceci en raisonnant par l'absurde. Soit $Q \in C\ell(P)$, $Q \neq P$ et $Q \rightsquigarrow P$ s'il existait $Q_1, \dots, Q_\ell \in C\ell(P) \setminus P$, non nécessairement distincts tels que $Q_1, \dots, Q \cap P \subseteq QP$, l'un des $Q_i$ serait contenu dans $Q$ et donc il existerait $Q_{i_o}$, $1 \leqslant i_o \leqslant \ell$, $Q_{i_o} = Q$ avec $Q_j \neq Q$ pour $j \subsetneq i_o$. On On aurait en posant $I = Q_1 \dots Q_{i_o - 1}$ et $J = Q_{i_o+1} \dots Q_\ell$, $IQJ \cap P \subseteq QP$ c'est-à-dire $I.\frac{Q \cap P}{QP}.J = (\bar{0})$ dans $\frac{Q \cap P}{QP}$. Donc $I$ serait contenu dans l'annulateur à gauche de $\frac{Q \cap P}{QP}.J$ ainsi que $Q$. Comme $Q$ est un idéal maximal, $\frac{Q \cap P}{QP}.J = (\bar{0})$, donc $J$ serait contenu dans l'annulateur à droite de $\frac{Q \cap P}{QP}$ qui est égal à $P$ car $Q \rightsquigarrow P$, ce qui n'est pas.

Evidemment, si $\mathcal{G}$ est nilpotente, pour tout idéal premier $P$ de $U(\mathcal{G})$, $\mathcal{E}(P)$ est une famille d'Artin Rees minimale.

Références


[1] G. Barou et M.P. Malliavin - Sur la résolution injective minimale de l'algèbre enveloppante d'une algèbre de Lie résoluble,Journal of Pure and Applied Algebra,37,(1985),1-25.

[2] A.D. Bell - Localization and ideal theory in iterated differential operator rings,J. of Algebra 106, 376-402,(1987).

[3] A.D. Bell - Localization and ideal theory in noetherian strongly group-graded rings, J. of Algebra, 105, 76-115 (1987).

[4] W. Borho, P. Gabriel et R. Rentschler - Primideale in Einhüllenden auflösbarer Lie algebren, Lecture Notes in Math 357, Springer-Verlag, 1973.

[5] K.A. Brown - The structure of modules over polycyclic groups, Math. Proc. Camb. Phil Soc. (1981), 89, 257-283.

[6] K.A. Brown - Localization, bimodules and injective modules for enveloping algebras of solvable Lie algebras, Bull. Sc. Math (2) 107 (1983) 225-251.

[7] K.A. Brown et R.B. Warfield Jr - The influence of ideal structure on representation theory - à paraître.

[8] K.A. Brown et Fokko du Cloux - On the representation theory of solvable Lie algebras - à paraître.

[9] J. Dixmier - Algèbres enveloppantes, Paris, Gauthier-Villars 1974 (Cahiers scientifiques, 37).

[10] Fokko du Cloux - Foncteurs dérivées des vecteurs $\mathcal{G}$-finis (preprint)

[11] A.V. Jategaonkar - Localization in Noetherian rings, London Math. Soc. Lecture Note Séries, n°98, Cambridge University Press, Cambridge 1986.

[12] T. Levasseur - L'enveloppe injective du module trivial sur une algèbre de Lie résoluble, Bull. Sc. Math. 2ème série 110, 1986, p.49-61.

[13] M.P. Malliavin - Représentations injectives d'algèbres de Lie résolubles , Communication in Algebra 14 (8) 1503-1513, 1986.

[14] B.J. Müller - Localization in non communicative Noetherian rings , Canad. J. Math. 28 (1976) 600-610.

[15] J.T. Stafford - The Goldie rank of a module, Proc. Conf. Noetherian rings, Oberwolfach 1983. Mathematical Surveys n°24 A.M.S. Providence Rhode Islande.

[16] P. Tauvel - Sur les quotients premiers de l'algèbre enveloppante d'une algèbre de Lie résoluble , Bull. Soc. Math. France 106, 1987, p.177-205.

[17] D. Wigner - Sur l'homologie relative des algèbres de Lie et une conjecture de Zuckermann, C.R. Acad. Sci. Paris t.305 série I, p.59-62, 1987.

# Noncommutative Valuation Rings

H.H. Brungs
Department of Mathematics
University of Alberta
Edmonton, Canada

Abstract. Various possible versions of noncommutative valuation rings and their applications are discussed.

## 1. Definitions and first results.

A valuation on a (commutative) field $K$ is a mapping $v$ from $K^*$ ($= K\backslash 0$) onto an ordered group $(G,+)$ with

1) $v(a\ b) = v(a) + v(b)$ and
2) $v(a+b) \geq \min\{v(a),v(b)\}$ for $a,b \in K^*$.

To such a valuation corresponds a valuation ring $V$ of $K$ with $V^* = \{a \in K|\ v(a) \geq v(1)\}$ which has the following property:

3) $a \in K\backslash V$ implies $a^{-1} \in V$.

Conversely, for any subring $R$ of $K$ with property 3) there exists a valuation on $K$ with $R$ as corresponding valuation ring ([12],[14]).

If $K$ is replaced by a skew field $D$ we can define as a valuation on $D$ any mapping $v$ from $D^*$ onto an ordered gorup $G$ satifying 1) and 2) and we say that this valuation is abelian if $G$ is abelian. The corresponding valuation ring $V$ still satisfies 3) and in addition



F. van Oystaeyen and L. Le Bruyn (eds.), Perspectives in Ring Theory, 105–115.

4) $d^{-1}Vd \subseteq V$ for all $d \in D^*$,

since $v(d^{-1}ad) \geq v(1) = e$, the identity of $G$, if $v(a) \geq e$.
As in the commutative case, we can define for a given subring $R$ of $D$ satisfying 3) and 4) an ordered group $G = \{sR \mid s \in D^*\}$ with $sRtR = stR$ and $sR > tR$ if and only if $sR \subset tR$ and a valuation $v$ from $D^*$ to $G$ with $v(s) = sR$. This is the valuation concept used in Schilling ([16]) and we will say the corresponding valuation ring $R$ is total, i.e. $x \in D^*\backslash R$ implies $x^{-1} \in R$ and invariant, i.e. 4) holds.

Before we discuss some results about total valuation rings that are not necessarily invariant we consider a one-sided generalization of valuation rings. We say a ring $R$ is a right chain ring if $aR \subseteq bR$ or $bR \subset aR$ holds for any $a,b$ in $R$. It follows immediately that the right ideals of such a ring are in a chain and that a total valuation ring is exactly a right and left chain domain.

We assume in the following that all rings $R$ have an identity and we denote with $J(R)$ the Jacobson radical of $R$ and with $U(R)$ the group of units of $R$. If $R$ is a right chain domain we write $Q(R) = \{ab^{-1} \mid a, 0 \neq b \in R\}$ for the skew field of quotients of $R$.

Lemma 1. A ring $R$ is a right chain ring if and only if $R$ is a right Bezout ring and $R/J(R)$ is a division ring.

That a right chain ring satisfies the two conditions in the Lemma follows from the definition. Conversely, let $a,b \in R$, not both equal zero and $aR + bR = dR$ for $d \neq 0$ in $R$ follows. Hence, $d = ax + by$, $a = da_1$, $b = db_1$ for $x,y,a_1,b_1 \in R$. We obtain $d(a_1x + b_1y - 1) = 0$ and $a_1,b_1 \in J(R)$ leads to the contradiction $d = 0$. For $a_1 \notin J(R)$ exists $r_1 \in R$ with $a_1r_1 = 1+j$ for some $j \in J(R)$ and $a_1r_1(1+j)^{-1} = 1$ which implies $aR = dR \supseteq bR$. The case $b_1 \notin J(R)$ is treated similarly.

We consider one instance where right chain rings appear to be the correct generalization of commutative valuation rings.

Prüfer domains are characterized by the equivalent properties that their lattice of ideals is distributive or that their localizations at maximal ideals are valuation rings. A corresponding result holds in the noncommutative case ([1]):

Theorem 1. An integral domain $R$ has a distributive lattice of right ideals if and only if $S = R\backslash N$ is a right Ore set in $R$ and $RS^{-1}$ is a right chain ring for every maximal right ideal $N$ or $R$.

Related results were proved by Gräter ([13]) and Tuganbaev ([17]).

Above it was pointed out how to regain the ordered group of values from a total invariant valuation ring $R$ in a division ring $D = Q(R)$. For right chain domains one obtains the following result ([2]):

Lemma 2. The set $H(R) = \{aR \mid a \in R^*\}$ of non-zero principal right ideals of the right chain domain $R$ forms a semigroup under right ideal multiplication if and only if $R$ is right invariant, i.e. every right ideal of $R$ is two-sided.

For a proof we only need to assume that $RaR = bR$ is again a principal right ideal for any $a \neq 0$ in $R$. Then $b = \Sigma\, r_i a s_i$ for certain elements $r_i$ and $s_i$ in $R$ and $a = bc$ for some $c$ in $R$. Hence, $b = \Sigma\, r_i a s_i = \Sigma\, r_i b c s_i = \Sigma\, b r_i' c s_i$ for some $r_i'$ in $R$ since $bR$ is a two-sided ideal. It follows that $c \notin J(R)$ and $bR = aR$.

The mapping $v$ from $R^*$, $R$ a right invariant chain domain, to $H(R)$ with $v(a) = aR$ satisfies conditions 1) and 2) but can not be extended to $Q(R)^*$ in general; in fact $H(R)$ is a totally ordered semigroup that is embeddable into a group only in exceptional cases.

We make the following observation:

Lemma 3. A right chain domain $R$ is right invariant if and only if $a^{-1}U(R)a \subseteq U(R)$ for all $a \in R^*$. If in addition $aU(R)a^{-1} \subseteq U(R)$ for all $a \in R^*$ than $R$ is an invariant and total valuation ring of $Q(R)$.

Proof. Let $R$ be a right chain domain. If $R$ is right invariant and $u \in U(R)$ then $ua = au'$ for some $u' \in U(R)$ and $a^{-1}U(R)a \subseteq U(R)$ follows. Conversely, if $aR$ is any right ideal and $a^{-1}U(R)a \subseteq U(R)$ holds we have $(1+j)a = av$ for $j \in J(R)$ and some $v \in U(R)$. Hence, $ja = a(v-1)$ and $R$ is right invariant. To prove the second part of the Lemma, we use the same argument to show that every left ideal is also a right ideal and $R$ is a total invariant valuation ring in $Q(R)$.

A right noetherian right chain domain $R$ is right invariant and $H(R) \cong \{\alpha \mid \alpha < \omega^I\}$ is isomorphic to the semigroup of ordianls less than a power of $\omega$, the order type of the natural nubmers. This result can then be applied to study right noetherian domains with a distributive lattice of right ideals, a noncommutative version of Dedekind domains ([1]).

Other results on $H(R)$, $R$ an invariant right chain domain, are obtained in [5]. In fact, one can associate a generalized semigroup (group) of values with any right chain (total valuation) ring ([2], [15]), however not many results are known in this general case.

We conclude this section by considering two instances in which right chain domains are also left chain domains.

Lemma 4. Let $R$ be a right chain domain with $D = Q(R)$ finite dimensional over its center $K$. Then $R$ is total in $D$.

Proof. Let $V = R \cap K$ and $a,b$ be elements in $V$. Then $a = br$ or $b = ar$ for some $r$ in $R$. In any case $r$ is in $K$ and hence in $V$, i.e. $V$ is a valuation ring. To show that $K = Q(V)$, let $\alpha \in K$ and $\alpha = st^{-1}$, $s,t \in R$. We obtain $s = \alpha t = t\alpha$ and $\alpha = t^{-1}s$ which equals $s_1$ if $s = ts_1$ and equals $t_1^{-1}$ if $t = st_1$ for some $s_1,t_1$ in $R$. This shows that either $\alpha$ or $\alpha^{-1}$ is in $V$.

We show next that $D = RS^{-1}$ for $S = V^*$. Let $e_1,\dots,e_n$ be a basis of $D$ over $K$ and we can write $e_i = r_i d^{-1}$ for $r_i$, $d \in R$. Then $\{r_1,\dots,r_n\}$ is also a basis for $D$ over $K$ and every element $ab^{-1}$ in $D$, $a, 0 \neq b \in R$ can be written as $ab^{-1} = rs^{-1}$ for $r \in R$, $s \in S$. We have either $r = sr_1 = r_1 s$ and $ab^{-1} = r_1$, $a = r_1 b$ follows, or $s = rs_1$ and $ab^{-1} = rs^{-1} = s^{-1}r = s_1^{-1}$ and $s_1 a = b$ where $r_1$ and $s_1$ respectively are elements in $R$. Hence, $R$ is also a left chain ring.

If a right chain domain $R$ is left Ore, then $R$ is also a left chain domain: Let $q = ab^{-1} \in Q(R)$ and $c, 0 \neq d$ exists in $R$ with $q = ab^{-1} = d^{-1}c$. But either $dR \subseteq cR$ or $cR \subseteq dR$ and $q^{-1} = c^{-1}d$ or $q = d^{-1}c \in R$; i.e. $Ra \subseteq Rb$ or $Rb \subseteq Ra$.

## 2. The extension problem.

Given any field extension $F \supseteq K$ of a field $K$ with a valuation $v$ on $K$. Then this valuation can be extended to $F$ or equivalently, there exists a valuation ring $V'$ of $F$ with $V' \cap K = V$ if $V$ is the valuation ring corresponding to $v$.

In the skew field $H$ of the quaternions over the rationals $Q$ there does not exist a total valuation ring $B$ of $H$ with

$B \cap Q = \mathbb{Z}_p$ for a prime $p \neq 2$ where $\mathbb{Z}_p = \{ab^{-1} \mid a,b \in \mathbb{Z},\ p \nmid b\}$ is the p-adic valuation ring of Q.

We consider the following problem: Let D be a division algebra, finite dimensional over its center K and V a valuation ring of K. Describe the set $\mathcal{B} = \{B \mid B \cap K = V\}$ of total valuation rings B of D with $B \cap K = V$; such a ring B is called a extension of V in D.

Wadsworth in [18], extending a result of Cohn ([7]), obtained the following theorem for invariant extensions of V in K.

Theorem 2: A valuation ring V of the center K of a finite dimensional division algebra D has an invariant extension B in D if and only if V has a unique extension in every subfield F with $K \subseteq F \subseteq D$.

This theorem implies that $|\mathcal{B}| = 1$ in case there exists an invariant extension B of V in D.

However, if V has a rank that is greater than 1, it is possible that V has extensions that are not invariant. In [3] the following result was proved:

Theorem 3: Let $[D:K] = n^2$ and V be a valuation ring of K. Then $|\mathcal{B}| \leq n$ and any two elements in $\mathcal{B}$ are conjugate in D.

Essential for the proof of this theorem is the fact that there exists in D an invariant and total valuation ring $R \neq D$ with $R \supseteq B$ for all B in $\mathcal{B}$ and a careful study of the maximal separable extension S of $\frac{R \cap K}{J(R) \cap K}$ in $Z(R/J(R))$, the center of $R/J(R)$.

The elements in D which are integral over V do in general not form a subring of D. The following result gives an exact condition under which this happens.

Theorem 4 ([3]). Let D be a division algebra finite dimensional over its center K and V a valuation ring of K. Then the integral closure T of V in D is a subring of D if and only if V has an extension in D. In that case: $T = \cap B$, $B \in \mathcal{B}$.

The existence of extensions of V is the exception, as was indicated by the example in the beginning; however, if $|\mathcal{B}| \geq 1$ then the second part of Theorem 4 generalizes the theorem about the integral closure of a valuation ring in finite field extensions.

In the next section a type of valuation ring is introduced so that any V in K has such an extension in D.

## 3. Dubrovin valuation rings.

So far we only considered integral domains and division rings. It is clear however that in the noncommutative situation one should study prime or prime Goldie rings and simple artinian rings. The following observation shows that the valuation rings introduced so far are not appropriate for this situation.

Lemma 5 ([6]) A prime Goldie right chain ring R is a left and right chain domain.

Proof. Since a right chain ring does not contain any nontrivial direct sums of right ideals we have Q(R) = D, a division ring. That R is also a left chain ring follows from the comment at the end of section 1.

More difficult to answer is the question whether prime right chain rings exist that are not domains - we will not discuss this problem here.

Dubrovin in [10] introduced a class of rings that share many properties with valuation rings, lead to a more general extension theorem than the results in Section 2 and are orders in artinian simple algebras.

Before we turn to this class of rings an even more general type of ring is introduced.
A subring $R$ of a ring $Q$ is called a right $n$-chain ring of $Q$ if for any $n+1$ elements $a_0,\dots,a_n$ in $Q$ there exists an $i$, $0 \leq i \leq n$ with $a_iR \subseteq \sum_{j\neq i} a_jR$.

A right chain ring $R$ is then just a 1-chain ring in $R$.

A special class of $n$-chain rings is formed by the following rings which will be called $Q$-valuation ring.

Definition. A subring $R$ of a simple artinian algebra $Q$ is called a $Q$-valuation ring if $R$ contains an ideal $M$ such that $R/M$ is simple artinian and for any element $q \in Q\backslash R$ exist $r_1, r_2 \in R$ with $r_1q\ , qr_2 \in R\backslash M$.

One can show that $M = J(R)$ in case $R$ is a $Q$-valuation ring.

As a motivation for this definition one can interpret $R$ as the set of elements with finite values under a (generalized) place mapping from $(Q,\infty)$ to $(R/M,\infty)$. The $Q$-valuation rings $R$ are the rings one obtains if one replaces the division rings $Q(R)$ and $R/J(R)$ in Lemma 1 by simple artinian rings; more precisely the following result ([10]) holds:

Theorem 5: A subring R of a simple artinian ring Q is a Q-valuation ring if and only if R is a Bezout order in Q and R/J(R) is simple artinian.

Every simple artinian ring Q is isomorphic to a matrix ring $(D)_n$ for a division ring D and some n. One can prove ([10],[4]) that a $(D_n)$-valuation ring R is conjugate in $(D)_n$ to a ring $(T)_n$ where T is a D-valuation ring. As the example $T = \mathbb{Z}_3 + \mathbb{Z}_3 i + \mathbb{Z}_3 j + \mathbb{Z}_3 k \subset H$, the quaternions over Q, shows, a D-valuation ring need not be total. However, $T \cap Q = \mathbb{Z}_3$. In fact, the class of D-valuation rings is large enough that every valuation ring V in the center K of a finite dimensional division algebra D can be extended to a D-valuation ring. More general, the following result was obtained in [11] and in [4] a somewhat different proof was given:

Theorem 6. Let Q be a simple algebra finite dimensional over its center K and V a valuation ring in K. Then there exists a Q-valuation ring R with $R \cap K = V$.

In [4] it was also proved that any two Q-valuation rings $R_1$, $R_2$ with $R_1 \cap K = R_2 \cap K = V$ are conjugate in Q if V has finite rank and the conditions of Theorem 6 hold.

Wadsworth has announced among other interesting results about Q-valuation rings that the condition on the rank is not needed for the above theorem.

P.M. Cohn ([9]), using matrix localizations has obtained conditions for the existence of real valuations on division rings, see also [8] where abelian extensions are treated.

References

[1] Brungs, H.H. Rings with a distributive lattice of right ideals, J . Algebra 40(1976), 392-400.

[2] Brungs, H.H. and Törner, G. Right chain rings and the generalized semigroup of divisibility; Pac. J. Math. 97(1981) 293-305.

[3] Brungs, H.H. and Gräter, J. Valuation rings in finite dimensional division algebra; to appear in J. Algebra.

[4] Brungs, H.H. and Gräter, J. Extensions of valuation rings in central simple algebras, 1987.

[5] Brungs, H.H. and Törner, G. A structure theorem for right invariant right holoids, 1987.

[6] Le Bruyn, L. and Van Oystaeyen, F. A note on noncommutative Krull domains; Comm. ALgebra 14(8)(1986) 1457-1472.

[7] Cohn, P.M. On extending valuations in division algebras; Studia Scient. Math. Hung. 16(1981) 65-70.

[8] Cohn, P.M. and Mahdavi-Hezavehi, M. Extensions of valuations on skew fields; Proc. Ring Theory Week, Antwerp 1980 (ed. F. Van Oystaeyen) Springer Lecture Notes in Math. 825, Springer-Verlag Berlin-New York 1980, 28-41.

[9] Cohn, P.M. The construction of valuations on skew fields Department of Mathematics, University of Alberta, Canada, 1986 46 pp.

[10] Dubrovin, N.I. Noncommutative valuation ring; Trudy Moskov Mat. Obshch. 45(1982), 265-289. Engl. Translation: Moscow Math. Soc. 1(45)(1984) 273-287.

[11] Dubrovin, N.I. Noncommutative valuation rings in simple finite-dimensional algebras over a field; Mat. Sbornik 123(165)(1984) Engl. Translation: Mat USSR Sbornik 51(1985) 493-505.

[12] Endler, O. Valuation Theory; Springer Verlag, Berlin-New York 1972.

[13] Gräter, J. Zur Theorie nicht kommutativer Prüferringe; Archiv der Math. 41(1983) 30-36.

[14] Krull, W. Allgemeine Bewertungstheorie; J. reine u.angew. Math. 167(1932) 160-196.

[15] Mathiak, K. Bewertungen nicht kommutativer Körper; J. Algebra 48(1977) 217-235.

[16] Schilling, O.F.G. The Theory of Valuations, Math. Surveys, No. 4. Amer. Math. Soc. Providence, R.I. 1950.

[17] Tuganbaev, A.A. Rings whose structure of right ideals is distributive; Jzvestiya VUZ Matematika, Vol. 30, No.2(1986) 44-49.

[18] Wadsworth, A.R. Extending valuations to finite dimensional division algebras; Proc. Amer. Math. Soc. 98(1986) 20-22.

## On Hopf Galois extensions of fields and number rings

Lindsay N. Childs
Department of Mathematics
State University of New York at Albany
Albany, NY 12222, USA

This paper is a summary of some results on the problem of determining the $H$-Galois extensions of a commutative ring $R$, where $H$ is a finite abelian Hopf $R$-algebra.

### Definitions.

Let $R$ be a commutative ring with unity. A Hopf $R$-algebra $H$ is an $R$-algebra, finitely generated and projective as $R$-module, which also has a comultiplication $\Delta : H \to H \oplus H$, counit $\epsilon : H \to R$, and antipode $\lambda : H \to H$, which make $H$ into a bialgebra with antipode (c.f. [Sw 69] for basic theory). We will call $H$ abelian if $H$ is both commutative and cocommutative.

An $R$-algebra $S$ is an $H$-module algebra if $S$ is an $H$-module such that $H$ measures $S$ to $S$ in the sense of Sweedler [Sw 69], that is, for $s, t$ in $S, h$ in $H$, $h(st) - \Sigma_{(h)} h_{(1)}(s) h_{(2)}(t)$, where $\Delta(h) = \Sigma_{(h)} h_{(1)} \otimes h_{(2)}$ in the usual Sweedler notation, and $h(1) = \epsilon(h)1$. See [Bg 85] for an explanation of the naturalness of this idea. We call $S$ an $H$-extension of $R$ if $S$ is an $H$-module algebra and $S^H = R$, where

$$S^H = \{s \text{ in } S | hs = \epsilon(h)s \text{ for all } h \text{ in } H\}$$

An $R$-algebra $S$ is a Galois extension of $R$ (e.g. [CS 69]) if $S$ is an $R$-algebra which is a finitely generated projective $R$-module and there is a Hopf $R$-algebra $H$ so that $S$ is a Galois $H$ extension of $R$. That is, $S$ is an $H$-module algebra, $S^H = R$ and the map $j : S \# H \to \mathrm{End}_R(S)$ is an isomorphism. $S/R$ is an abelian Galois extension if $H$ is abelian. If $S/R$ is $H$-Galois, then $\mathrm{rank}_R(H) = \mathrm{rank}_R(S)$.



*F. van Oystaeyen and L. Le Bruyn (eds.), Perspectives in Ring Theory, 117–128.*

**Consequences of being Galois.**

Knowing that $S$ is an $H$-Galois extension can be a useful fact. For example, suppose $S$ is an $H$-module algebra and $M$ is an $S$-module on which $H$ acts "semilinearly", that is, $M$ is an $S\#H$-module. If $S$ is an $H$-Galois extension, then $S\#H \cong \mathrm{End}_R(S)$, so Morita theory applies and gives that $M \cong S \otimes IM$ as $S\#H$-modules, where

$$I = \{\phi \text{ in } H | h\phi = \epsilon(h)\phi \text{ for all } h \text{ in } H\}$$

is the space of left integrals of $H$ and the $S\#H$-module structure on the right side is induced from that on $S$ ([CS69], Corollary 9.7).

If $S$ is an $H$-Galois extension of $R$ then $S$ is locally (with respect to primes of $R$) isomorphic to $H$ as $H$-module [KC 76]. In fact, if $R$ is a perfect field and $S$ is an $H$-module algebra with $H$ abelian and connected, then $S$ is isomorphic to $H$ as $H$-module iff $S$ is an $H$-Galois extension ([Hu84], Theorem 4.4.). This fact is of interest in Galois module theory; let $L/K$ be a Galois extension of local fields with valuation rings $S \supseteq R$ and abelian Galois group $G$. Let $\mathcal{U}$ be the order of $S$ in $KG$,

$$\mathcal{U} = \{\alpha \text{ in } KG | \alpha S \subseteq S\} = \mathrm{Hom}_{RG}(S, S)$$

(c.f. [Le59], [BF72], [Be79]). Then $\mathcal{U} \supseteq RG$. If $\mathcal{U}$ is a Hopf algebra with operations induced from those on $KG$, or as we shall say, a Hopf order in $KG$, and $\mathcal{U}$ is connected, then $S$ is an $\mathcal{U}$-Galois extension and is isomorphic to $\mathcal{U}$ as $\mathcal{U}$-module. This local information leads to an extension of the theorem of Noether which is a foundation of Galois module theory, as follows : if $L/K$ is a Galois extension of number fields with abelian Galois group $G$, and rings of integers $\mathcal{O}_L \supseteq \mathcal{O}_K$, then if the order $\mathcal{U}$ of $\mathcal{O}_L$ in $KG$ is a Hopf order in $KG$, then $\mathcal{O}_L$ is locally isomorphic to $\mathcal{U}$ as $\mathcal{U}$-module ([ChT87], Theorem 2.1). Noether's theorem is the case $\mathcal{U} = RG$. (Note that Noether's theorem holds for arbitrary $G$, not just $G$-abelian. Extending Noether's theorem tot non-commutative $H$ appears difficult, c.f. the example in Waterhouse [Wa88]).

Gamst and Hoechsmann [GH69] (c.f. [Ul86]) showed how one could use Galois extensions to produce Azumaya algebras, thereby generalizing the classical norm residue map whose properties were the subject of the Merkurjev-Suslin theorem. Namely, let $H$ be a Hopf $R$-algebra, finitely generated and projective as $R$-module, and let $S$, $T$ be Galois extensions of $R$ for $H$ and $H^* = \mathrm{Hom}_R(H, R)$, respectively.

Then the smash product $S\#T$ of $S$ and $T$ may be defined by forming the tensor product over $R$, $S \otimes T$, and twisting the multiplication via the coactions, as follows :

$$(s \otimes t)(s' \otimes t') = \Sigma s.s'_{(1)} < s'_{(2)}, t_{(2)} > \otimes t_{(1)} t'$$

where $\alpha : S \to S \otimes H^*, \alpha(s) = \Sigma s_{(1)} \otimes s_{(2)}$ is the comodule structure induced from the $H$-structure on $S$, etc.

If $H$ is abelian the set of isomorphism classes of Galois $H$-extensions forms an abelian group (e.g. [Ch86]), and Gamst and Hoechsmann [GH69] showed that the map $S \times T \mapsto S\#T$ induces a bimultiplicative map from $\mathrm{Gal}(H) \otimes \mathrm{Gal}(H*)$ to the Brauer group $\mathrm{Br}(R)$. In case $R$ is a field, $H = RG$, $G$ cyclic of order $n$, and $R$ contains $1/n$ and a primitive $n^{\text{th}}$ root of unity, then this map is the same as the map from $R^*/R^{*n} \times R^*/R^{*n} \to \mathrm{Br}(R)$ which induces the map from ${}_nK_2(R)$ to ${}_n\mathrm{Br}(R)$ which Merkurjev and Suslin showed was onto.

It is perhaps at least as interesting to consider the Gamst-Hoechsmann map for $R$ a commutative ring as for a field, since Azumaya algebras and Brauer classes are less easily obtained by classical constructions, e.g. crossed products, over commutative rings than over fields. Thus for example, the non-trivial class in $\mathrm{Br}(R)$, $R = \mathbb{Z}[\sqrt{2}]$, cannot be obtained as a classical crossed product, but can be obtained as the smash product of Galois extensions for the self-dual Hopf algebra $H = R[x]/(x^2 - \sqrt{2x})$, namely as $S\#S$, where $S = R[t]/(t^2 - \sqrt{2t} + 1)$. See [Ch89],[ChP87],[ChC87] for some results related to this construction over number rings.

**Examples.**

The three "classical" examples of extensions $S/R$ which are $H$-Galois for some Hopf-algebra $H$ are :

$H = RG$, in which case $S/R$ is Galois in the sense of Chase, Harrison, Rosenberg [CHR65]. For example, if $S, R$ are rings of integers in a Galois extension $L/K$ of number fields with Galois group $G$, then $G$ acts on $S, S^G = R$, and $S/R$ is Galois iff $S$ is unramified at all finite primes.

$H = RG^*, G$ a finite abelian group, in which case $S$ is a $G$-graded $R$-algebra (with the image of $R$ contained in the trivially graded component). Then $S$ is an $H$-Galois extension of $R$ iff $S$ is fully $G$-graded (c.f. [Da82]), that is, $S = \Sigma S_\sigma$ with $S_\sigma S_\tau = S_{\sigma\tau}$ for all $\sigma, \tau$ in $G$, and $S_1 = R$.

$H$ is the restricted universal enveloping algebra of the $p$-Lie algebra of derivations of an exponent one purely inseparable field extension $S/R$ (characteristic $p$). Then $S$ is $H$-Galois over $R$, and the Jacobson Galois theory applies ([J64], pp. 167 ff).

The problem of extension the Jacobson theory to higher exponent purely inseparable field extensions spurred the development of Hopf Galois theory during the decade centered around 1970 (culminating in [C76], c.f. [Ho73]). Perhaps the main point of this paper is to suggest that it is also fruitful to go beyond classical examples in characteristic zero.

*Describing Galois extensions.*

A natural question which has been studied since the development of the concept of Galois extension is that of attempting to classify the Galois extensions of a given commutative ring. For fixed abelian Hopf algebra $H$, this amounts to computing the Harrison group Gal($H$) of $H$-Galois extensions of $R$. There is a large literature on this question, of which some of the highlights are the homological classifications of Chase ([CS69], Theorem 16, 14) (cf. also [Sh69], [Wa71] and [CM74]) and Orzech [Or69], the work on quadratic extensions of C. Small [Sm72], the classification of Galois extensions with group $G$ cyclic of order $p$, prime, with or without $1/p$ or a primitive $p^{\text{th}}$ root of unity [Ch71], [Bo79], [Ch77], [KM84], [Mau84], [Wa87], and of order $p^n, n > 1$ [Ker83], [Gre88], [KM88], [Wy87]. Results on classifying $H$-Galois extensions where the Hopf algebra $H$ is not $RG$ (other than via cohomology) are more scarce. For fixed Hopf algebra $H$, there are classification of $H$-Galois extensions where $H = RG^*$, for an arbitrary finite group $G$ by Ulbrich [Ul81] (see also [Da82] and, related to this, the Galois theory of [GrH87]); quadratic Galois extensions for any $H$ of rank 2 are classified by Kreimer [Kr82]; and the group of $H$-Galois extensions with normal basis for $H$ a Hopf algebra of rank $p$, a prime, of the form $H_b = R[x]/(x^p - bx)$, where $b$ has a $p-1^{\text{st}}$ root in $R$ is computed by Hurley [Hu87].

The results we wish to discuss here are of a somewhat different character, in that the Hopf algebra is not necessarily specified in advance.

*Over fields of characteristic p.*

The earliest results along these lines are those of Chase [C76]. He proves that if $L/K$ is a finite extension of fields of characteristic $p$, then $L/K$ is $H$-Galois for some commutative Hopf $K$-algebra $H$ whose dual $H^*$ is a truncated polynomial algebra, (i.e. of the form $H^* = K[t_1, \ldots, t_n]$ with $t_j^{q_j} = 0, j = 1, \ldots, n$) iff $L/K$ is purely inseparable and modular. The Hopf algebra $H$ associated to $L/K$ is not unique, and Theorem 5.3. of [C76] in a certain sense classifies all such $H$.

*Over fields of characteristic* 0.

Greither and Pareigis [GP87] have recently answered the question, given a field extension $L/K$ of rank $n$, in characteristic 0, is there a Hopf $K$-algebra $H$ so that $L/K$ is $H$-Galois ? Suppose the normal closure $E$ of $L/K$ has Galois group $G$. If $\mathrm{Gal}(E/L) = G'$, then $L/K$ is a $H$-Galois iff there exists a regular subgroup $N$ in $\mathrm{Perm}(G/G')$ normalized by the image of $G$ in $\mathrm{Perm}(G/G')$. Using this criterion, they show that every cubic or quartic extension of $K$, normal or not, is $H$-Galois for some Hopf algebra $H$. (Every quadratic extension is $H$-Galois for $H = KG$, $G$ cyclic of order 2).

One can obtain from their work that if $K$ has characteristic zero and $p$ is prime, then a field extension $L/K$ of dimension $p$ is $H$-Galois for some $H$ iff the Galois group of (the normal closure of) $L/K$ is solvable.

A further interesting consequence of their work is that the non-uniqueness of the Hopf algebra which motivated Chase's work in purely inseparable Galois theory also occurs in characteristic zero. For example, if $L/K$ is $KG$-Galois for $G$ non-abelian, or even cyclic of order $p^n$, $p$ prime, $n > 1$, then $L/K$ is also $H$-Galois for some other $H$. The only situation $I$ know where if $L/K$ is $H$-Galois then $H$ is unique is if $L \supseteq K$ are fields of characteristic zero, amd $H = KG$, $G$ cyclic of order $p$, prime.

*Over complete dvrs.*

Let $K$ be a local field with valuation ring $R$, and $p$ be a prime number which is in the maximal ideal $p = (\pi)$ of the valuation ring $R$ of $K$.

Suppose $q = p^n$, and $K$ contains a primitive $q^{\text{th}}$ root of unity. Suppose $L$ is a field extension of $K$ which is $KG$-Galois for $G$ cyclic of order $q$ : that is, $L$ is a Kummer extension of $K$ with group $G$. Can we determine the Galois extensions of $R$ of rank $q$ which are orders in $L$ ?

Let $\mathcal{O}$ be the integral closure of $R$ in $L$. Given an order $S$ over $R$ in $L$ let $\mathcal{U} = \{\alpha \text{ in } KG | \alpha S \subseteq S\}$ be the associated order. Sometimes $\mathcal{U}$ is a Hopf order (in $KG$); often not. As observed earlier, it is of interest in Galois module theory to know when $\mathcal{U}$ is a Hopf order such that $S$ is $\mathcal{U}$-Galois. For example, let $\mathcal{U}_0$ be the order associated to $\mathcal{O}$. Then $\mathcal{U}_0$ always contains $RG; \mathcal{U}_0 = RG$ (hence, of cource, is a

Hopf algebra) iff (under our assumptions on $K$ and $G$) $L/K$ is unramified, in which case $\mathcal{O}/R$ is $\mathcal{U}_0$-Galois.

There are partially defined maps between orders over $R$ in $L$ and orders over $R$ in $KG$, as follows.

Given an order $S$ over $R$ in $L$, let

$$\mathcal{U}(S) = \{\alpha \in KG | \alpha S \subseteq S\}$$

be the associated order in $KG$.

Given a Hopf order $\mathcal{B}$ over $R$ in $KG$, let

$$\mathcal{O}(\mathcal{B}) = \{x \in L | \beta x \in \mathcal{O} \text{ for all } \beta \text{ in } \mathcal{B}\}$$

the largest $\mathcal{B}$-submodule of $\mathcal{O}$. If $\mathcal{B}$ is a Hopf order in $KG$, then $\mathcal{O}(\mathcal{B})$ is closed under multiplication and is an order over $R$ in $L$ (e.g. [T87], Lemma 3.3).

There are two "extreme" examples of these maps. One is for $\mathcal{B} = RG$, in which case $\mathcal{O}(\mathcal{B}) = \mathcal{O}$, the integral closure of $R$ in $L$, and $\mathcal{U}(\mathcal{O}) = \mathcal{U}_0$, which equals $\mathcal{B}$ iff $L/K$ is unramified.

The other is for $\mathcal{B} = \overline{RG}$, which under our hypothesis is isomorphic to $RG^*$, hence is a Hopf order. Then $\mathcal{O}(\mathcal{B}) = \tilde{\mathcal{O}}$, the Kummer order of $\mathcal{O}$ studied by Frohlich [Fr62] and defined as follows. Let $L = K[z], z^q = w$ in $K$. If $\sigma \epsilon G, \sigma(z) = \chi(\sigma)z$ for some character $\chi$ of $G$. For each $\psi$ in $\hat{G}$, set

$$L_\psi = \{s \text{ in } L | \sigma(s) = \psi(\sigma)s \text{ for all } \sigma \text{ in } G\}$$

Then $L = \Sigma L_\psi$. Let $\mathcal{O}_\psi = \mathcal{O} \cap L_\psi, \tilde{\mathcal{O}} = \Sigma \mathcal{O}_\psi$. Then $\tilde{\mathcal{O}}$ is the Kummer order over $R$ in $L$. Clearly $\mathcal{U}(\tilde{\mathcal{O}}) = \overline{RG}$.

The maps $S \mapsto \mathcal{U}(S), \mathcal{B} \mapsto \mathcal{O}(\mathcal{B})$ suggest the possibility of a Galois correspondence between certain containing $\mathcal{U}_\mathcal{O}$ in $KG$ and certain orders in $L$ containing $\tilde{\mathcal{O}}$. We obtain some of it : namely, if $\mathcal{B} = \mathcal{U}(S)$ is a Hopf algebra, then $\mathcal{U}(\mathcal{O}(\mathcal{B})) = \mathcal{B}$; while if $S$ is an order which is of the form $S = \mathcal{O}(\mathcal{B})$ then $S = \mathcal{O}(\mathcal{U}(S))$.

We also have the following results :

**Proposition 1.** If $S$ is an order in $L$ and $\mathcal{U}(S)$ is a Hopf order which is a local ring, then $S$ is $\mathcal{U}(S)$-Galois and $S = \mathcal{O}(\mathcal{U}(S)$.

**Proof.** Let $tr$ in $RG$ denote the sum of the elements of $G$. Suppose $tr(S) = eR$ for some $e$ in $R$. Then $(tr/e)(S) = R$, so $\phi = tr/e$ is in $\mathcal{U}(S)$ and $\phi S = R$. It is easy to see that $\phi$ is an integral of $\mathcal{U}(S)$, since $tr = \sum_{\sigma \in G} \sigma$ generates the space of integrals of $KG$. Since $\phi S = R$, $S$ is $\mathcal{U}(S)$-projective and hence $S$ is an invertible $\mathcal{U}(S)$-module (i.e. $S$ is a tame $\mathcal{U}(S)$-object in the sense of [CH86] or [Wa88]) and so $S$ is isomorphic to $\mathcal{U}(S)$ as $\mathcal{U}(S)$-module. But if $\mathcal{U}(S)$ is local, then modulo the maximal ideal of $R, \mathcal{U}(S)$ is connected and $S$ is tame; hence by [Hu84], Theorem 4.4. $S$ is $\mathcal{U}(S)$-Galois.

Now letting $\mathcal{U}(S) = \mathcal{B}, \mathcal{O}(\mathcal{B})$ is the largest order in $\mathcal{O}$ on which $\mathcal{B}$ acts, hence $S \subseteq \mathcal{O}(\mathcal{B})$. The inclusion map is an $R$-algebra, $\mathcal{B}$-module homomorphism, hence, since $S$ is Galois, $\mathcal{O}(\mathcal{B}) \cong S \otimes I\mathcal{O}(\mathcal{B}), I$ the space of integrals of $\mathcal{B}$. But $IS = R$, hence $I\mathcal{O}(\mathcal{B}) = R$ and $S = \mathcal{O}(\mathcal{B})$. □

The requirement that $\mathcal{B}$ be local is needed : if $\mathcal{B} = RG^*$, then $S$ can be tame, i.e. $\phi S = R$ for an integral $\phi$ of $\mathcal{B}$ but $S$ need not be Galois, for example if $\widetilde{\mathcal{O}} = R[z]$ and $S = R[cz]$ for a non-unit $c$ of $R$.

**Proposition 2.** If $\mathcal{B}$ is a Hopf order in $KG$ and $S$ is an order in $L$ which is $\mathcal{B}$-Galois, then $S = \mathcal{O}(\mathcal{B}), \mathcal{B} = \mathcal{U}(S)$ and $S$ is the unique $\mathcal{B}$-Galois extension which is an order in $L$.

**Proof.** That $S = \mathcal{O}(\mathcal{B})$ and $S$ is the unique $\mathcal{B}$-Galois extension was proved in the second paragraph of the last proof. To show $\mathcal{B} = \mathcal{U}(S)$, note that since $S$ is $\mathcal{B}$-Galois, $S \cong \mathcal{B}$ as $\mathcal{B}$-module, hence $S = \mathcal{B}s$ for some basis element $s$ in $S$. Now $\mathcal{B} \subseteq \mathcal{U}(S)$; but $\mathcal{B}s = \mathcal{U}(S)s$ since both sides equal $S$. Viewing both as submodules of $KGs$ and noting that since $s$ is an $\mathcal{B}$-basis of $S$, $s$ is also a $KG$-basis of $L$, it follows that $\mathcal{B} = \mathcal{U}(S)$. □

The uniqueness in Proposition 2 means that inside any Kummer extension $L/K$ with group $G$ cyclic or order $q$ live at most as many Galois extensions as there are Hopf orders in $KG$.

If $q = p$, prime, then one can describe all Galois extensions inside $L$ because the Hopf orders in $KG$ have been completely classified by Tate and Oort [TO70]. Namely, any Hopf order in $KG$ is, as algebra, of the form $H_b = R[x]/(x^p - bx)$ where $b$ is a $p-1^{\text{st}}$ power and $b$ divides $p$; $b$ is unique up to the $p-1^{\text{st}}$ power of a unit of $R$. Using this classifications and the correspondence above, we have

**Theorem** (ChT87], Theorem 14.1). If $\widetilde{\mathcal{O}}$ is not $\overline{RG}$-Galois, then there are no Galois extensions $S$ of $R$ which are orders over $R$ in $L$. If $\widetilde{\mathcal{O}}$ is $\overline{RG}$-Galois, then there are bijections between

$$\text{(Hopf orders over } R \text{ in } KG \text{ containing } \mathcal{U}_0)$$

{orders over $R$ in $L$ containing $\widetilde{\mathcal{O}}$ which are Galois extensions of $R$}, and

$$\{\text{ideals } J \text{ of } R,\ pR \subseteq J \subseteq \mathrm{tr}(\mathcal{O}),\ J \text{ a } p-1^{\mathrm{st}} \text{ power}\}$$

**Sketch of proof.** The correspondence between the first two sets have already been defined. The correspondence from orders $S$ in $L$ to ideals of $R$ is given by the map $S \mapsto \mathrm{tr}(S)$.

Suppose $\pi R$ is the maximal ideal of $R$ and $pR = \pi^{e(p-1)}R$. Suppose $L = K[z]$ with $z$ chosen so that $z^p = w = 1+\pi^{qp+r}u$, let $x = (z-1)/\pi^s$. for any $s, 0 < s \leq \min\{q, e\}$. Then $x$ is integral over $R$. Let $S = R[x]$. Let $H_b = R[\xi], b = \pi^{s(p-1)}$, acting on $S$ via $\xi x = 1+\pi^s x = z$. Then $S$ is an $H$-module algebra. One shows that $\phi = b-\xi^{p-1}$ generates the space of integrals of $H_b$, and $\phi(x^{p-1}) = 1$, hence $\phi S = R$, and $S$ is tame. Since $s > 0$, $S$ is $H_b$-Galois by the proof of Proposition 1. If $q = 0$ then the only possible Galois extension in $L$ is $\widetilde{\mathcal{O}} = R[z]$, and this is a Galois $H_1 \cong \overline{RG}$-extension iff $w$ is a unit of $R$. Propositions 1 and 2 and the classification of Tate and Oort then complete the bijection between Hopf orders and Galois extensions.

Since $\phi = b - \xi^{p-1} = \mathrm{tr}/\pi^{(e-s)(p-1)}$, we have $\mathrm{tr}(S) = pR/\pi^{s(p-1)}$. Thus $\mathrm{tr}(S)$ is a $p - 1^{\mathrm{st}}$ power which is contained in $\mathrm{tr}(\mathcal{O})$ and contains $pR$. □

This correspondence globalizes ([ChT87],17.5). Suppose $R$ is the ring of integers of a number field $K$ containing a primitive $p^{\mathrm{th}}$ root of unity. Let $L/K$ be a Kummer extension of order $p$. If for each prime $q$ of $R$ there exists an element $z$ of $L$ so that $L = K[z], z^p = w$ in $K$ and $w$ is a unit mod $q$, then there exist Galois extensions $S$ of $R$ which are orders over $R$ in $L$ : in particular the Kummer order of $L$. Otherwise there exist no Galois extensions. If Galois extensions exist, then they are in $1-1$ correspondence with ideals of $R$ which are $p - 1^{\mathrm{st}}$ powers and which contain $pR(\mathrm{tr}(\mathcal{O}_L))^{-1}$. In particular, if the Kummer order is a Galois extension, then $\mathcal{O}_L$ itself is a Galois extension iff its associated order $\mathcal{U}$ is a Hopf algebra, iff $tr(\mathcal{O}_L)$ is the $p - 1^{\mathrm{st}}$ power of an ideal of $R$.

All of this depends on the Tate-Oort classification of Hopf algebras of rank $p$, prime, over local and global number rings. To extend these results to $q$ a prime power, it would be desirable to have some kind of classification of rank $q$ abelian Hopf algebras over complete local dvrs. But to my knowledge the only general construction of abelian Hopf algebras of prime power rank over complete discrete valuation rings is via formal groups. Taylor [T87], [T88] has taken advantage of those constructions and shown that if $H$ is such a Hopf algebra and $S$ is its corresponding order inside a Kummer extension in the sense of formal groups, then $S$ is $H$-Galois. However, the rank $p$ Hopf algebras which arise from formal groups are only those of the form $H_b$ where $bR = p^s, pR = p^e$ and $s$ divides $e$, not just $s \leq e$. Thus a classification of abelian Hopf algebras of prime power rank over valuation rings of local fields appears not yet available.

Finally, as suggested earlier, these results have implications for the construction of Azumaya algebras and Brauer classes of a ring of integers $R$ via smash products of Galois extensions. For such that $R$, all Brauer classes are represented by maximal orders in certain quaternion algebras over the number field $K$, so of particular interest are rank 4 Azumaya $R$-algebras, obtainable as smash products of rank 2 Galois extansions of $R$.

The results above imply that to obtain all rank 2 Galois extensions of a number ring $R$, it suffices to determine all Galois extensions $\widetilde{S}$ for the Hopf algebra $\overline{RG} = RG^*$, $G$ cyclic or order 2, and then determine the Hopf algebras which lie between the order of $S$ and $\overline{RG}$, and their corresponding Galois extensions which are orders in $\widetilde{S} \otimes K$.

The determination of Gal($RG^*$) has been available for some time; assuming that known, it follows that the number of smash products is bounded by

$$|\text{Gal}(RG^*)|^2 . |\text{ Hopf algebras } H_B|$$

Both factors depend on the arithmetic of $R$.

By this approach, one can show [Ch89] that if $m$ is any prime congruent to 1 modulo 4, then the non-trivial class in $\text{Br}(\mathcal{O}_m), \mathcal{O}_m$ the ring of integers of $\mathbb{Q}(\sqrt{m})$, cannot be represented by the smash product of two rank 2 Galois extensions of $\mathcal{O}_m$, and that for totally real number fields $K$ of dimension $n$ over $\mathbb{Q}$, if the discriminant of $K/\mathbb{Q}$ is sufficiently large there will exist rank 4 Azumaya $\mathcal{O}_K$-algebras which are not smash products of Galois extensions of Galois extensions of $\mathcal{O}_K$.

**References.**


[Be78] A. Berge, *Arithmétique d'une extension Galoisienne a groupe d'inertie cyclique*, Ann. Inst. Fourier, Grenoble 28, 4(1978), 17-44.

[Bg85] G. Bergman, *Everybody knows what a Hopf algebra is*, AMS Contemporary Math. 43 (1985), 25-48.

[BF72] F. Bertrandias, M.J. Ferton, *Sur l'anneau des entiers d'une extension cyclique de degré premier d'un corps local*, C. R. Acad. Sc. Paris 274 (1972), A, 1330-1333.

[Bo79] Z. Borevich, *Kummer extensions of rings*, J. Soviet Math. 11, (1979), 514-534.

[C76] S. U. Chase, *Infinitesimal group scheme actions on finite field extensions*, Americal J. Math. 98 (1976), 441-480.

[CHR65] S. U. Chase, D. K. Harrison, A. Rosenberg, *Galois Theory and Galois Cohomology over a Commutative Ring*, Mem. Amer. Math. Soc. 52 (1965), 15-33.

[CS69] S. U. Chase and M. E. Sweedler, *Hopf Algebras and Galois Theory*, Springer Lecture Notes in Math, 97 (1969).

[Ch71] L. Childs, *Abelian Galois extensions of rings containing roots of unity*, Illinois J. Math. 15 (1971), 273-280.

[Ch77] L. Childs, *The group of unramified Kummer extensions of prime degree*, Proc. London Math. Soc 35 (1977), 407-422.

[Ch86] L. Childs, *Products of Galois objects and the Picard invariant map*, Math. J. Okayama Univ., 28 (1986), 29-36.

[ChP87] L. Childs, *Representing classes in the Brauer group of quadratic number rings as smash products*, Pacific J. Math. (to appear).

[ChT87] L. Childs, *Taming wild extensions by Hopf algebras*, Trans Amer. Math. Soc. (to appear).

[ChC87] L. Childs, *Non-isomorphic equivalent Azumaya Algebras*, Canad. Math. Bull. (in press.)

[Ch89] L. Childs, *Azumaya algebras which are not smash products*, Rocky Mountain J. Math. (to appear).

[CM74] L. Childs, A. Magid, *The Picard invariant of a principal homogeneous space*, J. Pure Appl. Algebra 4 (1974), 273-286.

[CH86] L. Childs and S. Hurley, *Tameness and local normal bases for objects of finite Hopf algebras*, Trans. Amer. Math. Soc. 298 (1986), 763-778.

[Da82] E. Dade, *The equivalence of various generalizations of group rings and modules*, Math. Z. 181 (1982), 335-344.

[Fr62] A. Frolich, *Module structure of Kummer extensions over Dedekind domains*, J. Reine Angew. math. 209 (1962), 39-53.

[GH69] J. Gamst and K. Hoechsmann, *Quaternions generalises*, C. R. Acad. Sci. Paris 269 (1969)A, 560-562.

[GrH87] C. Greither and D. Harrison, *A Galois correspondence for radical extensions of fields*, J. Pure Appl. Algebra 43 (1987), 257-270.

[GP87] C. Greither and B. Pareigis, *Hopf Galois theory for separable field extensions*, J. Algebra 106 (1987), 239-258.

[Gre88] C. Greither, *On Galois extensions of commutative rings with Galois Group* $Z/p^nZ$, preprint.

[Ho73] R. T, Hoobler, *Purely inseparable Galois theory*, in : Ring Theory, proc. Oklahoma Conference, Dekker, 1973, 207-240.

[Hu84] S. Hurley, *Tame and Galois Hopf objects with normal bases*, Thesis, SUNY at Albany, 1984.

[Hu87] S. Hurley, *Galois objects with normal bases for free Hopf Algebras of Prime Degree*, J. Algebra (to appear).

[J64] N. Jacobson, *Lectures in Abstract Algebra III : Theory of Fields and Galois Theory*, Van Nostrand, 1964.

[Ker83] I. Kersten, *Eine neue Kummertheorie für zyklische Galoiserweiterungen vom Grad* $p^2$, Algebra Berichte 45 (1983).

[KM84] I. Kersten and J. Michalichek, *Applications of Kummer theory without roots of unity*, in Methods of Ring Theory, F. Van Oystaeyen ed. D. Reidel, 1984, 201-205.

[KM88] I. Kersten and J. Michalichek, *Kummer theory without roots of unity*, J. Pure Appl. Algebra (to appear).

[KC76] H. F. Kreimer and P. Cook, *Galois theories and normal bases*, J. Algebra 43 (1976), 115-121.

[Kr82] H. F. Kreimer, *Quadratic Hopf algebras and Galois extensions*, Contemp. Math. 13(1982), 353-361,

[Le59] H. W. Leopoldt, *Uber die Hauptordnung der ganzen Elemente eines abelschen Zahlkorpers*, J. reine angew. Math. 201 (1959), 119-149.

[Mau84] D. Maurer, *Stickelberger's criterion, Galois algebras and tame ramification in algebraic number fields*, J. Pure Appl. Algebra 33, 1984, 281-293.

[Or69] M. Orzech, *A cohomological description of abelian Galois extensions*, Trans. Amer. Math. Soc. 137 (1969), 481-499.

[Sh69] S. Shatz, *Principal homogeneous spaces for finite group schemes*, proc. Amer. Math. Soc. 22 (1969), 678-680.

[Sm72] C. Small, *The group of quadratic extensions*, J. Pure Appl. Algebra 2(1972), 83-105, 395.

[Sw69] M. Sweedler, *Hopf algebras*, W. A. Benjamin, 1969.

[TO70] J. Tate and F. Oort, *Group schemes of prime order*, Ann. Sci. Ecole Norm. Sup. (4) 3 (1970), 1-21.

[T87] M. Taylor, *Hopf structure and the Kummer theory of formal groups*, J. reine angew. Math. 375/376 (1987), 1-11.

[T88] M. Taylor, *A note on Galois modules and group schemes*, preprint.

[Ul81] K. H. Ulbrich, *Vollgraduierte Algebren*, Abh. Math. Sem. Univ. Hamburg 51 (1981), 136-148.

[Ul86] K. H. Ulbrich, *Sur le smash-produit des algèbres galoisiennes*, C. R. Acad. Sci. Paris 303, Série 1 (1986), 769-772.

[Wa71] W. C. Waterhouse, *Principal homogeneous spaces and group scheme extensions*, Trans. Amer Math. Soc. 153 (1971), 181-189.

[Wa87] W. C. Waterhouse, *A unified Kummer-Artin-Schreier sequence*, Math. Ann. 277 (1987), 447-451.

[Wa88] W. C. Waterhouse, *Tame objects for finite commutative Hopf algebras*, preprint.

[Wy87] T. Wyler, *Torsors under abelian p-groups*, J. Pure Appl. Algebra 45 (1987), 273-286.

# The Brauer Long group of Z/2 dimodule algebras

Frank DeMeyer
Dept. of Math.
Colorado State University
Fort Collins Colorado 80523
U.S.A.

Beginning with the papers of G. Azumaya [2] and M. Auslander and O. Goldman [1] introducing the Brauer group B(R) of a commutative ring R, there have followed a series of extensions of the Brauer group of division algebras over a field to satisfy various needs. A. Grothendeick viewed the Brauer group of a commutative ring as the local part of a Brauer group of schemes and related the Brauer group to the second etale cohomology group of the scheme with values in the units scheaf [11]. At about the same time, C. T. C. Wall introduced in [14] a Brauer group of equivalence classes of **Z**/2 graded algebras over a field with multiplication induced by a twisted tensor product to study the Witt ring of quadratic forms. Wall's construction was extended to commutative rings R by H. Bass and C. Small [13], and this group is now called the Brauer Wall group of R and denoted BW(R). A Brauer group of algebras over a field graded by an arbitrary finite abelian group was introduced by M. Knus [9], and extended to arbitrary commutative rings (with a twisted multiplication) by L. Childs, G. Garfinkel and M. Orzech [5]. The Childs-Garfinkel-Orzech construction contained the Brauer Wall group as a special case. An "equivariant Brauer group" of algebras on which a fixed group acted as a group of automorphisms was constructed by O. Frolich and C. T. C. Wall. In his thesis [10], F. W. Long introduced a Brauer group of dimodule algebras on which a grading group G acted as a group of automorphisms which included the affine versions of all the previous extensions of the Brauer group as subgroups. Long's group is now called the Brauer Long group of R and is denoted BD(R,G). After Long introduced his group, a steady stream of papers have considered the properties and calculations of BD(R,G) and its siblings. Some of these are listed among the references at the end of this report.

Here I want to focus on the properties and calculation of BD(R,G) in the easiest and historically most important case, when the grading-action group G is **Z**/2. These results are the consequence of joint work with Tim Ford and the details of the calculations will appear in [6]. All other unexplained terminology and notation will be as in Long's original work [10].



*F. van Oystaeyen and L. Le Bruyn (eds.), Perspectives in Ring Theory, 129–133.*

Let $G = \langle \sigma \mid \sigma^2 = e \rangle = \mathbf{Z}/2 = \{0, 1\} = \{+, -\}$ be the cyclic group of order = 2 and let R be a commutative ring with exactly two idempotents 0 and 1 in which 2 is a unit. A quadratic extension of R is a Galois extension S of R with Galois group G. The action of G on S induces a $\mathbf{Z}/2$ grading of S where $S_0 = R$ and $S_1 = \{ s \in S \mid \sigma(s) = -s\}$. $S_1$ is a Rank = 1 projective R module. If $a \in$ Units(R) then $S = R[x] / (x^2 - a) = R\langle a\rangle$ is a quadratic extension of R and these observations give the maps in the exact sequence

$$1 \longrightarrow R^*/ R^{*2} \longrightarrow H^1(R, \mathbf{Z}/2) \longrightarrow {}_2\mathrm{Pic}(R) \longrightarrow 1$$

where $R^*$ = Units(R), ${}_2\mathrm{Pic}(R)$ is the subgroup of elements of order 1 or 2 in the group of rank = 1 projective R modules, and $H^1(R, \mathbf{Z}/2)$ is the etale cohomology group which classifies the Galois extensions of R with Galois group isomorphic to $\mathbf{Z}/2$.

Let A be a finitely generated projective separable R algebra which is graded by $\mathbf{Z}/2$ and on which $G = \mathbf{Z}/2$ acts as a group of algebra automorphisms which respect the grading. If A is an Azumaya R algebra in the ungraded sense (the center of A = R) then A is a + dimodule algebra. If A is not an Azumaya algebra and $A = A_0 + A_1$ is the decomposition of A into homogeneous elements, where $A_i$ denotes the elements of A of degree i (i = 1,2), then A is a - dimodule algebra when $A_0$ is an Azumaya algebra. In either case A is just a dimodule algebra. Notice that each quadratic extension S of R has a non trival $\mathbf{Z}/2$ grading induced by the action of G on S, and with respect to this action grading pair, S is a - dimodule algebra. Let M be a finitely generated projective R module which is graded by $\mathbf{Z}/2$ and on which G acts as a group of R automorphisms which respect the grading. Such an M is called a dimodule. Notice that each dimodule algebra is a dimodule. If $\mathrm{END}_R(M)$ is the R algebra of R endomorphisms of M which respect the grading of M then there is a natural $\mathbf{Z}/2$ grading and action on $\mathrm{END}_R(M)$ which turns $\mathrm{END}_R(M)$ into a dimodule algebra. If A and B are dimodule algebras then the smash product A # B is a dimodule algebra where A # B is the tensor product of A and B as R modules with multiplication on homogeneous elements given by

$$(a_i \# b_j)\,(c_k \# d_q) = a_i\sigma^j(c_k) \# b_j d_q$$

Two dimodule algebras A and B are called equivalent if there are dimodules M and N with

$$A \# \mathrm{END}(M) = B \# \mathrm{END}(N)$$

The Brauer Long group $BD(R,\mathbf{Z}/2)$ is the group whose elements are the equivalence classes of dimodule algebras with multiplication induced by # on representatives[10]. This is the group we are trying to understand.

Let A be a dimodule algebra. The $\mathbf{Z}/2$ grading of A induces a decomposition $A = A_0 + A_1$ where $A_i$ denotes the elements of A of degree i (i = 1,2). If A is trivially graded then $A_1 = \phi$ and A is associated to the trivial quadratic extension R<1>. Otherwise $A^{A_0} = T$ is a quadratic extension of R and the isomorphism class of T depends only on the class of A in $BD(R,\mathbf{Z}/2)$ [13]. The action of G on A induces a grading where $A_0 = \{a \varepsilon A \mid \sigma(a) = a\}$ and $A_1 = \{ a \varepsilon A \mid \sigma(a) = -a\}$. This grading determines as above a quadratic extension S of R. The + algebras determine a subgroup BAz(R) of BD(R,**Z**/2) of index = 2 and the non identity coset can be represented by either R<1> or R<-1>, for example. Thus each - algebra A is equivalent in BD(R) to R<-1> # D for some + algebra D. The class of A in BD(R,**Z**/2) determines and is determined by the class of D. These observations give the definition of a function

f: BD(R, **Z**/2) ---> Z/2 x $H^1$(R, **Z**/2) x $H^1$(R, **Z**/2) x B(R)

which turns out to be a bijection and is given on representatives of equivalence classes by

(1) f(A) = (+, S, T, A) if A is a + algebra
(2) f(A) = (-, S, T, (R<-1> # A) (R<-1> # T)) if A is a - algebra

The point of the next result is that with the multiplication rules it gives, f is an isomorphism of groups. Of course, we guessed the multiplication rules from the map f and the multiplication in BD(R,**Z**/2), so the proof of this result amounts to a long calculation, checking the guess was really the right one. One more preliminary observation is that if S and T are quadratic extensions of R with their non degenerate action gradings, then S # T represents an element in B(R) and this element coincides with the result of the natural cup product pairing $H^1$(R, **Z**/2) x $H^1$(R, **Z**/2) ---> B(R) given since 2 is a unit in R [8].

Theorem: Let R be a commutative ring with exactly two idempotents 0 and 1. Assume 2 is a unit in R. Then BD(R,**Z**/2) is isomorphic under the map f given above to the group of 4-tuples **Z**/2 x $H^1$(R, **Z**/2) x $H^1$(R, **Z**/2) x B(R) under the product rules

(1) (+,S,T,D) (+,S',T',D') = (+,SS',TT',DD' (S # T'))
(2) (-,S,T,D) (+,S',T',D') = (-,SS',TT',DD' (R<-1>S # T'))
(3) (+,S,T,D) (-,S',T',D') = (-,TS',ST',DD' (T # R<-1>STT'))
(4) (-,S,T,D) (-,S',T',D') = (+,R<-1>TS',R<-1>ST',DD'(T # STT'))

The idea of the Theorem and its proof is clear but the details are involved and can be found in [6]. Since the smash product of quadratic extensions is given by the cup product map, this theorem reduces the calculation of BD(R,**Z**/2) to the calculation of the groups $H^1$(R, **Z**/2), B(R) and the cup product map $H^1$(R, **Z**/2) x $H^1$(R, **Z**/2) ---> B(R). Next we list some corollaries which illustrate that once the Brauer Long group has been calculated, the other related groups given in the introduction can be calculated too.

For the following corollaries let w: **Z**/2 x **Z**/2 ---> $H^1$(R, **Z**/2) be the symmetric bilinear pairing defined by w(a, b) = R<$(-1)^{ab}$>. Let Galz(R, **Z**/2 x **Z**/2) = BD(R, **Z**/2) / B(R) be the group of **Z**/2 x **Z**/2 Galois extensions of R as defined by L. Childs in [4]. Let BC(R, **Z**/2) be the subgroup of classes in BD(R, **Z**/2) with trivial **Z**/2 action and let BM(R, **Z**/2) be the subgroup of classes with trivial **Z**/2 grading (see, for example [3]).

Corollary 1. (a) BW(R) is isomorphic to the subgroup of BD(R,**Z**/2) consisting of all 4-tuples of the form (+-,S,S,D).

(b) BW(R) is isomorphic to the group $\mathbf{Z}/2 \times H^1(R, \mathbf{Z}/2) \times B(R)$ with multiplication given by
$(a,S,D)\ (a',S',D') = (a+a', SS'w(a,a'), DD'(Sw(a,a+a') \# S'w(a+a',a'))$

Corollary 2. Galz(R, $\mathbf{Z}/2 \times \mathbf{Z}/2$) is isomorphic to the set $\mathbf{Z}/2 \times H^1(R, \mathbf{Z}/2) \times H^1(R, \mathbf{Z}/2)$ with multiplication given by
(1) $(a,S,T)\ (0,S',T') = (a,SS', TT')$
(2) $(a,S,T)\ (1,S',T') = (a+1, w(a,1)TS', w(a,1)ST')$

Corollary 3. BAz(R) corresponds to the subgroup $\{+\} \times H^1(R, \mathbf{Z}/2) \times H^1(R, \mathbf{Z}/2) \times B(R)$ of BD(R,$\mathbf{Z}/2$).

Corollory 4. (a) BC(R,$\mathbf{Z}/2$) corresponds to the subgroup $\mathbf{Z}/2 \times \{1\} \times H^1(R, \mathbf{Z}/2) \times B(R)$ of BD(R, $\mathbf{Z}/2$).
(b) BM(R, $\mathbf{Z}/2$) corresponds to the subgroup $\mathbf{Z}/2 \times H^1(R, \mathbf{Z}/2) \times \{1\} \times B(R)$ of BD(R, $\mathbf{Z}/2$).

If G is a group let $G_2$ be the elements of G of order is a power of 2.

Corollary 5. (a) $BD(R, \mathbf{Z}/2)_2 = \mathbf{Z}/2 \times H^1(R, \mathbf{Z}/2) \times H^1(R, \mathbf{Z}/2) \times B(R)_2$ is a subgroup of BD(R, $\mathbf{Z}/2$).
(b) $BD(R, \mathbf{Z}/2) = BD(R, \mathbf{Z}/2)_2 + U$ where U is the complement of $B(R)_2$ in B(R).

Example [7]. Let R be the coordinate ring of a non singular affine real curve X, and let s denote the number of real components of X in the Euclidean topology. Then ${}_2Pic(R) = \{1\}$ and $H^1(R, \mathbf{Z}/2) = R^*/R^{*2} = (\mathbf{Z}/2)^s$. If $a,b \in R^*$ then R<a> # R<b> is an Azumaya algebra which is non trivial precisely on those real components of X on which both a, b are negative. Thus, if s = 1 then BD(R, $\mathbf{Z}/2$) = BD($\mathbf{R}$, $\mathbf{Z}/2$) = $\mathbf{D}_8$ is the dihedral group of order 16. In general, BD(R, $\mathbf{Z}/2$) is a non abelian group of order $2^{3s+1}$ when s > 1. Now $B(R) = \langle D_1, \ldots, D_s \mid D_i^2 = e, D_iD_j = D_jD_i$ all $i,j\rangle$ where $D_i$ is the class which is non trivial on the i'th real component of X and trivial on the other components. In the same way, $H^1(R, \mathbf{Z}/2) = \langle T_1, \ldots T_s \mid T_i^2 = e, T_iT_j = T_jT_i\rangle$ where $T_i = R\langle a_i\rangle$ with $a_i$ negative on the i'th real component of X and positive on the other components. Let $x_i = (0,R,D_i)$ and $y_i = (1,T_i,R)$ in the notation of Corollary 1. Then one can check that $y_i^4 = \Pi x_j$ for all i and $y_i^{-1}x_jy_i = \Pi x_k$ where the product is over all k but k = j. This gives the group BW(R) by generators and relations.

I have restricted attention here to $\mathbf{Z}/2$ graded algebras, but M. Beattie and S. Caenepeel have given a description of BAz(R,G) along the same lines as the Theorem when G is a cyclic group of prime order. The index of BAz(R,G) in BD(R,G) is two in this case. Both Lindsay Childs and Tim Ford have been examining the cup product map on $H^1(R, \mathbf{Z}/2)$. Some of their results are contained in this volume and make it easy to give calculations of BD(R, **Z**/2) for other choices of R.

## REFERENCES

[1] M. Auslander and O. Goldman, The Brauer group of a commutative ring, Trans. Amer. Math. Soc. 97 (1960), 367-409.

[2] G. Azumaya, On maximally central algebras, Nagoya J. Math. 2 (1951), 119-150.

[3] M. Beattie, The subgroup structure of the Brauer group of RG-dimodule algebras, Ring Theory, Lecture Notes in Math. Vol. 1197, Springer-Verlag, Berlin (1986), 20-30.

[4] L.N. Childs, The Brauer group of graded Azumaya algebras II: graded Galois extensions, Trans. Amer. Math. Soc. 204 (1975), 137-160.

[5] L.N. Childs, G. Garfinkel and M. Orzech, The Brauer group of graded Azumaya algebras, Trans. Amer. Math. Soc. 175 (1973), 299-326.

[6] F. DeMeyer and Tim Ford, Computing the Brauer Long group of Z/2 dimodule algebras, J. Pure and Applied Algebra (to appear).

[7] F. DeMeyer and M. A. Knus, The Brauer group of a real curve, Proc. Amer. Math. Soc. 57 (1976), 227-232.

[8] J. Gamst, K. Hoechsmann, Quaternions generalises, C. R. Acad. Sci. Paris 269A (1969), 560-562.

[9] M. A. Knus, Algebras graded by a group; Category Theory, Homology Theory, and their applications II. Lecture Notes in Math. Vol. 92, Springer-Verlag, Berlin (1969), 117-133.

[10] F. W. Long, A generalization of the Brauer group of graded algebras, Proc. London Math. Soc. 29 (1974) 237-256.

[11] J. S. Milne, Etale cohomology, Princeton Math Series #33, Princeton Univ. Press, Princeton, NJ, 1980.

[12] M. Orzech, Brauer groups of graded algebras, Brauer Groups, Evanston 1975, Lecture Notes in Math. Vol. 549, Springer-Verlag, Berlin, 1976.

[13] C. Small, The Brauer-Wall group of a commutative ring, Trans. Amer. Math. Soc. 156 (1971), 455-491.

[14] C.T.C. Wall, Graded Brauer groups, J. Reine Agnew. Math. 213 (1963/64), 187-199.

# ON THE BRAUER GROUP AND THE CUP PRODUCT MAP

Timothy J. Ford (*)
Department of Mathematics
Florida Atlantic University
Boca Raton, Florida 33431

ABSTRACT. This article is concerned with the cup product map $\mu:H^1(X,Z/n)\otimes H^1(X,Z/n) \to {}_nB(X)$. Under certain conditions we describe the image and kernel of $\mu$ for the spectrum of $k[x_1, \ldots, x_\nu, f^{-1}]$ and for a fiber product space.

0. Throughout X will be a connected scheme over $Z[1/n][\omega]$ where n > 1 is an integer and $\omega$ is a primitive n-th root of unity. We denote by B(X) the Brauer group of X and by B'(X) the cohomological Brauer group of X [13]. For any abelian group A we let ${}_nA$ denote the subgroup of A annihilated by n. All cohomology and sheaves are for the étale topology. Let $G_m$ denote the sheaf of units on X and $\mu_n$ the

(*) This research was partially supported by the NSF.



*F. van Oystaeyen and L. Le Bruyn (eds.), Perspectives in Ring Theory, 135–145.*

sheaf of n-th roots of unity. The sequence

$$1 \to \mu_n \to G_m \xrightarrow{n} G_m \to 1 \tag{1}$$

is exact. Since $\Gamma(X,G_m)$ contains $\omega$, $\mu_n$ is (noncanonically) isomorphic to the constant sheaf $\mathbf{Z}/n$. The long exact sequence of cohomology associated to (1) is

$$\begin{aligned}1 \to \mu_n \to \Gamma(X,G_m) \xrightarrow{n} \Gamma(X,G_m) \to H^1(X,\mathbf{Z}/n) \to \text{Pic } X \xrightarrow{n} \text{Pic } X \\ \to H^2(X,\mathbf{Z}/n) \to B'(X) \xrightarrow{n} B'(X) \to \dots\end{aligned} \tag{2}$$

where we have identified $\text{Pic } X = H^1(X,G_m)$ and $B'(X) = \text{tors}(H^2(X,G_m))$. If X is affine, it is known that $B(X) = B'(X)$ under the natural map $B(X) \to H^2(X,G_m)$ [11], [17]. The cup product map [21, V.1.17] $H^1(X,\mathbf{Z}/n)\otimes H^1(X,\mathbf{Z}/n) \to H^2(X,\mathbf{Z}/n)$ followed by the homomorphism $H^2(X,\mathbf{Z}/n) \to {}_nB'(X)$ defines a homomorphism

$$\mu\colon H^1(X,\mathbf{Z}/n)\otimes H^1(X,\mathbf{Z}/n) \to {}_nB'(X) \tag{3}$$

which will also be called cup product.

This article is concerned with the study of the map $\mu$. If X is the spectrum of a field k this problem has been completely solved by Merkurjev [19], [22] if $n = 2$ and by Merkurjev and Suslin [20] for all $n > 1$. For Spec k, $\mu$ is always surjective and ker $\mu$ is the Steinberg relation group of k. In [3] L. Childs shows that if R is the ring of algebraic integers in a number field, then ${}_nB(R)$ is not always generated by im $\mu$.

The group $H^1(X,\mathbf{Z}/n)$ classifies Galois covers of X with group $\mathbf{Z}/n$. It is known that $\mu$ corresponds to taking the smash product of two cyclic Galois covers of X [12]. Since the smash product of cyclic Galois extensions is an Azumaya algebra, im $\mu \subseteq {}_nB(X)$. When n = 2 it is shown in [7] that $\mu$ is intimately connected to the group structure of the Brauer-Wall group BW(X) and the Brauer-Long group $BD(X,\mathbf{Z}/2)$. To compute $BD(X,\mathbf{Z}/2)$ it suffices to compute B(X), $H^1(X,\mathbf{Z}/2)$, and the cup product map $\mu$.

1. First we consider rings of the form $R = k[x_1, \ldots, x_\nu, f^{-1}]$. If f factors into linear polynomials, Theorem 1 shows $\mu$ is onto and ker $\mu$ is described. Examples 2 and 3 illustrate that this is not the case in general.

Let $Y_0, \ldots, Y_m$ be distinct hyperplanes in $\mathbf{P}^\nu$, $\nu > 1$. Let $Y = Y_0 \cup \ldots \cup Y_m$. Let P denote the singular set of Y, $P = \{Y_i \cap Y_j \mid i \neq j\}$. Write $P = p_1 \cup \ldots \cup p_s$ where the $p_i$ are the irreducible components of P. Each p is a linear subvariety of $\mathbf{P}^\nu$ of codimension 2, hence is isomorphic to $\mathbf{P}^{\nu-2}$. Define a graph $\Gamma$ associated to Y. The vertices of $\Gamma$ are the hyperplanes $Y_0, \ldots, Y_m$ and the varieties $p_1, \ldots, p_s$. There is an edge connecting $Y_i$ and $p_j$ if and only if $p_j$ is a subvariety of $Y_i$. The graph $\Gamma$ is bipartite and connected. We orient $\Gamma$ by taking the positive end of an edge E the $Y_i$ and the negative end the $p_j$. Let e be the number of edges.

Theorem 1. [10, Theorem 1] Let k be an algebraically closed field of characteristic p. Let $f_1, \ldots, f_m$ be linear polynomials in $k[x_1, \ldots, x_\nu]$ and $R = k[x_1, \ldots, x_\nu][f_1^{-1}, \ldots, f_m^{-1}]$. Let $Y_0$ be the hyperplane at infinity and $Y_1, \ldots, Y_m$ the complete hyperplanes in $P^\nu$ defined by $f_1, \ldots, f_m$. Assume that the $Y_i$ are distinct. Let $Y = Y_0 \cup \ldots \cup Y_m$ and $\Gamma$ the graph of Y. Then modulo p-groups $B(R) \simeq Q/Z^{(r)}$ where $r = e - m - s$ is the rank of the cycle space of $\Gamma$. The cup product map $\mu : H^1(R, Z/n) \otimes H^1(R, Z/n) \to {}_nB(R)$ is surjective for all n relatively prime to p and ker $\mu$ is generated by

$$\{f_i \otimes f_j \mid Y_i \cap Y_0 = Y_j \cap Y_0\} \cup$$
$$\{(f_i \otimes f_j)(f_i \otimes f_t)(f_j \otimes f_t)^{-1} \mid Y_i \cap Y_t = Y_j \cap Y_t\}.$$

Example 2. Let k = C be the field of complex numbers. Choose four points in the affine plane over k not all on a conic of the form $y = ax^2 + bx + c$ and no three on a line. Choose four conics A, B, C, D each with equation of the form $y = ax^2 + bx + c$, each passing through exactly three of the above points, no two conics containing the same three points. Let $R = k[x,y][\alpha^{-1}, \beta^{-1}, \gamma^{-1}, \delta^{-1}]$ where $\alpha$, $\beta$, $\gamma$, $\delta$ are the polynomials in k[x,y] corresponding to A, B, C, D. In [10] it is shown that $H^1(R, Z/2) = (Z/2)^{(4)}$, ${}_2B(R) = (Z/2)^{(8)}$ and im $\mu \simeq (Z/2)^{(5)}$. Thus $\mu$ is not surjective.

Example 3. Let k be an algebraically closed field and n relatively prime to the characteristic of k. Let $f = x^n - y^{n-1}z$ and $R = k[x, y, z,$

$f^{-1}]$. In [10] it is shown that $H^1(R,Z/n) \simeq Z/n$, $B(R) \simeq Z/n$ and $\mu$ is the zero map.

2. Now we consider fiber product spaces. In Example 4 we see that the Brauer group of a Laurent polynomial ring is generated by cup products and the Brauer group of the base ring. Corollary 6, a Künneth formula for the Brauer group, gives sufficient conditions for ${}_nB'(X\times Y)$ to be generated by ${}_nB'(X)$, ${}_nB'(Y)$ and cup products.

Example 4. Suppose R is a $Z[1/n][\omega]$- algebra and Spec R is connected. Let t be an indeterminate. In [9] it is shown

$$ {}_nB(R[t,1/t]) \simeq {}_nB(R) \oplus (H^1(R,Z/n) \;/\; (C/nC)) \tag{4} $$

where $C = \text{Pic } R[t,1/t] \;/\; \text{Pic } R$. The homomorphism $H^1(R,Z/n) \to {}_nB(R)$ is induced (noncanonically) by taking the smash product of a cyclic Galois extension L with the cyclic extension $R[t,1/t][t^{1/n}]$. Therefore ${}_nB(R[t,1/t])$ is generated by ${}_nB(R)$ and im $\mu$. If R contains an algebraically closed field this is a special case of Corollary 6.

Theorem 5. [21, VI.8.25] and [4, Th. finitude, 1.11] Let X and Y be schemes of finite type over the separably closed field k. Let n be relatively prime to the characteristic of k. Let F and G be sheaves of Z/n modules on X and Y respectively. The Künneth map

$$R\Gamma(X,F) \otimes^L R\Gamma(Y,G) \to R\Gamma(X\times Y, F\boxtimes^L G) \qquad (5)$$

is a quasi-isomorphism.

Corollary 6. Let X and Y be connected schemes of finite type over the algebraically closed field k. Let $n > 1$ be relatively prime to the characteristic of k. Suppose $H^i(X,Z/n)$ is a free $Z/n$- module for $i \geq 0$. Then the following sequences are exact, where C is defined by the first sequence.

$$0 \to \mathrm{Pic}\, X \oplus \mathrm{Pic}\, Y \to \mathrm{Pic}\, X\times Y \to C \to 0 \qquad (6)$$

$$0 \to C/nC \to H^1(X,Z/n)\otimes H^1(Y,Z/n) \xrightarrow{\mu} {}_nB'(X\times Y) \to {}_nB'(X) \oplus {}_nB'(Y) \to 0 \qquad (7)$$

Proof: Because $H^i(X,Z/n)$ is flat for $i \geq 0$ Theorem 5 gives

$$H^2(X\times Y,Z/n) \simeq \oplus_{p+q=2} H^p(X,Z/n)\otimes H^q(Y,Z/n) \qquad (8)$$

Since k is algebraically closed the natural projections $X\times Y \to X$, $X\times Y \to Y$ admit sections. Therefore, the natural maps $B'(X) \oplus B'(Y) \to B'(X\times Y)$ and $\mathrm{Pic}\, X \oplus \mathrm{Pic}\, Y \to \mathrm{Pic}\, X\times Y$ split. Let C be defined by (6). Then $\mathrm{Pic}\, X\times Y \simeq \mathrm{Pic}\, X \oplus \mathrm{Pic}\, Y \oplus C$ and $\mathrm{Pic}(X\times Y) / n\mathrm{Pic}(X\times Y) \simeq \mathrm{Pic}\, X / n\mathrm{Pic}\, X \oplus \mathrm{Pic}\, Y / n\mathrm{Pic}\, Y \oplus C / nC$. Kummer theory (2) gives a commutative diagram

$$\begin{array}{ccccccccc}
0 \to & \mathrm{Pic}\,Y / n\mathrm{Pic}\,Y & \to & H^2(Y,\mathbf{Z}/n) & \to & {}_nB'(Y) & \to 0 \\
& \downarrow & & \downarrow & & \downarrow & \\
0 \to & \mathrm{Pic}\,X{\times}Y / n\mathrm{Pic}\,X{\times}Y & \to & H^2(X{\times}Y,\mathbf{Z}/n) & \to & {}_nB'(X{\times}Y) & \to 0 \\
& \uparrow & & \uparrow & & \uparrow & \\
0 \to & \mathrm{Pic}\,X / n\mathrm{Pic}\,X & \to & H^2(X,\mathbf{Z}/n) & \to & {}_nB'(X) & \to 0
\end{array} \tag{9}$$

with split vertical arrows and exact rows. From (9) we have the exact sequence:

$$0 \to \frac{\mathrm{Pic}(X{\times}Y)\otimes\mathbf{Z}/n}{(\mathrm{Pic}\,X \oplus \mathrm{Pic}\,Y)\otimes\mathbf{Z}/n} \to \frac{H^2(X{\times}Y,\mathbf{Z}/n)}{H^2(X,\mathbf{Z}/n)\oplus H^2(Y,\mathbf{Z}/n)}$$

$$\to \frac{{}_nB'(X{\times}Y)}{{}_nB'(X)\oplus {}_nB'(Y)} \to 0 \tag{10}$$

Combining (10) and (8) yields (7).

<u>Corollary 7.</u> Let X and Y be smooth curves over the algebraically closed field k of characteristic p. If n is relatively prime to p, then there are exact sequences

$$0 \to \mathrm{Pic}\,X \oplus \mathrm{Pic}\,Y \to \mathrm{Pic}\,X{\times}Y \to C \to 0 \tag{11}$$

$$0 \to C/nC \to H^1(X,\mathbf{Z}/n)\otimes H^1(Y,\mathbf{Z}/n) \xrightarrow{\mu} {}_nB(X{\times}Y) \to 0 \tag{12}$$

where C is defined by the first sequence.

Proof: For the smooth surface $X \times Y$, $B(X \times Y) = B'(X \times Y)$. For smooth curves the groups $H^i(X, Z/n)$ are free $Z/n$- modules, $i \geq 0$. Over k the Brauer group of a curve is trivial.

Corollary 8. Let X and Y be projective nonsingular varieties over the algebraically closed field k of characteristic p. Assume n is relatively prime to p and either

a. X and Y are both curves, or

b. n is a prime.

If $C = \text{Pic } X \times Y / (\text{Pic } X \oplus \text{Pic } Y)$ and

$$\mu: K(X)^*/K(X)^{*n} \otimes K(Y)^*/K(Y)^{*n} \to B(K(X \times Y)) \qquad (13)$$

then $C/nC \simeq \ker \mu$.

Proof: There is a natural injection $H^1(X, Z/n) \otimes H^1(Y, Z/n) \to K(X)^*/K(X)^{*n} \otimes K(Y)^*/K(Y)^{*n}$. Choose arbitrary open subsets U and V of X and Y respectively. Let $D = \text{Pic } U \times V / (\text{Pic } U \oplus \text{Pic } V)$. The diagram

$$\begin{array}{ccccccccc} 0 & \to & \text{Pic } X \oplus \text{Pic } Y & \to & \text{Pic } X \times Y & \to & C & \to & 0 \\ & & \downarrow\alpha & & \downarrow\beta & & \downarrow\phi & & \\ 0 & \to & \text{Pic } U \oplus \text{Pic } V & \to & \text{Pic } U \times V & \to & D & \to & 0 \end{array} \qquad (14)$$

commutes. Since $\alpha$ and $\beta$ are surjective $\phi$ is surjective. The diagram

$$\begin{array}{ccccccc} 0 \to C/nC & \to & H^1(X,\mathbf{Z}/n)\otimes H^1(Y,\mathbf{Z}/n) & \to & {}_nB(X\times Y) \\ \downarrow\sigma & & \downarrow\tau & & \downarrow\gamma \\ 0 \to D/nD & \to & H^1(U,\mathbf{Z}/n)\otimes H^1(V,\mathbf{Z}/n) & \to & {}_nB(U\times V) \end{array} \tag{15}$$

commutes and $\tau$ and $\gamma$ are one-to-one. Therefore $\sigma$ is one-to-one, hence $C/nC \simeq D/nD$. Taking the limit of $D/nD$ over all U and V gives ker $\mu$.

Example 9. Let X be a projective nonsingular elliptic curve over the algebraically closed field k. Suppose char k = p, $p \neq 2$. With notation taken from [16, IV.4], say $\tau = i$, $j = 1728$. Then Pic X $\simeq \mathbf{Z} \oplus (\mathbf{R}/\mathbf{Z})^2$. If C is as in (11), then $C = \mathrm{End}(X,P_0) \simeq \mathbf{Z}[i]$. Thus $C/nC \simeq (\mathbf{Z}/n)^2$. $H^1(X,\mathbf{Z}/n) \simeq {}_n\mathrm{Pic}\ X \simeq (\mathbf{Z}/n)^2$. Applying Corollary 7 we get ${}_nB(X\times X) \simeq (\mathbf{Z}/n)^2$. Modulo p-groups, $B(X\times X) \simeq (\mathbf{Q}/\mathbf{Z})^2$.

Example 10. Let k be the complex number field. Let X be the complement of the curve $x^n = y^{n-1}z$ in the projective plane $P^2$. Let Y be the affine nodal cubic curve $y^2 = x^2(x+1)$. In [8] it was shown that $B(X\times Y) \simeq \mathbf{Z}/n$. One can compute $B(X) = B(Y) = (0)$, Pic X = $\mathbf{Z}/n$, $H^1(X,\mathbf{Z}/n) = \mathbf{Z}/n$, $H^2(X,\mathbf{Z}/n) = \mathbf{Z}/n$, $H^1(Y,\mathbf{Z}/n) = \mathbf{Z}/n$ and in (6) $C = (0)$. Applying Corollary 6 we see that the generator of $B(X\times Y)$ is a cup product.

Corollary 11. Let X and Y be connected schemes of finite type over the algebraically closed field of characteristic p. Let $n > 1$ be relatively prime to p. Suppose $H^i(X,\mathbf{Z}/n)$ is a free $\mathbf{Z}/n$- module for $i \geq 0$. If ${}_nB(X) = {}_nB'(X)$ and ${}_nB(Y) = {}_nB'(Y)$, then ${}_nB(X\times Y) = {}_nB'(X\times Y)$.

Proof: From Corollary 6, ${}_nB'(X\times Y)$ is generated by ${}_nB(X)$, ${}_nB(Y)$ and im $\mu$. But these groups are subgroups of ${}_nB(X\times Y)$.

## REFERENCES


1. M. Artin and D. Mumford, Some elementary examples of unirational varieties which are not rational, Proc. London Math. Soc. 25 (1972), 75-95.

2. M. Auslander and O. Goldman, The Brauer group of a commutative ring, Trans. Amer. Math. Soc. 97 (1960), 367-409.

3. L. N. Childs, Azumaya algebras which are not smash products, preprint.

4. P. Deligne et al., SGA 4 1/2, Cohomology étale, "Lecture Notes in Math. No. 569," Springer-Verlag, New York/Berlin, 1977.

5. F. DeMeyer and T. Ford, On the Brauer group of surfaces, J. Algebra 86 (1984), 259-271.

6. F. DeMeyer and T. Ford, On the Brauer group of surfaces and subrings of k[x,y], Brauer Groups in Ring Theory and Algebraic Geometry, in "Lecture Notes in Math. No. 917," pp. 211-221, Springer-Verlag, New York, 1982.

7. F. DeMeyer and T. Ford, Computing the Brauer-Long group of Z/2-dimodule algebras, to appear in J. Pure and Applied Algebra.

8. T. Ford, Every finite abelian group is the Brauer group of a ring, Proc. Amer. Math. Soc. 82 (1981), 315-321.

9. T. Ford, On the Brauer group of a Laurent polynomial ring, to appear in J. Pure and Applied Algebra.

10. T. Ford, On the Brauer group of $k[x_1, \ldots, x_n, 1/f]$, preprint.

11. O. Gabber, Some theorems on Azumaya algebras, Le Groupe de Brauer, in "Lecture Notes in Math. No. 844," pp. 129-209, Springer-Verlag, New York/Berlin, 1981.

12. J. Gamst and K. Hoechsmann, Quaterions generalises, C. R. Acad. Sci. Paris 269 (1969), 560-562.

13. A. Grothendieck, Le Groupe de Brauer I, II, III, Dix Exposes sur la Cohomologie des Schemas, North-Holland, Amsterdam; Masson, Paris, 1968, 46-188.

14. A. Grothendieck with M. Artin and J.-L. Verdier, SGA 4, Theorie des Topos et Cohomologie étale des Schemas (1963-64), "Lecture Notes in Math. Nos. 269, 270, 305," Springer-Verlag, New York/Heidelberg, 1972-73.

15. R. Hartshorne, Residues and duality, "Lecture Notes in Math. No. 20," Springer-Verlag, New York/Berlin, 1966.

16. R. Hartshorne, Algebraic Geometry, Springer-Verlag, New York, 1977.

17. R. Hoobler, When is Br(X) = Br'(X)?, Brauer groups in Ring Theory and Algebraic Geometry, in "Lecture Notes in Math. No. 917," pp. 231-245, Springer-Verlag, New York, 1982.

18. R. Hoobler, Functors of graded rings, Methods in Ring Theory, F. van Oystaeyen, ed., NATO ASI Series, D. Reidel, Dordrecht, 1984, 161-170.

19. A. Merkurjev, On the norm residue symbol of degree 2, Dokladi Akad. Nauk. SSSR, 261 (1981), 542-547; (English trans.) Soviet Math. Doklady, 24 (1981), 1546-551.

20. A. Merkurjev and A. Suslin, K-cohomology of Severi-Brauer varieties and the norm residue homomorphism, Math. USSR Izv. 21 (1983), 307-340.

21. J. Milne, Etale cohomology, Princeton Univ. Press, Princeton, N.J., 1980.

22. A. Wadsworth, Merkurjev's elementary proof of Merkurjev's theorem, Contemporary Mathematics 55, Part II (1986), 741-776.

# Group Actions on Module-finite Rings

S. JØNDRUP

In this note $R$ is a ring, which we assume is a finite module over its center, and $G$ is a finite group of automorphisms of $R$. We try to find conditions either on $R$ or $G$, which ensure that $R^G$ (the fixed ring) is a finite module over its center.

A similar famous question is: Let $R$ be $C$-affine (where $C$ is a subring of the center) and $G$ is a finite group of $C$-automorphims: Will $R^G$ be $C$-affine? ([5] and [6] for non commutative results).

**1. The Noetherian case.** We first consider the case where $R$ is not only a finite module over its center, but also a finite $C$-algebra and $G$ is a finite group of $C$-automorphism.

Let us suppose $R$ is generated by $a_1, \quad, a_m$ as a $Z(R)$-module ($Z(R)$ denotes the center of $R$), and $R$ is generated by $b_1, \quad, b_n$ as a $C$-algebra. We then have for suitable $z_{ij}, z_{jlk}$ in $Z(R)$

$$b_i = \sum_j z_{ij} a_j \; 1 \leq i \leq n,\, 1 \leq j \leq m \quad a_j a_l = \sum z_{jlk} a_k \quad 1 \leq j, l, k \leq m.$$

By $T$ we denote the $C$-algebra generated by $z_{ij}^g$, $z_{jlk}^g$, $g \in G$, $1 \leq i \leq n$, $1 \leq j, l, k \leq m$.

$T$ is clearly a finite $C$-algebra and a subring of $Z(R)$, moreover $R$ is a $T$-module.

$T$ is by construction $G$-stable, it is quite easy to prove that $T$ is a finite module over $T^G$, but we need a slightly better result.

So for short let $T$ be generated as a $C$-algebra by $w_1, \quad, w_s$, where $w_i^g \in \{w_1, \quad, w_s\}$. If $|G| = r$ we consider the $r$ symmetric polynomials in $w_j^g$, $g \in G$, $1 \leq j \leq s$ and the $C$-algebra generated by these is denoted by $T_0$, $T_0$ is by construction a finite $C$-algebra, each $w_i$ is integral over $T_0$ ($w_i$ is root in the polynomial $\Pi_g(X - w_i^g)$ with coefficients in $T_0$). Since $T$ is commutative, $T$ is a finite $T_0$-module. Putting these observations together we get that there exists a $C$-affine algebra $T_0$, a subring of $Z(R)^G$ such that $R$ is a finite $T_0$-module.

From above we now get

PROPOSITION 1.1. *Let $R$ be a finite module over its center and a finite $C$-algebra, where $C$ is a subring of the center of $R$. If $G$ is a finite group of $C$-automorphisms of $R$ with $|G|^{-1}$ in $R$, then $R^G$ is $C$-affine and $R^G$ is a finite module over its center.*

PROOF: Both claims follows since we have an $R^G$-homomorphism from $R$ onto $R^G$.

The first result is [3, Proposition 1.1].



*F. van Oystaeyen and L. Le Bruyn (eds.), Perspectives in Ring Theory, 147–152.*

PROPOSITION 1.2. *Let $R$ be a (left) noetherian ring, which is a finite module over its center. If $R$ is $C$-affine, where $C$ is a subring of $Z(R)$, and $G$ is a finite group of $C$-automorphisms, then $R^G$ is a finite module over its center and $C$-affine.*

PROOF: Using earlier notation and remarks we see that $R$ is finite $T_0$-module, thus $T_0$ is noetherian [2]. $R^G$, being a $T_0$-submodule of the finitely generated $T_0$-module $R$, is therefore a finite $T_0$-module. Since $T_0$ is a subring of $Z(R^G)$ and $T_0$ is $C$-affine, the result follows.

A similar argument shows

PROPOSITION 1.3. *[cf. 6, Corollary 3]. Let $R$ be a $C$-affine algebra over a commutative Noetherian ring $C$. If $R$ is a finite module over its center and $G$ a finite group of $C$-automorphisms. Then $R^G$ is $C$-affine and a finite module over its center.*

The following example shows that the assumption that $R$ is $C$-affine is essential for Proposition 1.1.

Example 1. Let $K$ be a field of characteristic not 2 and $R$ the polynomial ring in countably many indeterminates over $K$, $R = K[(x_i)_{i\in \mathbf{N}}]$.

If $I$ denotes the ideal generated by $X_1$ and $X_2$ we let $A$ be the ring

$$\begin{pmatrix} R & R \\ I & R \end{pmatrix}.$$

The automorphism $\sigma$ of $A$ induced by mapping $x_i$ to $-x_i$ has order 2 and fixed point ring

$$\begin{pmatrix} R^\sigma & R^\sigma \\ I^\sigma & R^\sigma \end{pmatrix},$$

where $I^\sigma$ is generated by $X_1X_j$, $j \in \mathbf{N}$ and $X_2X_j$, $j \in \mathbf{N}$. Clearly $A$ is a finite module over its center, but $A^{<\sigma>}$ is not.

Let us also note the following:

PROPOSITION 1.4. *Let $R$ be a noetherian ring, which is a finite module over its center and $G$ a finite group of automorphisms of $R$ with $|G|^{-1}$ in $R$. Then $R^G$ is a finite module over its center.*

PROOF: $Z$ is noetherian [2] and a finite $Z^G$-module [4, Corollary 5.9], so $R$ is a finite $Z^G$-module and $Z^G$ is noetherian. Thus $R^G$ is a finite $Z^G$-module and hence a finite module over its center.

We finish this section by an example showing that in Proposition 1.4 one cannot omit the assumption "$|G|^{-1}$ in $R$".

Example 2. The example is based on the example of Nagarajan (cf. [4, Example 5.5]).

Let us recall Nagarajan's construction.

$S = F[[X,Y]]$, the ring of powerseries in 2 indeterminates over a field of characteristic 2.

$\sigma$ is an automorphism of order 2, $\sigma(x) = x$, $\sigma(y) = y$.

$S$ is not a finite module over $S^{<\sigma>}$ where $< \sigma >$ denotes the group generated by $\sigma$.

Clearly $S$ is a complete, local Noetherian ring, $S^{<\sigma>}$ is local with maximal ideal $(x,y)S \cap S^{<\sigma>} = \eta$. It is easy to see that $S^{<\sigma>}$ is complete with respect to the powers of $\eta$.

[7, Corollary 3.17] shows that $\eta$ is not a finite $S^{<\sigma>}$ ideal.

Let $R = \begin{pmatrix} S & S \\ (x,y) & S \end{pmatrix}$ and let $\sigma$ act componentwise on $R$, this gives an automorphism of $R$, say $\sigma_0$.

$R$ is clearly a finite module over its center and $(x,y)S \cap S^{<\sigma>}$ is not finitely generated over $S^{<\sigma>}$, $R^{<\sigma_0>}$ is not a finite module over its center.

If one repeats the argument of Chuang and Lee and the construction above one gets an example of a Noetherian ring $R$, which is a finite module over its center, and a group $G$ such that $R$ has no $|G|$-torsion and such that $R^G$ is not a finite module over its center.

**2. Von Neumann regular rings.** We start this section by an example, which Proposition 1.1 and 1.4 should be compared with.

Example 3. The ring $A$ constructed below has the following properties:

(i) $A$ is von Neumann regular.
(ii) $A$ is a finite module over its center.
(iii) $G$ is a finite group of automorphisms of $A$ with $|G|^{-1}$ in $A$.
(iv) $A^G$ is not a finite module over its center.

$A$ is the ring of all sequences of $(2 \times 2)$-matrices over the complex numbers ultimately constant and real.

An automorphism $\sigma$ of $M_2(\mathrm{C})$, C the complex numbers is defined by sending

$$\begin{pmatrix} \lambda_1 & \lambda_2 \\ \lambda_3 & \lambda_4 \end{pmatrix} \text{ to } \begin{pmatrix} \overline{\lambda}_4 & -\overline{\lambda}_3 \\ -\overline{\lambda}_2 & \overline{\lambda}_1 \end{pmatrix}$$

where $^-$ denotes complex conjugation.

[In fact $\sigma$ the inner automorphism determined by $\begin{pmatrix} 0 & -i \\ 1 & 0 \end{pmatrix}$ followed by componentwise complex conjugation]. $\sigma$ has order 2 and induces an sutomorphims $\widehat{\sigma}$ on $A$ also of order 2.

The fixed ring $A^{<\widehat{\sigma}>}$ is the ring of also sequences of the form

$$\begin{pmatrix} a_n & b_n \\ -\overline{b}_n & \overline{a}_n \end{pmatrix} \text{ ultimately constant and real.}$$

The center of this ring is the ring of sequences of real numbers ultimately constant. It is now easy to see that the condition (i)-(iv) will hold.

In case $G$ is a group of inner automorphisms with $|G|^{-1}$ in $A$, then clearly $A^G$ is a finite module over its center (cf. Proposition 1.1). But in this case the assumption $|G|^{-1}$ is in $A$ can in fact be removed.

Before we prove this result we need a result which might be of independent interest.

PROPOSITION 2.1. *Let the von Neumann regular ring $A$ be a finite module over its center $Z$. Then $A$ is a projective $Z$-module.*

PROOF: We use the Pierce decomposition of $Z$. For $x$ in $\operatorname{Spec} B(Z)$, $A_x$ is von Neumann regular and since $A$ is finite $Z$-module it is easy to see that the center of $A_x$ is $Z_x$, a field, so $A_x$ is by Kaplansky's Theorem on P.I. rings a central simple algebra of dimension $n_x^2$ of $Z_x$.

In this case Amitsur [1] has constructed a central polynomial

$$h_x(x_1, \quad , x_{n_x^2}, y_1, \quad , y_m),$$

which is linear and homogeneous in the $x$'s.

If we fix a choice $(y_1)_x, \quad , (y_m)_x$ of elements in $A_x$ such that $h_x(x_1, \quad , x_{n_x^2}, (y_1)_x, \quad , (y_m)_x)$ is non vanishing on $A_x$, then $h_x(x_1, \quad , x_{n_x^2}, y_1, \quad , y_m)$ is central in some neightbourhood of $x$, since $A$ is a finite $Z$-module. In this neighbourhood $h_x(x_1, \quad , x_{n_x^2}, y_1, \quad , y_m)$ is also non-vanishing in the support of $h_x(\widehat{x}_1 \quad , \widehat{x}_{n_x^2}, y_1, \quad , y_m)$, where $\widehat{x}_1, \quad , \widehat{x}_{n_x^2}$ is a choice making $h_x(\widehat{x}_1, \quad , \widehat{x}_{n_x^2}, y_1, \quad , y_m)$ non-zero at $x$. Since $Z$ is semihereditary the support of an element is an open-closed subset of $\operatorname{Spec} B(Z)$, hence $h_x(x_1, \quad , x_{n_x^2}, y_1, \quad , y_m)$ is also non-vanishing in some neighbourhood of $x$.

If $x_1, \quad , x_{n^2}$ generate $A$ over $Z$ at $x$, then since $A$ is a finite module over $Z$, $x_1, \quad , x_{n^2}$ will generate $A$ over $Z$ in some neighbourhood of $x$.

The following well-known lemma from linear algebra shows that if $x_1, \quad , x_t$ is a base for $A_x$ over $Z_x$, then $x_1, \quad , x_t$ is a base for $A_y$ over $Z_y$ for $y$ in some neighbourhood of $x$.

LEMMA. *Let $V$ be a vectorspace over a field $K$, $k = \dim_F V$ and $f : V^k \to K$ is non-zero multilinear alternating map.*

*$f(v_1, \quad , v_k) = 0$ if and only if $v_1, \quad , v_k$ are linearly independent.*

Standard sheaf technique now shows that $A$ is a finite direct sum of rings, each of which a finite free module over its center and the result follows.

Notice that we in fact have proved:

A von Neumann regular module finite ring is a finite direct sum of rings each of which as a free module over its center.

PROPOSITION 2.2. *Let $A$ be a von Neumann regular ring and $G$ a finite group of inner automorphism. If $A$ is a finite module over its center $Z$ so is $A^G$.*

PROOF: By the result and the remark above we may without loss of generality assume that $A$ is a free module over its center, i.e. $A = Za_1 \oplus \quad \oplus Za_t$.

Write $a_i a_j = \sum_k z_{ijk} a_k \quad 1 \le i, j \le t$. For $u = \sum u_j a_j$ and $a = \sum \lambda_j a_j$ we get

$$ua = \sum_{i,j,k} u_i \lambda_j z_{ijk} a_k \quad \text{and}$$

$$au = \sum_{i,j,k} u_i \lambda_j z_{jik} a_k.$$

Thus $ua = au$ if and only if

$$\sum_{i,j} (u_i \lambda_j)(z_{ijk} - z_{jik}) = 0 \quad \text{for all} k.$$

But $\sum_{i,j}(u_i\lambda_j)(z_{ijk} - z_{jik}) = \sum_j \lambda_j (\sum_i u_i(z_{ijk} - z_{jik}))$. Denote by $\delta^u_{jk}$, $\sum_i u_i(z_{ijk} - z_{jik})$, we then get

$$ua = au \quad \text{if and only if}$$
$$(\lambda_-)(\delta^u_{jk}) = 0,$$

where $\lambda_- = (\lambda_1, \ldots, \lambda_t)$.

We have a $Z$-homomorphism from $Z^t$ to $Z^t$ given by the matrix $(\delta^u_{jk})$ such that the submodule of $A$ consisting of the elements $\{a \mid ua = au\}$ as a $Z$-module is isomorphic to the kernel of this homomorphism.

Consequently $A^G$ is isomorphic as a $Z$-module to the kernel of a $Z$-homomorphism $\delta$ from $Z^t$ to a free $Z$-module.

Because $Z$ is semihereditary, $\operatorname{Im} \delta$ is projective and therefore $\operatorname{Ker} \delta$ is a direct summand of $Z^t$.

Remarks. It is easy to see that if a P.I. von Neumann regular ring is a free module over its center, then it is a finite module over its center.

The ring of $(2 \times 2)$-matrices over a field ultimately constant and scalar, shows that a von neumann regular ring can be a projective module over its center without being finitely generated.

## References


1. S. A. Amitsur, *Identities and linear dependence*, Israel J. Math. **22** (1975), 127-137.
2. D. Eisenbud, *Subrings of Artinian and Noetherian Rings*, Math. Ann. **185** (1970), 247-249.
3. S. Jøndrup, *Fixed point rings of some non noetherian rings*, Comm. in Algebra **14** (1986), 109-124.
4. S. Montgomery, *Fixed rings of finite automorphism groups of associative rings*, Lecture Notes in Mathematics **818** (1980).
5. S. Montgomery and L. W. Small, *Fixed Rings of Noetheiran Rings*, Bull. London Math. Soc. **13** (1981), 33-38.
6. S. Montgomery and L. W. Small, *Some Remarks on Affine Rings*, Proc. Amer. Math. Soc. **98** (1986), 537-544.
7. M. Nagata, *Local Rings*, New York, Interscience Publishers (1960).

# Algebraic Conjugacy Classes and Skew Polynomial Rings

T. Y. Lam[(*)]
University of California
Berkeley, California 94720
U. S. A.

André Leroy
Université de l'État à Mons
B-7000 Mons
Belgium

Abstract. The goal of this paper is to develop further the theory of skew polynomial rings over division rings, using as our main tools the notions of invariant and semi-invariant polynomials. These notions arise naturally when one tries to study the algebraic conjugacy classes (in a suitably generalized sense) of the underlying division ring. A substantial part of our effort will also be devoted to the investigation of the properties and the characterizations of algebraic derivations, algebraic endomorphisms, and their respective minimal polynomials. This investigation is made possible by the discovery of the relationship between polynomial equations and differential equations, and the relationship between polynomial dependence and linear dependence. Applications of these results to the study of non-commutative Hilbert 90-type theorems will be presented in a forthcoming work $[LL_2]$.

## §1. Introduction

Let $K$ be a division ring equipped with a given endomorphism $S: K \longrightarrow K$. By an S-derivation on $K$, we mean an additive map $D : K \longrightarrow K$ with the property that $D(ab) = S(a)D(b) + D(a)b$ for all $a, b \in K$. For a given indeterminate $t$, let $R = K[t,S,D]$ denote the skew polynomial ring with respect to the triple $(K,S,D)$, consisting of all left polynomials $\sum a_i t^i$ $(a_i \in K)$ which are added in the

(*)Supported in part by N.S.F.



F. van Oystaeyen and L. Le Bruyn (eds.), Perspectives in Ring Theory, 153–203.

usual way and multiplied according to the rule $ta = S(a)t + D(a)$ for any $a \in K$. This definition of skew polynomial rings was first introduced by Ore [O], who combined earlier ideas of Hilbert (in the case $D = 0$) and Schlessinger (in the case $S = I$). Ore lay a firm foundation for the study of $R = K[t,S,D]$ by establishing the unique factorization property of $R$, and using this, he studied, among other things, the problem of finding the greatest common divisors and the least common multiples of pairs of skew polynomials. Ever since the appearance of Ore's fundamental paper [O], the skew polynomial rings $K[t,S,D]$ (and their generalizations) have played an important role in non-commutative ring theory. About 15 years after Ore's paper appeared, Amitsur [A] made a basic contribution to the study of $K[t,S,D]$ by proving a generalization of a theorem on linear differential equations in a purely algebraic setting. Through this paper of Amitsur, the interesting role played by the so-called algebraic derivations (D is called <u>algebraic</u> if it satisfies a monic equation $\sum_{i=1}^{n} a_i D^i = 0$ over K) came to light. In [A'], Amitsur also studied, in the special case when $S = I$, the structure of the 2-sided ideals in $K[t,S,D]$; this work has been recently extended to the general case by Cauchon [C] and Lemonnier [Lem] (see also [Ca]).

Our present work is, in many ways, a continuation of the work on skew polynomial rings cited above. The point of departure is the introduction of the notion of "evaluation" of skew polynomials $f \in R$ on the constants $a \in K$. Surprisingly, the discovery of the right definition of $f(a)$ came rather late in the game: two pertinent references are [Sm] and [Le], but even in these references, the fact that $f(a)$ amounts to the "evaluation" of $f$ at $a$ was not explicitly pointed out. In [$LL_1$], we rectified this by initiating the notation $f(a)$ for evaluation, and proved the all-important Product Theorem [$LL_1$ : (2.7)] for the evaluation of a product of two polynomials at $a \in K$. This notion of the evaluations of polynomials at constants enabled us to generalize the theory of Vandermonde and Wronskian matrices to the non-commutative setting, as in [$LL_1$].

The main goal of the present paper is to study the algebraic conjugacy classes in a division ring $K$ equipped with $(S,D)$. (We shall often write $(K,S,D)$ to refer to this setting.) Recall from [$LL_1$] that two elements $a, b \in K$ are said to be $(S,D)$-<u>conjugate</u> if there exists an element $c \in K^*$ such that $b = a^c := S(c)ac^{-1} + D(c)c^{-1}$. $(S,D)$-conjugacy being an equivalence relation, we shall write $\Delta^{S,D}(a) := \{a^c : c \in K^*\}$ for the $(S,D)$-conjugacy class determined by $a$. This class is said to be $(S,D)$-<u>algebraic</u> (or algebraic for short) if there is a nonzero $f \in R$ which vanishes on all of $\Delta^{S,D}(a)$. The (unique) monic $f$ of the least degree with this property is said to be the <u>minimal polynomial</u> of $\Delta^{S,D}(a)$. Such a polynomial $f$ is always <u>right invariant</u>, in the sense that $f \cdot R \subseteq R \cdot f$. Therefore, the study of algebraic conjugacy classes is closely tied to the study of right invariant polynomials, which is, in turn, tied to the study of the 2-sided ideal structure of $R$.

This paper is organized as follows. In §2, we first study right invariant polynomials in $R$, along with the right <u>semi</u>-invariant polynomials. (We say that $g \in R$ is <u>right semi-invariant</u> if $g \cdot K \subseteq K \cdot g$.) We recall Cauchon's result on the classification of right invariant polynomials, and obtain (in the special case when $S$ is an automorphism) a parallel result on the classification of right semi-invariant polynomials. In §3, we study Lemonnier's notion of <u>quasi-algebraic derivations</u>, and characterize in different ways the least possible degree of the non-constant right semi-invariant polynomials (if they exist). In §4, we fix our attention on a single $(S,D)$-conjugacy class $\Delta^{S,D}(a)$ and study the "polynomial dependence" (or P-dependence) among elements of $\Delta^{S,D}(a)$. It turns out that the P-dependence among elements of $\Delta^{S,D}(a)$ is "controlled" by the linear dependence of elements of $K$ viewed as a right vector space over the division subring $C^{S,D}(a) := \{0\} \cup \{c \in K^* : a^c = a\}$. This fact is most succintly expressed by saying that there is a one-one correspondence between the lattice of "full" $(S,D)$-algebraic subsets of $\Delta^{S,D}(a)$ <u>and</u> the lattice of finite dimensional right $C^{S,D}(a)$-subspaces of $K$ (cf. Theorem 4.5). This one-one correspondence is essentially given by the process of "exponentiation".

In §5 (which is perhaps the heart of this paper), we take up in earnest the study of the $(S,D)$-algebraic conjugacy classes in $K$. These classes are characterized in various ways, and their minimal polynomials are linked to the minimal polynomials of certain algebraic derivations. This tie between the two kinds of minimal polynomials is made possible by Proposition 5.8 which establishes the basic relationship between polynomial equations and differential equations. One easily stated result is Corollary (5.12) which says that K has at least one (S,D)-algebraic class iff D is the sum of an inner S-derivation and an algebraic S-derivation. Among the many ramifications of our results characterizing the (S,D)-algebraic classes, one finds an interesting relationship between such classes and the notion of primitive rings: by Corollary 5.23, R is a left primitive ring unless all (S,D)-conjugacy classes are algebraic. Toward the end of the paper, we analyze the algebraic classes of K according as S is an automorphism of finite inner order or otherwise. In the latter case, we show that there is at most one (S,D)-algebraic class (Theorem (5.25)), while in the former case, we show that, with possibly one exception, the minimal polynomials of the algebraic classes are scalar multiples of central polynomials, and their degrees are all divisible by the inner order of S (Theorem (5.28)). Further results on the criteria for an (S,D)-conjugacy class to be algebraic, and for two elements in K to be (S,D)-conjugate (proved by using a certain "Composite Function Theorem") will be presented in a forthcoming work [$LL_2$].

Since this paper is largely a continuation of our earlier work [L] and [$LL_1$], the notations and terminology in these two papers will be used rather freely. However, the crucial definitions are recalled for the convenience of the reader whenever possible. The definition of the evaluation of a polynomial $f \in R$ at $a \in K$ is needed only to the extent that $f(a)$ is the unique constant $c$ such that $f(t) \in R \cdot (t-a) + c$. Whereas this is no doubt the best conceptual way to understand $f(a)$, we would be remiss if we do not mention at least once the "computational" definition of $f(a)$: if $f(t) = \sum b_i t^i$, then $f(a) := \sum b_i N_i(a)$, where the $N_i$ 's are defined inductively by: $N_o(a) = 1$, $N_{i+1}(a) = S(N_i(a))a + D(N_i(a))$. Note, however, that these formulas apply only to the evaluation of $f$ <u>on constants</u>. An expression such as $f(D)$ (resp. $f(S)$) shall still have its usual meaning, namely, it stands for the operator $\sum b_i D^i$ (resp. $\sum b_i S^i$). The minimal polynomial of $D$ (in case $D$ is algebraic) is the monic polynomial $f \in R$ of the least degree such that $f(D) = 0$ (and similarly for $S$).

Often, we shall have occasion to specialize to the case $S = I$ (resp. the case $D = 0$). When we do this, we shall drop $S$ (resp. $D$) from our notations. Thus, we shall write $K[t,D]$ to mean $K[t,I,D]$, and write $K[t,S]$ to mean $K[t,S,0]$. The same conventions will also apply to $\Delta^{S,D}(a)$ and $C^{S,D}(a)$. In any case, the abbreviated notations shall always be clear from the context.

We wish to thank Professor S. Amitsur for pointing out to us that his theorem on linear differential equations in [A] can be proved by using the Density Theorem. Our presentations in the second half of §5 have taken his insightful comments into account.

## §2. Invariant and Semi-invariant Polynomials

In this beginning section, we shall introduce the notions of right invariant and right semi-invariant polynomials and discuss their basic properties and characterizations. The important roles played by these two kinds of polynomials will be clear in the later sections when we take up the study of algebraic conjugacy classes in division rings. Throughout this section (and in fact the whole paper), we assume that the data $(K,S,D)$ are given and fixed, where $K$ is a division ring, $S$ is an endomorphism of $K$, and $D$ is an $S$-derivation on $K$. We shall always write $R$ for the associated skew polynomial ring $K[t,S,D]$, and write $K^*$ for the multiplicative group $K\setminus\{0\}$ of $K$.

<u>Definition 2.1</u>. A polynomial $f(t) \in R$ is called <u>right invariant</u> if $f\cdot R \subseteq R\cdot f$. (This means that the left ideal $R\cdot f$ is a 2-sided ideal of $R$.) A polynomial $g(t) \in R$ is called right <u>semi-invariant</u> if $g\cdot K \subseteq K\cdot g$. Left invariant and left semi-invariant polynomials are defined analogously.

The term "right invariant" is fairly standard in ring theory. Our choice of the new term "right semi-invariant" is based on the following rationale: Since $R$ is generated as a ring by $K$ and $t$, it follows that $f \in R$ is right invariant iff $f$ is right semi-invariant and in addition $f\cdot t \subseteq R\cdot f$. This says that right semi-invariance amounts to "half" of the condition for right invariance. Also, note that the nonzero right (semi-) invariant polynomials are closed under multiplication, so they form a semigroup. In particular, if $a \in K^*$, then f is right (semi-) invariant iff $a \cdot f$ is right (semi-) invariant. Because of this, it is generally sufficient to focus our study of right (semi-) invariant polynomials on the monic ones.

Lemma 2.2. *For a monic polynomial* $g(t) = \sum_{i=o}^{n} a_i t^i \in R$ *of degree* $n$, *the following are equivalent:*

(1) $g$ *is right semi-invariant;*

(2) $g(t)c = S^n(c)g(t)$ *for every* $c \in K$;

(3) $S^n(c)a_j = \sum_{i=j}^{n} a_i f_j^i(c)$ *for every* $j$ *and every* $c \in K$, *where the operators* $\{f_j^i\}$ *are defined as at the beginning of* §2 *of* $[LL_1]$.

Proof. (1)$\Longleftrightarrow$(2) follows by observing that, as a left polynomial, the leading coefficient of $g(t)c$ is $S^n(c)$. (2)$\Longleftrightarrow$(3) follows by comparing the coefficients of $S^n(c)g(t)$ with those of

$$\begin{aligned} g(t)c &= \sum_{i=o}^{n} a_i t^i c \\ &= \sum_{i=o}^{n} a_i \sum_{j=o}^{i} f_j^i(c)t^j \\ &= \sum_{j=o}^{n} \left( \sum_{i=j}^{n} a_i f_j^i(c) \right) t^j . \end{aligned}$$

Q.E.D.

Note that if $\Delta_{n+1}(c)$ denotes the $(n+1)\times(n+1)$ lower triangular matrix whose $(i,j)$-entry is $f_{j-1}^{i-1}(c)$ (cf. $[LL_1: (6.8)]$ ), then the condition (3) above can be expressed succintly in the matrix form:

$$(a_o, a_1, \ldots , a_n)\Delta_{n+1}(c) = S^n(c)(a_o, a_1, \ldots , a_n).$$

This says that $(a_o, a_1, \ldots , a_n)$ is a left "eigenvector" for the matrix $\Delta_{n+1}(c)$ with "eigenvalue" $S^n(c)$.

In the classical case $(S, D) = (I, 0)$, we see immediately that the right invariant and right semi-invariant polynomials are just the polynomials of the form $a\cdot\sum a_i t^i$, where $a \in K$ and all $a_i$ 's belong to the center $Z(K)$ of $K$. In order to get a good perspective on the general case, we shall work out below the classes of right invariant and right semi-invariant polynomials in the cases when $D = 0$ and when $S = I$. Throughout this paper, we shall write $I_a$ for the inner automorphism $x \mapsto axa^{-1}$ on $K$ associated with $a \in K^*$. Also, we shall write $K^S = \{y \in K : S(y) = y\}$ and $K_D = \{y \in K : D(y) = 0\}$.

Proposition 2.3. *Assume that* $D = 0$, *and let* $f(t) = \sum_{i=o}^{n} a_i t^i \in K[t,S]$ *be monic of degree* $n$. *Then*

(1) $f$ *is right semi-invariant iff, for any* $j$ *such that* $a_j \neq 0$, *we have* $S^n = I_{a_j} \circ S^j$.

(2) $f$ *is right invariant iff* $f$ *satisfies the condition above and in addition* $a_j \in K^S$ *for all* $j$.

Proof. (1) Since $D = 0$, we have $f_j^i = 0$ whenever $i > j$. Thus, the condition in (2.2)(2) simplifies to $S^n(c)a_j = a_j S^j(c)$ $(\forall c \in K)$. If $a_j \neq 0$, this amounts to

$$S^n(c) = a_j S^j(c) a_j^{-1} = \left(I_{a_j} \circ S^j\right)(c) \qquad (\forall c \in K),$$

i.e. $S^n = I_{a_j} \circ S^j$. (Note. We cannot rewrite this as $S^{n-j} = I_{a_j}$ in general, since $S$ is not assumed to be an automorphism. In the case when $S$ is an automorphism, we can prove a much more precise result: see (2.12) below.)

(2) We need to work out here the condition for $f(t)t \in R\cdot f(t)$, i.e. for $f(t)t$ to be equal to $(t+c)f(t)$ for some $c \in K$. Since

$$(t+c)f(t) = t^{n+1} + \sum_{i=1}^{n} (S(a_{i-1}) + ca_i)t^i + ca_o ,$$

the conditions on $c$ are that $a_i = S(a_i) + ca_{i+1}$ $(0 \leq i < n)$, and $ca_o=0$. If $c \neq 0$, it follows by induction on $i$ that all $a_i$ 's are zero. This is not the case as $a_n = 1$. Therefore, we must have $c = 0$ and the conditions above boil down to $a_i \in K^S$ for all $i$. Q.E.D.

Proposition 2.4. *Assume that* $S = I$ *and let* $f(t) = \sum_{i=o}^{n} a_i t^i \in K[t,D]$ *be monic of degree* $n$. *Then*

(1) $f(t)$ *is right semi-invariant iff* $ca_j = \sum_{i=j}^{n} \binom{i}{j} a_i D^{i-j}(c)$ *for all* $c \in K$ *and all* $j \geq 0$.

(2) $f(t)$ *is right invariant iff* $f$ *satisfies the condition above and in addition* $a_j \in K_D$ *for all* $j$. *Such a polynomial in fact belongs to the center of* $K[t,D]$ .

Proof. (1) Since $S = I$, $f_j^i$ boils down to $\binom{i}{j} D^{i-j}$ for $i \geq j$. Thus the condition in (2.2)(2) simplifies to the one in (1).

(2) Again, we need to work out here the condition for $f(t)t$ to be equal to $(t+c)f(t)$ for some $c \in K$. Since

$$(t+c)f(t) = t^{n+1} + \sum_{i=1}^{n} \left(a_{i-1} + ca_i + Da_i\right) t^i + (ca_o+Da_o),$$

the conditions on $c$ are that $ca_i + Da_i = 0$ for $0 \leq i \leq n$. Since $a_n = 1$, this amounts to $c = 0$ and $Da_i = 0$ for all $i$. We have then $f(t)t = tf(t)$, and since $f(t)c = cf(t)$ also, $f(t)$ belongs to the center of $K[t,D]$ . Q.E.D.

By the above, we expect that there exist many examples of right semi-invariant polynomials which are <u>not</u> right invariant. Let us now record some such examples below.

Examples 2.5.

(a) Let $D = 0$, and let $S$ be an automorphism of order 2. Then by (2.3), $t^2+a$ is right semi-invariant for any $a \in Z(K)$, but such a polynomial is right invariant only if our $a$ is also fixed by $S$.

(b) Let $S = I$, $D$ be a derivation with $D^2 = 0$, and assume that char $K = 2$. Then by (2.4) (or by an explicit calculation), $t^2+a$ is right semi-invariant for any $a \in Z(K)$, but such a polynomial is right invariant only if our $a$ is also a constant of $D$.

The fact that we have to work with quadratic polynomials above has a good reason. In fact, the result below shows that, in the linear case, "right invariance" and "right semi-invariance" become synonymous terms.

Example 2.6. Here, we determine, in the general $(S,D)$-setting, all the the (monic) <u>linear</u> right invariant and right semi-invariant polynomials. Let $f(t) = t - b$, where $b \in K$. Then the following are equivalent:

(1) $f(t)$ is right invariant;

(2) $f(t)$ is right semi-invariant;

(3) $b \in Z^{S,D}(K) := \{ a \in K : a^c = a \quad \forall c \in K^* \}$.

We need only show (2) $\Longrightarrow$ (3) $\Longrightarrow$ (1). Assume $f$ is right semi-invariant. Then, for any $c \in K^*$,

$$S(c)(t - b) = (t - b)c = S(c)t + D(c) - bc.$$

Thus, $-S(c)b = D(c) - bc$ and hence $b = S(c)bc^{-1} + D(c)c^{-1} = b^c$, i.e. $b \in Z^{S,D}(K)$. Now assume $b \in Z^{S,D}(K)$. By reversing the above argument, we see that $f(t)$ is right semi-invariant. We finish by showing that $(t - b)t \in R\cdot(t - b)$. Assume, for the moment, that $b \neq 0$. From the equation $b^b = b$, we have $S(b)bb^{-1} + D(b)b^{-1} = b$, and so $D(b) = b'b$ for $b' = b - S(b)$. Of course, $D(b) = b'b$ also holds for $b = 0$. Thus, in any case,

$$\begin{aligned}(t - b)t &= t^2 - (S(b) + b')t + b'b - D(b)\\ &= t^2 - (S(b)t + D(b)) - b'(t - b)\\ &= t(t - b) - b'(t - b)\\ &= (t - b')(t - b) \in R\cdot(t - b),\end{aligned}$$

so $t - b$ is, in fact, right invariant. Q.E.D.

We have observed earlier that, if $g(t)$ and $h(t)$ in $R$ are both right (semi-) invariant, then so is $g(t)h(t)$. In the case when $S$ is an automorphism of $K$, we can prove some variations of this fact, as in part (3) of the following result.

Proposition 2.7. *Let* $S \in \mathrm{Aut}(K)$. *Then*

(1) $f(t) \in R$ *is right (semi-) invariant iff* $f$ *is left (semi-) invariant* (cf. [Co': pp.296-297]);

(2) *If* $f$ *is right invariant, then* $R\cdot f = f\cdot R$ ; *if* $f$ *is right semi-invariant, then* $K\cdot f = f\cdot K$.

(3) <u>Let</u> $f(t) = g(t)h(t) \neq 0$ <u>in</u> R <u>be right (semi-) invariant</u>. <u>Then</u> $g(t)$ <u>is right (semi-) invariant iff</u> $h(t)$ <u>is</u>.

<u>Proof</u>. (1) By symmetry, it is sufficient to prove the "only if" parts. Assume f is right invariant. For any $p(t) \in R$, we can write $p(t)f(t) = f(t)q(t) + r(t)$ for some $q(t), r(t) \in R$ such that $\deg r(t) < \deg f(t)$. (This is possible since S is assumed to be an automorphism.) Since f is right invariant, $f(t)q(t) = q'(t)f(t)$ for some $q' \in R$. Thus, $(p(t) - q'(t))f(t) = r(t)$. By degree consideration, this implies that $q'(t) = p(t)$ and $r(t) = 0$, and so $p(t)f(t) = f(t)q(t) \in f(t)\cdot R$, i.e. $f(t)$ is <u>left</u> invariant. If $f(t)$ is right semi-invariant, the same argument for $p(t)$ a scalar shows that f is also <u>left</u> semi-invariant.

(2) This is already covered by the argument above.

(3) Assume $f(t)$ and $h(t)$ are both right invariant. Then, for any $p(t) \in R$, $ph = hp'$ for some $p' \in R$ (since h is also left invariant). Thus, $gph = ghp' = fp' = p''f$ for some $p'' \in R$, since f is right invariant. Cancelling h on the right, we have $gp = p''g$. Since $p \in R$ is arbitrary, this shows that g is right invariant. Similarly, if $f(t)$ and $g(t)$ are both right invariant, we can show that $h(t)$ is also right invariant. The arguments for the case of semi-invariance are almost the same as those given above; we shall therefore leave them to the reader.

In [C], Cauchon has determined the structure of the right invariant polynomials in $R = K[t,S,D]$, generalizing earlier work of

Amitsur [A'] in the case $S = I$. For the convenience of the reader, we shall recall Cauchon's result, which will be exploited in §5. Cauchon's result holds more generally for any artinian simple ring $K$, but we shall only be concerned with the case when $K$ is a division ring here. Another pertinent reference for the result below is [Ca].

Theorem 2.8. (Cauchon) *Let* $q(t)$ *be a (monic) nonconstant right invariant polynomial of the least degree (if it exists). Then any right invariant polynomial in* $R$ *has the form* $\alpha \cdot h(t)q(t)^r$ *where* $\alpha \in K$, $r \geq 0$ *and* $h(t)$ *is a polynomial in* $Z(R)$, *the center of* $R$. *Moreover, let* $h_o(t)$ *be a nonconstant polynomial in* $Z(R)$ *of the least degree (if it exists); then* $h_o(t) = \lambda \cdot q(t)^s$ *for some* $\lambda \in K^*$ *and* $s \geq 1$, *and* $Z(R) = Z(K)_{S,D}[h_o(t)]$ *where* $Z(K)_{S,D} = Z(K) \cap K^S \cap K_D$.

Prompted by this result, we shall try to determine also the structure of all right semi-invariant polynomials in $R$. Our methods below will lead to such a complete determination in the case when $S$ is an automorphism of $K$. (The case when $S$ is not an automorphism seems to be much more difficult, and will not be attempted here.) The first step in this analysis is the following.

Proposition 2.9. *Assume* $S \in \mathrm{Aut}(K)$, *and let* $p(t)$ *be a (monic) non-constant right semi-invariant polynomial in* $R$ *of the least degree (if it exists). Then any right semi-invariant polynomial* $f(t)$ *lies in* $K[p(t)] = \{\sum a_i p(t)^i : a_i \in K\}$ *(the subring of* $R$ *generated by* $K$ *and* $p(t)$). *In particular,* $\deg f$ *must be a multiple of* $\deg p$.

Proof. Let $\deg p(t) = m > 0$ and $\deg f(t) = n$. We shall prove the Proposition by induction on $n$, the case $n = 0$ being clear. For $n > 0$, write $f(t) = q(t)p(t) + r(t)$, where $\deg r(t) < m$ (or $r(t) = 0$). Let $c \in K$. Assuming without loss of generality that $f$ is monic, we have $f(t)c = S^n(c)f(t)$ for any $c \in K$, and so

$$\begin{aligned} S^n(c)q(t)p(t) + S^n(c)r(t) &= q(t)p(t)c + r(t)c \\ &= q(t)S^m(c)p(t) + r(t)c. \end{aligned}$$

Transposition yields

$$\left[S^n(c)q(t) - q(t)S^m(c)\right] p(t) = r(t)c - S^n(c)r(t).$$

By degree consideration, we must have

$$(*) \qquad r(t)c = S^n(c)r(t) \qquad \text{and}$$

$$(**) \qquad q(t)S^m(c) = S^n(c)q(t).$$

Replacing $c$ by $S^{-m}(c)$ (the fact that $S \in \mathrm{Aut}(K)$ is needed here), (**) shows that $q(t)\cdot K \subseteq K\cdot q(t)$, i.e. $q(t)$ is right semi-invariant. Using the inductive hypothesis, we have then $q(t) \in K[p(t)]$. From (*), we see also that $r(t)$ is right semi-invariant. Since $\deg r(t) < m$, $r(t)$ must then be a constant. Thus, we have

$$f(t) = q(t)p(t) + r(t) \in K[p(t)]\cdot p(t) + K \subseteq K[p(t)]. \quad \text{Q.E.D.}$$

Proposition 2.10. *Assume that the above* $p(t)$ *exists (but not assuming* $S$ *to be an automorphism). Then* $p(t)$ *is unique up to an additive constant. Moreover, for* $a \in K^*$, $p(t) + a$ *is right semi-invariant* *iff* $S^m = I_a$ *(where* $m = \deg p$*). In particular,* $p(t)$ *is unique iff* $S^m$ *is not an inner automorphism.*

*Proof.* Suppose $p'(t)$ is another candidate. Then $\deg p' = \deg p = m$, from which we see easily that $p'(t) - p(t)$ is right semi-invariant. Therefore, $p'(t) - p(t) = a \in K$. For $p(t) + a$ $(a \in K^*)$ to be actually right semi-invariant, we need

$$S^m(c)(p(t) + a) = (p(t) + a)c$$
$$= S^m(c)(p(t) + a) + (ac - S^m(c)a)$$

for all $c \in K$, i.e. $S^m(c) = aca^{-1}$. Thus the necessary and sufficient condition is that $S^m = I_a$. The last statement of the Corollary now follows immediately from this. Q.E.D.

We are now in a position to determine the set of all right semi-invariant polynomials in $R$, in case $S$ is an automorphism. Letting $\mathrm{Inn}(K)$ denote the group of inner automorphisms of $K$, the order of $S$ in $\mathrm{Aut}(K)/\mathrm{Inn}(K)$ is called the *inner order* of $S$. It turns out that it is this inner order which holds the key to the structure of the set of right semi-invariant polynomials.

*Theorem 2.11. Assume* $S \in \mathrm{Aut}(K)$*, and let* $p(t)$ *be a (monic) non-constant right semi-invariant polynomial in* $R$ *of the least degree, say* $m$.

(1) *If* $S$ *has infinite inner order, then the right semi-invariant polynomials in* $R$ *are precisely those of the form* $a \cdot p(t)^r$ *where* $a \in K$ *and* $r \geq 0$.

(2) *Let* S *be of finite inner order* k, *say* $S^k = I_u$ $(u \in K^*)$. *Let* $d = \gcd(k, m)$ *and write* $k = dk'$, $m = dm'$. *Then the right semi-invariant polynomials in* R *are precisely those of the form*

$$(*) \qquad a \cdot \sum_{\substack{0 \leq i \leq r \\ i \equiv r \pmod{k'}}} \varepsilon_i \, u^{(r-i)m'/k'} p(t)^i ,$$

*where* $a \in K$, $\varepsilon_r = 1$ *and* $\varepsilon_i \in Z(K)$.

*Proof*. Let $f(t) \in R$ be a monic right semi-invariant polynomial. By (2.9), $f(t)$ is expressible in the form $\sum_{i=0}^{r} a_i p(t)^i$, with $n = \deg f = mr$, and $a_n = 1$. The right semi-invariance condition $f(t)c = S^n(c)f(t)$ $(\forall c \in K)$ now becomes

$$\sum S^n(c)a_i p(t)^i = \sum a_i p(t)^i c = \sum a_i S^{mi}(c)p(t)^i ,$$

i.e. $S^n(c)a_i = a_i S^{mi}(c)$ for $0 \leq i \leq r$. Replacing c by $S^{-mi}(c)$, this amounts to $S^{n-mi}(c)a_i = a_i c$ (for all $c \in K$). Therefore, in Case (1), all $a_i$ 's must be zero for $i < r$, and we get $f(t) = p(t)^r$. Conversely, of course, all $a \cdot p(t)^r$ are right semi-invariant. Now assume we are in Case (2), and use the notations there. Then, whenever $a_i \neq 0$, we have $S^{n-mi} = I_{a_i}$ and therefore the inner order k of S divides $n-mi = m(r-i)$, and so $k'$ divides $r-i$. Moreover,

$$I_{a_i} = S^{n-mi} = S^{k(r-i)m'/k'} = I_u^{(r-i)m'/k'}$$

implies that $a_i = \varepsilon_i u^{(r-i)m'/k'}$ for some $\varepsilon_i \in Z(K)$. Therefore, $f(t)$ has the form (*) (with $a = 1$). Conversely, consider any summand $f_i(t) = \varepsilon_i u^{(r-i)m'/k'} p(t)^i$ of (*) where $0 \leq i \leq r$, $i \equiv r \pmod{k'}$ and $\varepsilon_i \in Z(K)$. Writing $r-i = k'j$, we have $mi = mr - kjm'$, so for any $c \in K$:

$$\begin{aligned} f_i(t)c &= \varepsilon_i u^{jm'} \cdot S^{mi}(c)\, p(t)^i \\ &= \varepsilon_i u^{jm'} (S^{-k})^{jm'} \left(S^{mr}(c)\right) p(t)^i \\ &= \varepsilon_i u^{jm'} \cdot I_{u^{-1}}^{jm'} \left(S^{mr}(c)\right) p(t)^i \\ &= \varepsilon_i S^{mr}(c)\, u^{jm'}\, p(t)^i \\ &= S^{mr}(c)\, f_i(t) . \end{aligned}$$

It follows that any $a \cdot \sum f_i(t)$ as in (*) is right semi-invariant.

Q.E.D.

It is worthwhile to record the simplest manifestation of the theorem above, in the special case when $D = 0$. Note that in this case we can choose $p(t) = t$.

<u>Corollary 2.12</u>. *Assume that* $S \in \text{Aut}(K)$ *and* $D = 0$.

(1) *If* $S$ *has infinite inner order, then* $a \cdot t^r$ $(a \in K,\ r \geq 0)$ *are all the right semi-invariant polynomials in* $R = K[t,S]$.

(2) *Let* $S$ *be of finite inner order* $k$, *say* $S^k = I_u$. *Then the right semi-invariant polynomials in* $R$ *are precisely those of the form* $a \cdot \sum_{j \geq 0} c_j u^j t^{r-kj}$, *where* $a \in K$, $c_0 = 1$, *and* $c_j \in Z(K)$.

## §3. Quasi-algebraic derivations

The material in the second half of §2 calls to attention the important question: *When does there exist a (monic) non-constant right semi-invariant polynomial in* $R = K[t,S,D]$ ? In order to give an answer to this question, we recall the following definition which was first introduced in the 1984 thesis of B. Lemonnier [Lem].

*Definition 3.1.* An S-derivation D is called *quasi-algebraic* if there exist $a_1, \ldots, a_n \in K$ with $a_n = 1$ such that $\sum_{i=1}^{n} a_i D^i$ is an inner derivation with respect to the endomorphism $S^n$. (For instance, any S-inner derivation is quasi-algebraic, and so is any algebraic S-derivation.)

With this defintion, we have the following answer to the question raised at the beginning of this section, without any assumptions imposed on S. The equivalence (1)$\Longleftrightarrow$(3) herein is due to Lemonnier [Lem: Th. (9.21)].

*Theorem 3.2. The following are equivalent for* $R = K[t,S,D]$ :

(1) *There exists a (monic) non-constant right semi-invariant polynomial in* R ;

(2) *There exists a polynomial* $g(t) = \sum_{i=o}^{n} a_i t^i \in R$ *with* $n \geqslant 1$ *and* $a_n = 1$ *such that for any* $c \in K$ :

(*) $\qquad g(t)c \equiv S^n(c)g(t) \pmod{R \cdot t}$;

(3) *The S-derivation D is quasi-algebraic.*

Because of the intrinsic interest of this result, and because of the fact that Lemonnier's proof of (1)$\Longleftrightarrow$(3) is not easily accessible, we shall offer a complete and direct proof of the Theorem below.

Proof of (3.2). (1)$\Longrightarrow$(2) is obvious (see (2.2)(2)).

(2)$\Longrightarrow$(3). (*) means that $g(t)c$ and $S^n(c)g(t)$ have the same constant term when both are written out as left polynomials. Therefore, going through the proof of (2)$\Longleftrightarrow$(3) for Lemma 2.2, we can still compare the constant terms, and thereby ascertain the conclusion (2.2)(3) for $j = 0$, i.e. we'll have for all $c \in K$:

$$S^n(c)a_o = \sum_{i=o}^{n} a_i f_o^i(c) = \sum_{i=o}^{n} a_i D^i(c).$$

Therefore,

$$\text{(3.3)} \qquad \sum_{i=1}^{n} a_i D^i(c) = S^n(c)a_o - a_o c = D_{-a_o, S^n}(c) \qquad (\forall c \in K),$$

which means, by defintion, that $D$ is quasi-algebraic.

(3)$\Longrightarrow$(1). Suppose $D$ is quasi-algebraic, say with (3.3) holding for suitable constants $a_o, \ldots, a_n$, with $n \geqslant 1$, $a_n = 1$. Let $g(t) := \sum_{i=o}^{n} a_i t^i \in R$. Then, for any $x \in K$, we have

$$\text{(3.4)} \qquad g(D)(x) = \left(D_{-a_o, S^n} + a_o I\right)(x) = S^n(x)a_o .$$

Replacing $x$ by $cx$ (where $c \in K$), we then have

$$g(D)(cx) = S^n(c)S^n(x)a_o = S^n(c)g(D)(x).$$

Therefore, we have an operator equation

(3.5) $$g(D)c = S^n(c)g(D) \qquad (\forall c \in K),$$

where the constants are thought of as left multiplication operators on K. This equation holds in the image of the natural ring homomorphism

(3.6) $$\mathcal{E} : K[t,S,D] \longrightarrow \text{End}(K, +)$$

which sends t to D and sends the constants in K to their left multiplication operators. We now go into the following two cases.

Case A. $\ker \mathcal{E} = 0$ (i.e. D is not algebraic). Here, $\mathcal{E}$ is injective, so the equation (3.5) pulls back to a polynomial equation $g(t)c = S^n(c)g(t)$ in $K[t,S,D]$, and hence $g(t)$ is right semi-invariant.

Case B. $\ker \mathcal{E} \neq 0$ (i.e. D is algebraic). Let $\ker \mathcal{E} = R \cdot h$ (using the fact that R is a left PID). Then, since $\ker \mathcal{E}$ is a 2-sided ideal, $h \neq 0$ is right invariant, in particular right semi-invariant. Q.E.D.

We have also the following supplement to the theorem above.

Theorem 3.7. *Assume that* R *has a (monic) non-constant right semi-invariant polynomial.*

(1) *Let* $p(t)$ *be such a monic polynomial of the least degree, say* m.

(2) *Let* $p'(t)$ *be a monic non-constant polynomial of the least degree, say* $m'$, *such that* $p'(t)c \equiv S^{m'}(c)p'(t) \pmod{R\cdot t}$ $(\forall c \in K)$.

(3) *Let* $a_o,\dots,a_n \in K$ *be such that* $n \geqslant 1$, $a_n = 1$, *and*

$\sum_{i=1}^{n} a_i D^i = D_{-a_o, S^n}$, *where* $n$ *is chosen to be as small as possible;* *let* $g(t) = \sum_{i=o}^{n} a_i t^i$.

*Then we have* $m = m' = n$, $p'(t)$, $g(t)$ *are both right semi-invariant, and* $p(t) \equiv p'(t) \equiv g(t) \pmod K$. *Moreover,* $p(t) = p'(t) = g(t)$ *unless* $S^m$ *is an inner automorphism. If* $D$ *happens to be algebraic and* $S$ *is an automorphism, then the degree of the minimal polynomial for* $D$ *is always a multiple of* $m$.

*Proof.* From the proof of $(2) \Longrightarrow (3)$ in the theorem, we have $n \leq m'$, and of course also $m' \leq m$. As before, we shall distinguish the following two cases depending on the behavior of the homomorphism $\varepsilon$ in (3.6) :

*Case A.* $\ker \varepsilon = 0$. From the "pullback" argument used before, we see that $g(t)$ and $p'(t)$ are in fact right semi-invariant. Therefore $m \leq n$ and hence $n \leq m' \leq m$ now become equalities. By (2.10), we conclude further that $p(t) \equiv p'(t) \equiv g(t) \pmod K$.

*Case B.* $\ker \varepsilon \neq 0$. Write $\ker \varepsilon = R \cdot h$ where $h$ is the minimal polynomial of the algebraic derivation $D$. For the polynomial $g(t)$, recall that we have the operator equation

$$g(D)c = S^n(c)g(D) \qquad \text{(cf. (3.5))}$$

holding for every $c \in K$. Lifting this equation back to $R$ using the homomorphism $\varepsilon$, we have

$$g(t)c - S^n(c)g(t) = q_c(t)h(t), \tag{3.8}$$

where $q_c(t) \in R$ depends on $c$. Since $h(t)$ is right invariant, $\deg h \geq m$. If there exists a $c \in K$ such that the LHS above is not zero, then, for this $c$, we would have

$$n > \deg\,(\text{LHS}) \geq \deg h(t) \geq m,$$

a contradiction. Therefore, the LHS of (3.8) is zero for all $c \in K$, i.e. $g$ is right semi-invariant. Replacing $g$ by $p'$, we see similarly that $p'$ is right semi-invariant. Therefore, we can finish the argument exactly as in Case A.

The uniqueness statement in the Theorem now follows from Prop. (2.10). In fact, we are free to change any one of $p(t)$, $p'(t)$ and $g(t)$ by an additive constant $a \in K$ iff $S^m = I_a$. Finally, if $D$ is algebraic (Case B above), its minimal polynomial $h(t)$, being right invariant, will be expressible as a (left) polynomial in $p(t)$, in case $S$ is an automorphism (see (2.9)). It follows then that $\deg h$ is divisible by $\deg p = m$. Q.E.D.

<u>Corollary 3.9</u>. <u>Assume $S$ is an automorphism, and that $D$ is algebraic with a minimal polynomial $h$ of prime degree $\ell$. If $D$ is not an inner $S$-derivation, then no $D^r + b_{r-1}D^{r-1} + \ldots + b_1 D$ with $r < \ell$ can be an inner $S^r$-derivation, and $h(t)$ has the least degree among all non-constant right semi-invariant polynomials.</u>

<u>Proof</u>. Keeping the notations in Th. (3.7), we have $\deg g \mid \deg h$. Since $\deg h = \ell$ is prime and $n = \deg g \neq 1$, we must have $m = n = \ell$, which gives the two desired conclusions. Q.E.D.

Having looked at conditions for the existence of non-constant right semi-invariant polynomials, it is natural to look for conditions for the existence of non-constant right invariant polynomials (i.e. for the non-simplicity of R). There seems to be some evidence for the following

(3.10) Conjecture. *R has a non-constant right invariant polynomial iff it has a non-constant right semi-invariant polynomial. Or equivalently (in view of* Th. (3.2)), *R is non-simple iff D is quasi-algebraic*.

(Needless to say, the weight of the Conjecture is in the "if" part.)

Most remarkably, Lemonnier has proved the truth of this Conjecture in the case when S is an automorphism [Lem]. This result has provided the strongest evidence for the Conjecture so far. Recall also (from (2.6)) that if there is a linear right semi-invariant polynomial f, then f must be automatically right invariant. Such a polynomial f exists iff D is an inner S-derivation (cf. [$LL_1$: (3.4)(1)]): in this case, the Conjecture is, therefore, trivially true.

In view of Theorem (3.7), it will be of interest to give as much information as possible on "the" polynomial p(t), if it exists. We shall content ourselves here by dealing with the two key cases (a) D = 0, and (b) S = I. In case (a), we can just take p(t) = t and (3.7) gives all the desired information. In case (b), we have the following fairly precise description of $p(t)$, largely inspired by ideas in [A'] and in [LM].

<u>Theorem 3.11</u>. $(S = I)$ *Let* $p(t) = \sum a_i t^i$ *be a monic non-constant right semi-invariant polynomial of the least degree, say* $m$. *(We are assuming that* $p(t)$ *exists.) Then:*

(1) $D(a_i) = 0$ *for* $i = 1,2,\ldots,m$, *and* $D(a_o) \in Z(K)$, *the center of* $K$.

(2) *If* $\operatorname{char} K = p > 0$, *then* $p(t)$ *has the form* $\sum_{i=o}^{k} c_i t^{p^i} + a_o$, *where* $c_i \in Z(K) \cap K_D$ *and* $D(a_o) \in Z(K)$. *Moreover,* $\sum_{i=o}^{k} c_i D^{p^i} = D_{-a_o}$.

(3) *If* $\operatorname{char} K = 0$, *then* $p(t) = t + a_o$ *and* $D = D_{-a_o}$.

<u>Proof</u>. Since $S = I$, to say that a polynomial is right semi-invariant simply means that it commutes with constants (see(2.2)(2)). To prove (1), the crucial observation is that $tp(t) - p(t)t$ commutes with constants. In fact, for any $c \in K$,

$$\begin{aligned}[tp(t) - p(t)t]\,c &= tcp(t) - p(t)(ct + D(c)) \\ &= ctp(t) + D(c)p(t) - p(t)ct - p(t)D(c) \qquad (3.12) \\ &= c\,[tp(t) - p(t)t]\,.\end{aligned}$$

On the other hand,

$$\begin{aligned}tp(t) - p(t)t &= t^{m+1} + a_{m-1}t^m + (D(a_{m-1}) + a_{m-2})t^{m-1} + \ldots. \\ &\quad - t^{m+1} - a_{m-1}t^m - \ldots. \\ &= \sum_{i=o}^{m-1} D(a_i)t^i\end{aligned}$$

has degree $< m$. Thus, (by the minimal choice of $m$), $tp(t) - p(t)t$ must be a constant (namely, $D(a_o)$). This gives $D(a_i) = 0$ for $i \geq 1$, and (3.12) gives $D(a_o) \in Z(K)$.

(2) By (2.4)(1), we have

$$(3.13) \qquad ca_j = \sum_{i=j}^{m} \binom{i}{j} a_i D^{i-j}(c) \quad \text{for any } c \in K, \text{ and } j \geq 0.$$

For $j \geq 1$, define the polynomials

$$(3.14) \qquad \begin{cases} p_j(t) := \sum_{i=j}^{m} \binom{i}{j} a_i t^{i-j} \in K[t,D] & \text{with} \\ \deg p_j(t) \leq m - j < m. \end{cases}$$

Calculating as in [LM: pp.1255-1256], we can check via (3.13) that each $p_j(t)$ $(1 \leq j \leq m)$ commutes with constants. By the minimal choice of $m$, we must have then $p_j(t) = a_j \in Z(K)$ (in addition to $a_j \in K_D$ which we proved in (1)), and from (3.14), we see that

$$(3.15) \qquad \binom{i}{j} a_i = 0 \quad \forall\ i, j \text{ such that } 1 \leq j < i \leq m.$$

Exactly as in [LM], we conclude from this that, when $\operatorname{char} K = p > 0$, $a_j \neq 0$ $(j \geq 1)$ can occur <u>only when</u> $j$ is a power of $p$. Therefore, $p(t)$ has the form $\sum_{i=o}^{k} c_i t^{p^i} + a_o$, and it follows from (3.3) that $\sum_{i=o}^{k} c_i D^{p^i}$ is equal to the inner derivation $D_{-a_o}$. (Aside from the constant term $a_o$, $p(t)$ is a "p-polynomial" in the sense of Ore [O'].)

(3) Now assume $\operatorname{char} K = 0$. In this case, setting $i = m$ in (3.15), we have $\binom{m}{j} a_m = 0$ whenever $1 \leq j < m$. Since $a_m = 1$, the only way for this to be possible is when $m = 1$. Thus, $p(t) = t + a_o$, and by (3.3) (for instance), we have $D = D_{-a_o}$. Q.E.D.

## §4. P-dependence and linear dependence

We begin by recalling some basic notions from [L]. These notions were introduced in [L] in the case $D = 0$, but they are equally meaningful in the general case. A set $\Delta \subseteq K$ is called (S,D)-algebraic (or just algebraic if (S,D) is clear from the context) if there exists a nonzero polynomial $f(t) \in R = K[t,S,D]$ such that $f(\Delta) = 0$. In this case, the monic $f$ of the least degree with $f(\Delta) = 0$ is called the minimal polynomial of $\Delta$, and the rank of $\Delta$ is defined to be the degree of such an $f$. The basic properties developed in [L] for minimal polynomials carry over without change to the (S,D)-setting. In particular, the minimal polynomial $f$ for an (S,D)-algebraic set $\Delta$ has always a complete factorization $(t-a_1)\ldots(t-a_n)$ in $K[t,S,D]$ where each $a_i$ is (S,D)-conjugate to some element of $\Delta$, and any zero of $f$ is also (S,D)-conjugate to some element of $\Delta$. If $\Delta$ is (S,D)-algebraic, an element $b \in K$ is said to be P-dependent (or polynomially dependent) on $\Delta$ if every polynomial vanishing on $\Delta$ also vanishes on $b$ (or, equivalently, the minimal polynomial of $\Delta$ vanishes on $b$). By what we said above, such an element $b$ must be (S,D)-conjugate to some element of $\Delta$.

In this section, we shall focus our attention on subsets of a fixed (S,D)-conjugacy class $\Delta^{S,D}(a)$. Let $C$ denote the (S,D)-centralizer $C^{S,D}(a) = \{0\} \cup \{c \in K^* : a^c = a\}$ of $a$. Then $C$ is a division subring of $K$ (see [$LL_1$]), and $K$ may be viewed as a right vector space over $C$. In this section, we shall show that there is a very close relationship between P-dependence for elements in $\Delta^{S,D}(a)$ and right C-linear dependence for elements of K. The basic tool needed to establish this relationship is the idea of an "exponential space" introduced (though without such a name) in [$LL_1$]: for any polynomial $f(t) \in K[t,S,D]$, let

$$(4.1) \qquad E(f, a) = \{0\} \cup \{y \in K^* : f(a^y) = 0\}.$$

This is easily seen to be a right C-vector space, henceforth called the exponential space of f at a. In [$LL_1$: Th. (4.2)], we have proved the basic inequality $\dim_C E(f, a) \leq \deg f$ (which, in the special case $a = 0$, boils down to Amitsur's Theorem in [A]). We shall now explore some consequences of this important inequality.

Proposition 4.2. *Let Y be any subset of $K^*$, and let $a^Y$ denote $\{a^y : y \in Y\}$. Then $a^Y$ is (S,D)-algebraic iff span(Y) is finite dimensional over $C := C^{S,D}(a)$. (Here, span(Y) denotes the right C-vector space of K spanned by Y.) Furthermore, in this case, $\operatorname{rank}(a^Y) = \dim_C \operatorname{span}(Y)$.*

Proof. For the "only if" part, let $f \in R = K[t,S,D]$ be the minimal polynomial of $a^Y$. Then $Y \subseteq E(f, a)$ and so

$$\dim_C \operatorname{span}(Y) \leq \deg f = \operatorname{rank}(a^Y) < \infty .$$

Conversely, suppose span(Y) has right C-dimension $n < \infty$ and let $y_1,\dots,y_n \in Y$ form a C-basis for span(Y). Fix a polynomial $g \in R \setminus \{0\}$ of degree $\leq n$ such that $g(a^{y_i}) = 0$ for $i = 1,2,\dots,n$

(see [L : Prop. 6]). Then $y_i \in E(g, a)$ for all $i$ implies that $Y \subseteq E(g, a)$ and so $g(a^Y) = 0$. This shows that $a^Y$ is (S,D)-algebraic with rank $(a^Y) \leq \deg g \leq n = \dim_C \operatorname{span}(Y)$. Q.E.D.

Proposition 4.3. *Let $Y \subseteq K^*$ be such that $a^Y \subseteq \Delta^{S,D}(a)$ is (S,D)-algebraic. Then, for any $x \in K^*$, $a^x$ is P-dependent on $a^Y$ iff $x \in \operatorname{span}(Y)$.*

Proof. First assume $x \in \operatorname{span}(Y)$. Consider any $f \in R$ such that $f(a^Y) = 0$. Then $Y \subseteq E(f, a)$ implies that $\operatorname{span}(Y) \subseteq E(f, a)$. Therefore, $x \in E(f, a)$, i.e. $f(a^x) = 0$. This shows that $a^x$ is P-dependent on $a^Y$. Conversely, assume $a^x$ is P-dependent on $a^Y$. Then rank $\{a^Y, a^x\}$ = rank $(a^Y)$ and so, by the Proposition above, we have $\dim_C \operatorname{span}\{Y, x\} = \dim_C \operatorname{span}(Y)$. This clearly implies that $x \in \operatorname{span}(Y)$. Q.E.D.

Definition 4.4. An (S,D)-algebraic set $\Delta$ is said to be *full* if every $x \in K$ which is P-dependent on $\Delta$ actually belongs to $\Delta$.

From what we said earlier about minimal polynomials, it follows readily that *an (S,D)-algebraic set $\Delta$ is full iff $\Delta$ consists of all the zeros of its minimal polynomial in K.*

Theorem 4.5. *Let $a \in K$ be given and let $C = C^{S,D}(a)$. Then there is a one-one correspondence between the full (S,D)-algebraic subsets of $\Delta^{S,D}(a)$ and the finite dimensional right C-linear subspaces of K. Moreover, this one-one correspondence preserves inclusion and rank.*

Proof. For a finite dimensional right C-subspace $Y \dot{\cup} \{0\} \subseteq K$, we associate the (S,D)-algebraic subset $a^Y$ of $\Delta^{S,D}(a)$. We claim that $a^Y$ is full. Indeed, let $z$ be an element of $K$ which is P-dependent on $a^Y$. Then $z$ is a zero of the minimal polynomial of $a^Y$, and so $z \in \Delta^{S,D}(a)$. Write $z = a^x$ where $x \in K^*$. By Proposition (4.3), we must have $x \in \text{span}(Y) = Y \dot{\cup} \{0\}$ and so $a^x \in a^Y$, as desired. Next, we have to show that $Y \dot{\cup} \{0\} \longmapsto a^Y$ gives the desired one-one correspondence. First, suppose $a^Y = a^{Y'}$, where $Y \dot{\cup} \{0\}$, $Y' \dot{\cup} \{0\}$ are both finite dimensional right C-subspaces of K. Then for any $y \in Y$, we have $a^y = a^{y'}$ for some $y' \in Y'$. But then, conjugating by $y'^{-1}$, we get $a = (a^y)^{y'^{-1}} = a^{y'^{-1}y}$ (see [$LL_1$: (2.6)]), so $y'^{-1}y \in C$, i.e. $y \in y'C \subseteq Y' \dot{\cup} \{0\}$. This shows that $Y \subseteq Y'$ and so by symmetry we must have $Y = Y'$. Finally, let $\Delta$ be any full (S,D)-algebraic subset of $\Delta^{S,D}(a)$, say with minimal polynomial $f$. Then $\Delta$ consists of all zeros of $f$. Let $Y := \{ y \in K^* : a^y \in \Delta \}$. To show that $Y \dot{\cup} \{0\}$ is a right C-subspace of K, let $y_1, y_2 \in Y$ and $c_1, c_2 \in C$. We have $f(a^{y_1}) = f(a^{y_2}) = 0$ so $y_1, y_2 \in E(f, a)$, which implies that $y_1c_1+y_2c_2 \in E(f, a)$. If $y_1c_1+y_2c_2 \neq 0$, we'll have $f(a^{y_1c_1+y_2c_2}) = 0$ and so $a^{y_1c_1+y_2c_2} \in \Delta$. By the definition of Y, we have then $y_1c_1+y_2c_2 \in Y$. We have clearly $a^Y = \Delta$ and by Proposition (4.2), $\dim_C Y \dot{\cup} \{0\} = \text{rank}\, \Delta < \infty$. The proof is now complete.

In [L], it was shown that many of the key facts on linear dependence and bases in linear algebra have valid analogues for P-dependence and P-bases. (All arguments in [L] extend without change to the (S,D)-setting.) The above results giving the <u>explicit</u> relationship between P-dependence and linear dependence have now explained why such a close analogy should exist. Actually, this relationship has already been exploited in [$LL_1$: Th. (4.4)] in our computation of the rank of an (S,D)-Vandermonde matrix. The work we did in this section gives a fuller treatment of the ideas involved, and makes explicit the one-one correspondence in Theorem 4.5 above.

## §5. Algebraic Conjugacy Classes

We begin with a few basic observations.

**Lemma 5.1.** *Let* $f \in R = K[t,S,D]$ *be a right semi-invariant polynomial. If* $a \in K$ *is such that* $f(a) = 0$, *then* $f(\Delta^{S,D}(a)) = 0$.

*Proof.* For any $c \in K^*$, we have $f(t)c = c'f(t)$ for some $c' \in K$ depending on $c$. Using the Product Theorem [LL$_1$: (2.7)] to evaluate the two sides of this equation at $a$, we get $f(a^c)c = c'f(a) = 0$, and so $f(a^c) = 0$ for every $c \in K^*$. Q.E.D.

**Lemma 5.2.** *Let* $\Delta$ *be an* (S,D)-*algebraic subset of* K *which is closed under* (S,D)-*conjugation. (This means that* $\Delta$ *is the union of a finite number of* (S,D)-*conjugacy classes of* K.) *Then the minimal polynomial* $f(t) \in R$ *of* $\Delta$ *is a right invariant polynomial.*

*Proof.* Consider any $h(t) \in R$. If we can show that $f(t)h(t)$ vanishes on $\Delta$, then we will have $f \cdot h \in R \cdot f$ as desired. Let $b$ be any element of $\Delta$. By the Product Theorem again, we have

$$(fh)(b) = \begin{cases} 0 & \text{if } h(b) = 0, \\ f(b^c)c & \text{if } c := h(b) \neq 0. \end{cases}$$

In the second case, since $b \in \Delta$ implies that $b^c \in \Delta$, $f(b^c)$ is also zero, and so $fh$ vanishes on all of $\Delta$. Q.E.D.

**Proposition 5.3.** *Let* $\Delta$ *be a full* (S,D)-*algebraic set (in the sense of* (4.4)) *with minimal polynomial* $f \in R$. *Then the following are equivalent:*

In this case, the Corollary asserts that the "bound" of $\bigcap R\cdot(t-a_i)$ is given by $R\cdot f$, and that this is also the product of the bounds of $R\cdot(t-a_i)$ $(1\leq i\leq n)$. ]

Proof of (5.4). $(1)\Longleftrightarrow(1')$ follows from the fact that the union of a finite number of algebraic sets is algebraic. But we can also avoid using this fact by proving $(1')\Longrightarrow(2)$ and $(3)\Longrightarrow(1)$, for then we'll have a complete cycle of implications

$$(1)\Longrightarrow(1')\Longrightarrow(2)\Longrightarrow(3)\Longrightarrow(1).$$

$(1')\Longrightarrow(2)$ Let $f_i$ be the minimal polynomial of $\Delta^{S,D}(a_i)$. By (5.3), each $f_i$ is right invariant. It follows that $f_1 \cdots f_n$ is right invariant and is a left multiple of each of $f_i$. Since $f_i$ is a left multiple of $t-a_i$, $f_1 \cdots f_n$ gives the candidate for (2).

$(3)\Longrightarrow(1)$ Let $g(t)$ be a right semi-invariant polynomial as in (3). Then $g(a_i) = 0$ $(i\leq 1\leq n)$ implies that $g(\Delta^{S,D}(a_i)) = 0$ by (5.1), and so $g(\Delta) = 0$.

The above arguments also suffice to show that the monic $f(t)$ of the least degree as in (3) (or (2)) is exactly the minimal polynomial of $\Delta$, and that such an $f$ is a right factor of $f_1 \cdots f_n$. Since by the Union Theorem in [L] (which extends verbatim to the (S,D)-setting) $\operatorname{rank}\Delta = \sum_{i=1}^{n} \operatorname{rank} \Delta^{S,D}(a_i) = \sum_{i=1}^{n} \deg f_i$, it follows that $f = f_1 \cdots f_n$. Applying this in the case $n = 2$, we conclude further that $f_i f_j = f_j f_i$ whenever $i \neq j$, since both sides of the equation give the minimal polynomial of $\Delta^{S,D}(a_i) \cup \Delta^{S,D}(a_j)$. Q.E.D.

(1) $\Delta$ _is closed under_ $(S,D)$-_conjugation_;

(2) $f$ _is right invariant_;

(3) $f$ _is right semi-invariant_.

_Proof_. (1) $\Longrightarrow$ (2) is given by the preceding lemma, and (2) $\Longrightarrow$ (3) is obvious. For (3) $\Longrightarrow$ (1), let $a \in \Delta$ and consider any conjugate $a^c$ of $a$. Since we assume $f$ is right semi-invariant, $f(a) = 0$ implies that $f(a^c) = 0$ by Lemma 5.1. The fact that $\Delta$ is full means that $\Delta$ consists of all the zeros of $f$. Therefore, we have $a^c \in \Delta$. Q.E.D.

_Proposition 5.4_. _Let_ $\Delta$ _be a finite disjoint union_ $\bigcup_{i=1}^{n} \Delta^{S,D}(a_i)$. _Then the following are equivalent_:

(1) $\Delta$ _is_ $(S,D)$-_algebraic_;

(1') _Each_ $\Delta^{S,D}(a_i)$ $(1 \leq i \leq n)$ _is_ $(S,D)$-_algebraic_;

(2) _There is a nonzero right invariant polynomial which is a common left multiple of all_ $t - a_i$ $(1 \leq i \leq n)$;

(3) _There is a nonzero right semi-invariant polynomial which is a common left multiple of all_ $t - a_i$ $(1 \leq i \leq n)$.

_If these conditions hold, the monic_ $f(t)$ _of the least degree as in_ (2) (_or_ (3)) _is exactly the minimal polynomial of_ $\Delta$. _Furthermore, the minimal polynomials_ $f_i$ _of_ $\Delta^{S,D}(a_i)$ _pairwise commute, and we have_ $f(t) = f_1(t) \cdots f_n(t)$.

[_Note_. In the standard terminology of [J: p.38], the condition on the $a_i$'s in (2) is that the left ideal $\bigcap_{i=1}^{n} R\cdot(t-a_i)$ be _bounded_.

Before we proceed further with our treatment of algebraic conjugacy classes, let us recall some facts from $[LL_1]$ in a form most suitable for applications in this section.

Lemma 5.5 Let $p(t) \in K[t,S,D]$ and $y \in K^*$. Then

(1) $p(D)(y) = p(0^y)y$ ;

(2) If $D = 0$, then $p(S)(y) = p(1^y)y$. (Here, $1^y = S(y)y^{-1}$.)

Proof. (1) has been shown in the proof of Cor. 4.3 of $[LL_1]$. For (2), first note that $N_i(1) = 1$ for all $i \geqslant 0$. (The $N_i$ 's are the generalized "power functions" with respect to $(S,D)$: see $[LL_1 : \S 2]$.) From $[LL_1$: Prop. (2.9)(1)$]$ applied to the special case $D = 0$, we have then $N_i(1^y)y = S^i(y)$ for all $i \geqslant 0$. (This can also be checked by a direct calculation.) Thus, if $p(t) = \sum a_i t^i$ , we have

$$p(S)(y) = \sum a_i S^i(y) = \sum a_i N_i(1^y)y = p(1^y)y. \quad \text{Q.E.D.}$$

Let $a \in K$ be a fixed element. Then, for any $c \in K$,

$$\begin{aligned}(t-a)c &= S(c)t + D(c) - ac \\ &= S(c)(t-a) + S(c)a - ac + D(c) \\ &= S(c)(t-a) + D'(c),\end{aligned}$$

where $D' := D - D_{a,S}$ . ($D_{a,S}$ denotes the S-inner derivation of K sending y to $ay - S(y)a$.) Therefore, we have a well-defined ring homomorphism

$$(5.6) \qquad \Lambda : K[t',S,D'] \longrightarrow K[t,S,D]$$

which is the identity on $K$ and sends $t'$ to $t-a$. Clearly $\Lambda$ is an isomorphism of rings (with the inverse isomorphism sending $t$ to $t'+a$). Let $g(t')$ be any polynomial in $K[t',S,D']$, and let $f(t) = \Lambda(g(t')) \in K[t,S,D]$. Then

$$(5.7) \qquad \begin{cases} f(t) = \Lambda(g(t')) = g(\Lambda(t')) = g(t-a), \text{ and} \\ g(t') = \Lambda^{-1}(f(t)) = f(\Lambda^{-1}(t)) = f(t'+a). \end{cases}$$

As we have observed in [$LL_1$: §2], the division ring of constants of $D'$ is just $C := C^{S,D}(a)$. The following lemma provides the basic link between the solutions of polynomial equations and the solutions of differential equations in $K$.

Proposition 5.8. *For $f \in K[t,S,D]$ as above, and any $y \in K^*$, we have $g(D')(y) = f(a^y)y$. In particular, the exponential space $E(f, a)$ is exactly the right $C$-vector space of solutions (in $y$) of the differential equation $g(D')(y) = 0$.*

Proof. We first show that, for any $b \in K$, we have $f(b) = g(b-a)$. (This is a special case of a more general result called the "Composite Function Theorem" in [$LL_2$].) In fact, write

$$f(t) = q(t)(t-b) + f(b) \qquad \text{where } q(t) \in K[t,S,D].$$

Applying the inverse isomorphism $\Lambda^{-1}$, we get

$$g(t') = q(t'+a)(t'-(b-a)) + f(b) \qquad \text{in } K[t',S,D'].$$

Therefore, by the Remainder Theorem (applied to $K[t',S,D']$),

$f(b) = g(b-a)$. Now, let $b = a^y$, where $y \in K^*$. Since

$$D'(y)y^{-1} = \left[D(y) - (ay-S(y)a\right] y^{-1}$$
$$= D(y)y^{-1} - a + S(y)ay^{-1}$$
$$= a^y - a,$$

we get $f(a^y) = g(a^y-a) = g(D'(y)y^{-1})$. Therefore, by (5.5)(1) (applied to $g(t') \in K[t',S,D']$), we have

$$f(a^y)y = g(D'(y)y^{-1})y = g(D')(y). \qquad \text{Q.E.D.}$$

<u>Remark 5.9</u>. S. Amitsur [A] has shown that the solutions of the differential equation $g(D')(y) = 0$ form a right $C$-vector space of dimension $\leqslant \deg g$. (Amitsur's original arguments worked only in the case when $S$ is an automorphism. A more general argument establishing the result for any endomorphism $S$ can be found in [Co: p.65].) Assuming this result, the above Proposition leads to another proof of the fact that, as a right $C$-vector space, $E(f, a)$ has $C$-dimension $\leqslant \deg f$ (=deg g). This proof is somewhat different in spirit from the proof we gave earlier in [$LL_1$: Th. (4.2)].

Combining the preceding results with those of §4, we can now give some additional criteria (to (5.4)) for a given $(S,D)$-conjugacy class $\Delta^{S,D}(a)$ to be algebraic. Since $C^{S,D}(a)$ is just the division ring of the constants of the $S$-derivation $D' := D - D_{a,S}$, the equivalence of (2) and (3) below is well-known (dating from the work of Amitsur [A]). However, we'll prove this afresh as our arguments will also yield the exact information relating the minimal polynomials of algebraic conjugacy classes and those of algebraic derivations.

Theorem 5.10. *For* $a \in K$, *the following are equivalent:*

(1) $\Delta^{S,D}(a)$ *is* $(S,D)$-*algebraic;*

(2) $[K : C^{S,D}(a)]_{rt} < \infty$ ;

(3) $D' := D - D_{a,S}$ *is an algebraic* $S$-*derivation.*

*If these conditions hold, then*

$$\text{rank } \Delta^{S,D}(a) = [K : C^{S,D}(a)]_{rt} = \text{deg (min. poly. of } D').$$

*Moreover, if* $f(t) \in K[t,S,D]$ *is the minimal polynomial of* $\Delta^{S,D}(a)$, *then the minimal polynomial of* $D'$ *in* $K[t',S,D']$ *is given by* $f(t' + a)$.

Proof. Let $Y = K^*$. Then $\Delta^{S,D}(a)$ is just $a^Y$. Since $\text{span}(Y) = K$ as a right $C^{S,D}(a)$-vector space, $(1) \Longleftrightarrow (2)$ follows from Prop. (4.2). The last part of this Proposition also gives the equality $\text{rank } \Delta^{S,D}(a) = [K : C^{S,D}(a)]_{rt}$ .

$(1) \Longrightarrow (3)$ For the minimal polynomial $f(t)$ of $\Delta^{S,D}(a)$, we have $E(f, a) = K$. Let $g(t') = f(t'+a) \in K[t',S,D']$ so $f$ and $g$ are related as in (5.7). Then Proposition (5.8) gives $g(D') = 0$, so $D'$ is algebraic.

$(3) \Longrightarrow (1)$ Let $g_o(t') \in K[t',S,D']$ be the minimal polynomial of $D'$. Then by (5.8) again, we have $E(f_o, a) = K$ where $f_o(t) := g_o(t-a) \in K[t,S,D]$. This means that $f_o(a^y) = 0$ for all $y \in K^*$, so $\Delta^{S,D}(a)$ is $(S,D)$-algebraic.

Combining the arguments in the last two paragraphs, it is now clear that $g_o = g \in K[t',S,D']$. Q.E.D.

Let us now record a few consequences of the Theorem. As we have already pointed out, the following consequence (corresponding to the case $a = 0$) is largely classical: cf. [A] and [Le].

Corollary 5.11. *D is algebraic iff $[K : K_D] < \infty$ iff the logarithmic derivatives with respect to D form an (S,D)-algebraic class $\Delta^{S,D}(0)$. In this case, the minimal polynomial $g(t) \in K[t,S,D]$ of D is equal to the minimal polynomial of $\Delta^{S,D}(0)$. In particular, g splits completely in $K[t,S,D]$, and an element $b \in K$ is a logarithmic derivative with respect to D iff $g(b) = 0$.*

Next, note that the Theorem yields a criterion for the existence of an (S,D)-algebraic class:

Corollary 5.12. *K has an (S,D)-algebraic conjugacy class iff D is the sum of an inner S-derivation and an algebraic S-derivation.*

[Remark. If D is inner, say $D = D_{a,S}$, then $\Delta^{S,D}(a) = \{a\}$ is obviously an algebraic class. On the other hand, if D is algebraic, then $\Delta^{S,D}(0)$ is an algebraic class (see (5.11)). In general, if $D = D_{a,S} + D'$ where D' is an algebraic S-derivation, then (by the Theorem) $\Delta^{S,D}(a)$ is an algebraic class; moreover, an easy calculation shows that $\Delta^{S,D}(a) = a + \Delta^{S,D'}(0)$.]

If K is a field and $S = I$, then $D_{a,S} = 0$ for every $a \in K$. In this case, the Theorem gives the following:

Corollary 5.13. *Let* K *be a field, and* S = I. *Then a derivation* D *on* K *is algebraic iff one (or all) of the* (I,D)-*conjugacy classes is (are) algebraic.*

To put Theorem (5.10) in perspective, note that, in the classical case when (S,D) = (I,0), the condition that $\Delta^{S,D}(a)$ be algebraic means simply that a is algebraic over the center F of K, and rank $\Delta^{S,D}(a)$ is then given by the field extension degree $[F(a): F]$ (see [L: p.207]). In this case, the equality $[K : C(a)] = [F(a): F]$ is well-known, and is usually stated as a part of the Double Centralizer Theorem. Thus, the Theorem we proved above may be viewed as an extension of some of the consequences of the Double Centralizer Theorem to the (S,D)-setting.

Example. Let K be the division ring of the real quaternions and let S = I, D = 0, and a = -i. Then $\Delta^{S,D}(a)$ is (S,D)-algebraic (by the above) with minimal polynomial $f(t) = t^2+1 \in K[t]$. From (5.10), it follows that the minimal polynomial for the inner derivation $D' = 0 - D_{-i,I} = D_{i,I}$ is

$$\begin{aligned} g(t') = f(t'-i) = (t'-i)^2 + 1 &= t'^2 - it' - t'i \\ &= t'^2 - it' - (it' + D'i) \\ &= t'^2 - 2it' \in K[t',D']. \end{aligned}$$

Next, we shall obtain an analogue (and supplement) to Th. (5.10), in the case when D = 0. *Let* $a \in K$ *be given, and assume* $a \neq 0$. Since we now assume D = 0, we have

$$(a^{-1}t)c = a^{-1}S(c)t = a^{-1}S(c)a\cdot a^{-1}t \qquad (\text{for any } c \in K^*).$$

Writing $I_{a^{-1}}$ for the inner automorphism which sends $x$ to $a^{-1}xa$, we have then $(a^{-1}t)c = \tilde{S}(c)t$ where $\tilde{S} = I_{a^{-1}} \circ S$. Therefore, we have a well-defined ring homomorphism

$$(5.14) \qquad \Gamma : K[\tilde{t},\tilde{S}] \longrightarrow K[t,S]$$

which is the identity on $K$, and sends $\tilde{t}$ to $a^{-1}t$. As was the case for $\Lambda$ in (5.6), $\Gamma$ is an isomorphism, with its inverse sending $t$ to $a\tilde{t}$. For any $h(\tilde{t}) \in K[\tilde{t},\tilde{S}]$, define

$$(5.15) \qquad \begin{cases} f(t) := \Gamma(h(\tilde{t})) = h(a^{-1}t), & \text{so that} \\ h(\tilde{t}) = \Gamma^{-1}(f(t)) = f(a\tilde{t}). \end{cases}$$

We have now the following analogue of Prop. (5.8).

<u>Proposition 5.16</u>. $(D = 0)$ <u>With the above notations, we have</u> $f(a^y)y = h(\tilde{S})(y)$ <u>for any</u> $y \in K^*$. <u>In particular, the exponential space</u> $E(f, a)$ <u>is exactly the right</u> $C^S(a)$-<u>vector space of the solutions (in</u> $y$) <u>of the equation</u> $h(\tilde{S})(y) = 0$. [<u>Note</u>. <u>By</u> (5.8), $E(f, a)$ <u>is also the solution space of the differential equation</u> $g(-D_{a,S})(y) = 0$, <u>where</u> $g(t') = f(t'+a)$.]

<u>Proof</u>. Proceeding as in the proof of (5.8), we can show that $f(t) = h(a^{-1}t) \Longrightarrow f(b) = h(a^{-1}b)$ for every $b \in K$. (Again, this is a special case of the Composite Function Theorem in $[LL_2]$.) Now applying Lemma (5.5)(2) to $h(\tilde{t}) \in K[\tilde{t},\tilde{S}]$, we have, for every $y \in K^*$:

$$\begin{aligned} h(\tilde{S})(y) &= h(\tilde{S}(y)y^{-1})y \\ &= f(a\tilde{S}(y)y^{-1})y \\ &= f(a\cdot a^{-1}S(y)ay^{-1})y \\ &= f(a^y)y, \end{aligned}$$

as desired. Q.E.D.

We have now the following refinement of Th. (5.10) in the case $D = 0$ (and $S$ any endomorphism of $K$).

Theorem 5.17. ($D = 0$) *For $a \in K^*$, the following are equivalent:*

(1) *$\Delta^S(a) = \{S(c)ac^{-1} : c \in K^*\}$ is $S$-algebraic;*

(2) *$[K : C^S(a)]_{rt} < \infty$ (where $C^S(a) = \{c \in K : S(c)a = ac)\}$);*

(3) *$D_{-a,S}$ is an algebraic $S$-derivation;*

(4) *The endomorphism $\tilde{S} = I_{a^{-1}} \circ S$ is algebraic.*

*If these conditions hold, and $f(t) \in K[t,S]$ is the minimal polynomial of $\Delta^S(a)$, then the minimal polynomial of $D_{-a,S}$ in $K[t',S,D_{-a,S}]$ is $f(t'+a)$, and the minimal polynomial of $\tilde{S}$ in $K[\tilde{t},\tilde{S}]$ is $f(a\tilde{t})$; moreover, $S$ and $\tilde{S}$ must be automorphisms of finite inner order.*

Proof. The equivalence of (1), (2), (3) and the relation between the minimal polynomials of $\Delta^S(a)$ and $D_{-a,S}$ follow by specializing (5.10) to the case $D = 0$. The equivalence of (1) with (4) and the relation between the minimal polynomials of $\Delta^S(a)$ and $\tilde{S}$ now follow similarly by applying Proposition (5.16). By [L: Lemma 5], $f(t)$ has the form $(t-a_1)\cdots\cdots(t-a_n) \in K[t,S]$ for suitable $a_i \in \Delta^S(a)$.

Therefore, $f(t)$ has constant term $b = a_1 \dots a_n \neq 0$. By (2.3)(1), it follows that $S^n = I_b$, so $S$ is an automorphism of finite inner order dividing $n$. Since $\tilde{S}$ has the same inner class as $S$, the same holds for $\tilde{S}$. Q.E.D.

Remark 5.18. It follows easily from the above that the $S$-inner derivation $D_{-a,S}$ satisfies a polynomial $g(t') \in K[t', S, D_{-a,S}]$ iff the endomorphism $\tilde{S} = I_{a^{-1}} \circ S$ satisfies the polynomial $h(\tilde{t}) = g(a(\tilde{t}-1)) \in K[\tilde{t}, \tilde{S}]$.

Letting $a = 1$ in Theorem (5.17), we get:

Corollary 5.19. $(D = 0)$ *The set* $\{S(c)c^{-1} : c \in K^*\}$ *is* $S$-*algebraic iff* $[K : K^S]_{rt} < \infty$, *iff* $S - I$ *is an algebraic* $S$-*derivation, iff* $S$ *is an algebraic endomorphism. For any of these conditions to hold,* $S$ *must be an automorphism of finite inner order.*

(If $S$ is assumed to be an automorphism of $K$ to begin with, the implication that $[K : K^S]_{rt} < \infty \Longrightarrow S$ has finite inner order is, of course, a well-known fact in the Galois theory of division rings (see, e.g. [Co: p.47]).

Going back to the general $(S,D)$-setting, let us now give some simple characterizations for the minimal polynomials of the $(S,D)$-algebraic conjugacy classes in $K$.

Theorem 5.20. *For a monic non-constant polynomial* $f(t) \in R$, *the following are equivalent:*

(1) f(t) <u>is the minimal polynomial of an</u> (S,D)-<u>algebraic conjugacy class</u> $\Delta^{S,D}(a)$;

(2) f(t) <u>is right invariant, has a zero in</u> K, <u>and has no proper left or right factor which is right invariant</u>;

(3) f(t) <u>is right semi-invariant, has a zero in</u> K, <u>and has no proper right factor which is right invariant</u>.

<u>Proof</u>. (2) $\Longrightarrow$ (3) is a tautology.

(3) $\Longrightarrow$ (1) Let $a \in K$ be a root of f. Then by (5.1), $f(\Delta^{S,D}(a)) = 0$ and so f is right divisible by $f_o$, the minimal polynomial of $\Delta^{S,D}(a)$. Since $f_o$ is right invariant, (3) implies that $f = f_o$.

(1) $\Longrightarrow$ (2) We already know that f is right invariant, and that f has a root (namely, a ). Consider the simple left R-module $V = R/R\cdot(t-a)$. We shall identify V with K via the correspondence $\overline{g(t)} \longmapsto g(a)$. Viewing K as a left R-module through this identification, the action of a polynomial $g(t) \in R$ on an element $c \in K^*$ is given by

$$g(t) * c = g(t)c\,\Big]_{t=a} = g(a^c)c, \tag{5.21}$$

by the Product Theorem in $[LL_1$: (2.7)]. In particular,

$$\begin{aligned} \operatorname{ann}_R V &= \{ g(t) \in R : g(a^c) = 0 \quad \forall c \in K^* \} \\ &= \{ g(t) \in R : g(\Delta^{S,D}(a)) = 0 \} \\ &= R\cdot f . \end{aligned} \tag{5.22}$$

Therefore, V is a <u>faithful</u> simple left $R/R\cdot f$-module. Since $R/R\cdot f$ is artinian, it follows that $R/R\cdot f$ is a simple ring. This means that $R\cdot f$ is maximal as a 2-sided ideal in $R$, and so $f$ has no proper right factor which is right invariant. It follows from this that $f$ has also no proper left factor which is right invariant. Q.E.D.

The observation we made in the proof above about the simple left $R$-module $R/R\cdot(t-a)$ (for any $a$) have also some other consequences which are worth recording. If $a$ is such that $\Delta^{S,D}(a)$ is not $(S,D)$-algebraic, the first two equalities in (5.22) would show that $R/R\cdot(t-a)$ is a faithful simple left $R$-module. Therefore, we have:

Corollary 5.23. $R = K[t,S,D]$ is a left primitive ring unless all $(S,D)$-conjugacy classes are algebraic. (In particular, $K[t]$ is a left primitive ring unless $K$ is algebraic over its center.)

Also, whether $\Delta^{S,D}(a)$ is algebraic or not, for any polynomial $g(t) \in R$, the annihilator of $g$ on $K$ identified with $V = R/R\cdot(t-a)$ as a left $R$-module is (by (5.21)) exactly

$$\{0\} \cup \{c \in K^* : g(a^c) = 0\} = E(g, a).$$

This gives a very interesting new interpretation for the exponential space $E(g, a)$, which was pointed out to us by Professor S. Amitsur during the Conference. In fact, as Professor Amitsur further pointed out, the fact that $\dim_C E(g, a) \leq \deg g$ for $C = C^{S,D}(a)$ can be deduced from the Jacobson-Chevalley Density Theorem, upon noting that $C$ is isomorphic to $\mathrm{End}_R V$ as a ring of right operators on $V$. This deduction is a rather illuminating exercise which we shall leave to the reader. Let us now make two additional remarks. First, in the case when $\Delta^{S,D}(a)$ is algebraic with minimal polynomial $f$, the Artin-Wedderburn Theorem implies that

$$R/R\cdot f \cong \mathrm{End}_C\ (R/R\cdot(t-a)) \cong \mathbb{M}_n(C),$$

where $n = [K : C]_{rt} = \deg f$. Secondly, whether $\Delta^{S,D}(a)$ is algebraic or not, it is also possible to derive from the Density Theorem some of the key facts in §4 relating the P-dependence of elements in $\Delta^{S,D}(a)$ to the right C-linear dependence of the elements of K. However, we do not feel justified to include the details of these alternative proofs here.

To conclude this section, we shall now combine Theorem (5.20) with Cauchon's result (Theorem 2.8) to derive some more interesting information on the minimal polynomials of (S,D)-algebraic classes. We first make the following easy observation on Z(R) (the center of $R = K[t,S,D]$) which is essentially well-known:

<u>Lemma 5.24</u>. *If* $Z(R)$ *contains a polynomial* $h(t) = \alpha^{-1}t^n + \ldots.$ *of degree* $n \geq 1$, *then* $S^n = I_\alpha$ *and* $S(\alpha) = \alpha$.

*Proof*. For any $c \in K$, we have $(\alpha^{-1}t^n + \ldots.)c = c(\alpha^{-1}t^n + \ldots.)$. Comparing leading coefficients, we get $\alpha^{-1}S^n(c) = c\alpha^{-1}$, so $S^n = I_\alpha$. Similarly, $(\alpha^{-1}t^n + \ldots.)t = t(\alpha^{-1}t^n + \ldots.)$ leads to $S(\alpha^{-1}) = \alpha^{-1}$. Q.E.D.

In case $D = 0$ and $S$ is not an automorphism of finite inner order, we have shown in Theorem (5.17) that K has only one S-algebraic conjugacy class, namely, $\Delta^S(0) = \{0\}$, with minimal polynomial $t$. Using Cauchon's Theorem (2.8) and Theorem (5.20), we can now generalize this fact to the case where $D$ need not be zero.

Theorem 5.25. *Assume that $S$ is not an automorphism of finite inner order. (This includes the case when $S$ is not onto.) Then there is at most one $(S,D)$-algebraic class $\Delta^{S,D}(a)$, and if such a class exists, its minimal polynomial $f(t)$ is a non-constant right invariant polynomial of the least degree.*

Proof. Look at an algebraic class $\Delta^{S,D}(a)$ with minimal polynomial $f(t)$, and let $q(t)$ be a monic non-constant right invariant polynomial of the least degree, as in (2.8). Since by (5.24) there is no non-constant central polynomial, (2.8) and (5.20) imply that $f(t) = q(t)$; in particular, $\Delta^{S,D}(a)$ is unique (if it exists). Q.E.D.

Remark 5.26. It can be shown that the $f(t)$ above has, in fact, minimal degree among all non-constant right *semi-invariant* polynomials. We shall not prove this fact here since it does not follow directly from the techniques developed in this paper.

Finally, we treat the case when $S$ is an automorphism of finite inner order.

Theorem 5.27. *Let $S$ be an automorphism of finite inner order $k$. Then, for all $(S,D)$-algebraic classes $\Delta^{S,D}(a)$ with possibly one exception, the following holds:*

(1) *The minimal polynomial* $f(t)$ *of* $\Delta^{S,D}(a)$ *lies in* $K \cdot Z(R)$;

(2) $f(t)$ *commutes with all monic right semi-invariant polynomial; and*

(3) *The rank of* $\Delta^{S,D}(a)$ *is divisible by* $k$.

*Finally, the rank of any* (S,D)-*algebraic class is divisible by* $\deg q(t)$, *where* $q(t)$ *is a monic non-constant right invariant polynomial of the least degree.*

*Proof*. By (2.8), the minimal polynomial $f(t)$ of any algebraic class $\Delta^{S,D}(a)$ has the form $\alpha \cdot h(t)q(t)^r$, where $\alpha \in K^*$, $h(t) \in Z(R)$ and $r \geq 0$. Bringing (5.20) to bear, we see that $f(t)$ is either equal to $q(t)$ or $\alpha \cdot h(t)$. Thus, with the possible exception of one class, $\Delta^{S,D}(a)$ has minimal polynomial $f(t) = \alpha \cdot h(t) = \alpha(\alpha^{-1}t^n + \ldots)$, where $n = \deg h$. By (5.24), we must have $S^n = I_\alpha$ and $S(\alpha) = \alpha$. The former implies that $n = \text{rank } \Delta^{S,D}(a)$ is a multiple of the inner order $k$ of $S$. To prove the property (2), let $g(t)$ be any monic right semi-invariant polynomial, say of degree $m$. Then we have

$$\begin{aligned} g(t)f(t) &= g(t)\alpha h(t) = S^m(\alpha)g(t)h(t) \\ &= \alpha h(t)g(t) = f(t)g(t), \end{aligned}$$

as claimed. Finally, since in any case $f(t)$ is either $\alpha \cdot h(t)$ or $q(t)$, the last part of Cauchon's Theorem (2.8) implies that $n = \deg f$ is always divisible by $\deg q$. Q.E.D.

Note that part (2) above strengthens the fact, first proved in Proposition (5.4), that the minimal polynomials of two distinct (S,D)-

algebraic classes always commute. This fact is now an obvious consequence of (5.27)(2) when $S$ is an automorphism of finite inner order, and, in view of (5.25), is vacuous when $S$ is not an automorphism of finite inner order.

Of course, in Theorem (5.27), an "exceptional" (S,D)-algebraic class may indeed exist, and it would behave somewhat differently from the other algebraic classes. For instance, let $K$ be a field with an automorphism $S$ of order $k$, and let $F = K^S$. Then, in the notation of (2.8) (with $D = 0$), we have $q(t) = t$, $h_o(t) = t^k$ in $R = K[t,S]$, and $Z(R) = F[t^k]$. Any class $\Delta^S(a)$ with $a \in K^*$ has rank $k$ and minimal polynomial $t^k - N_{K/F}(a)$ (see [L: p.208]), but the "exceptional" class $\Delta^S(0) = \{0\}$ has rank $1$ and minimal polynomial $q(t) = t$. The former kind of minimal polynomials clearly all belong to $Z(R)$, but $q(t) = t$ does not. In fact, $t$ fails to commute with all monic right semi-invariant polynomials: for instance, $t^k + a$ $(a \in K)$ is always right semi-invariant, but $t$ does not commute with it unless $a \in F$.

## References

[A] S. Amitsur, A generalization of a theorem on linear differential equations, Bull. Amer. Math. Soc. 54(1948), 937-941.

[A'] S. Amitsur, Derivations in simple rings, Proc. London Math. Soc. 7(1957), 87-112.

[C] G. Cauchon, Les T-anneaux et les anneaux à identités polynomiales noethériens, Thèse, Orsay, 1977.

[Ca] J. Carcanague, Idéaux bilatères d'un anneau de polynômes non commutatifs sur un corps, J. Algebra 18(1971), 1-18.

[Co] P. M. Cohn, Skew Field Constructions, London Math. Soc. Lecture Notes Series, Vol. 27, Cambridge University Press, 1977.

[Co'] P. M. Cohn, Free Rings and Their Relations, Academic Press, New York, 1971.

[J] N. Jacobson, The Theory of Rings, Mathematical Surveys, No.2, Amer. Math. Soc., Providence, R.I., 1943.

[L] T. Y. Lam, A general theory of Vandermonde matrices, Expositiones Mathematicae 4(1986), 193-215.

[Le] A. Leroy, Dérivées logarithmiques pour une S-dérivation algébrique, Communications in Algebra 13(1985), 85-99.

[Lem] B. Lemonnier, Dimension de Krull et codéviations, quelques applications en théorie des modules, Thèse, Poitiers, 1984.

[$LL_1$] T. Y. Lam and A. Leroy, Vandermonde and Wronskian matrices over division rings, to appear in Journal of Algebra.

[$LL_2$] T. Y. Lam and A. Leroy, Hilbert 90-type theorems for division rings, in preparation.

[LM] A. Leroy and J. Matczuk, Dérivations et automorphismes algébriques d'anneaux premiers, Communications in Algebra 13(1986), 1245-1266.

[O] O. Ore, Theory of non-commutative polynomials, Annals of Math. 34(1933), 480-508.

[O'] O. Ore, On a special class of polynomials, Trans. Amer. Math. Soc. 35(1933), 559-584.

[Sm] T. H. M. Smits, Nilpotent S-derivations, Indagationes Math. 30(1968), 72-86.

# On the Gelfand-Kirillov Dimension of Normal Localizations and Twisted Polynomial Rings.

A. Leroy (U. de Mons, Belgium)
J. Matczuk (U. de Warsaw, Poland)
J. Okninski (U. de Warsaw, Poland)

It is well known that, if $S$ is a central subset consisting of regular elements of an associative algebra $A$ over a field $K$, then the algebras $A$ and $AS^{-1}$ have the same Gelfand-Kirillov dimension. This also holds if $S$ is a commutative set of regular elements determining locally nilpotent inner derivations of $A$, cf. [2], Chapter 4. Some other positive results are concerned with the localizations of the enveloping algebras of finite dimensional Lie algebras over a field of characteristic zero, [1]. On the other hand, there are known examples showing that in general the Gelfand-Kirillov dimension of a localization may be very far from that of the original algebra, [2], Chapter 4.

In this note we are concerned with the question of determining the behaviour of the Gelfand-Kirillov dimension under localizations with respect to the multiplicatively closed subsets of regular elements $s \in A$ such that $sA = As$. Elements of this type, called **normal** elements, provide an important generalization of central elements and are often encountered in the theory of associative algebras. We show that, in some cases, the Gelfand-Kirillov dimension does not increase under localizations with respect to normal elements. The special case of the localization of the twisted polynomial ring $A[t;\sigma] \subset A[t,t^{-1};\sigma]$ appears to be of special interest and is connected to the original general problem. Examples of a pathological behaviour of Gelfand-Kirillov dimension under constructions of the above types are also given.

For the basic material concerning the Gelfand-Kirillov dimension we refer to [2].



*F. van Oystaeyen and L. Le Bruyn (eds.), Perspectives in Ring Theory, 205–214.*

Throughout the paper $K$ will be a fixed field and $A$ an algebra with unity over $K$. The Gelfand-Kirillov dimension of $A$ is denoted by $GK$ dim $A$. If $S$ is a multiplicatively closed subset of regular elements of $A$, which is an Ore subset, then we write $AS^{-1}$ for the loclaization with respect to $S$. If $s \in A$ is a normal element and $S = \{s^n | n \in N\}$, then $AS^{-1}$ is denoted by $A\{s^{-1}\}$. In this case $s$ determined an automorphism $\sigma$ of $A$ such that $\sigma(a)s = sa$ for $a \in A$. On the other hand, it is clear that for any automorphism $\tau$ of $A$, $t$ is a normal element of the twisted polynomial ring and we have $A[t;\tau]\{t^{-1}\} = A[t,t^{-1};\tau]$- the twisted Laurent polynomial ring. Let us start with the following simple observations.

**Lemma 1.** Let $\sigma$ be an automorphism of $A$. Then

**1.** $GK.\mathrm{dim}A[t;\sigma] \geq GK.\mathrm{dim}A + 1$;

**2.** If a power of $\sigma$ is an inner automorphism of $A$, then $GK.\mathrm{dim}A[t,t^{-1};\sigma] = GK\mathrm{dim}A[t;\sigma] = GK.\mathrm{dim}A + 1$.

**Proof. 1.** Is standard, cf. [2], lemma 3.4.

**2.** Assume that for some $n \geq 1, \sigma^n$ is an inner automorphism determined by an invertible element $u \in A$. Since $A[t,t^{-1};\sigma]$ is a finitely generated right $A[t^n,t^{-n};\sigma]$-module, then from [2], proposition 5.5, it follows that $GK.\mathrm{dim}\ A[t,t^{-1};\sigma] = GK.\mathrm{dim}A[t^n,t^{-1};\sigma^n]$. Now, $A[t^n,t^{-n};\sigma^n] = A[u^{-1}t^n, ut^{-n}]$ and $u^{-1}t^n$ is a central element of this algebra. Consequently, $GK.\mathrm{dim}A[t,t^{-1},\sigma] = GK.\mathrm{dim}A + 1$ and the result follows by **1.**

It appears that there is a connection between the Gelfand-Kirillov dimension of the localization with respect to a normal element and that of the corresponding twisted polynomial ring.

**Lemma 2.** Let $\sigma$ be an automorphism of $A$, determined by a normal element $s \in A$. Then :

**1.** $GK.\mathrm{dim}A\{s^{-1}\} + 1 = GK.\mathrm{dim}A[t,t^{-1}\sigma] = GK.\mathrm{dim}A[t,\sigma^{-1}]$

**2.** $GK.\mathrm{dim}A[t,\sigma] = GK.\mathrm{dim}A + 1$.

**Proof. 1.** Since $\sigma$ becomes inner in $A\{s^{-1}\}$ and $A[t,t^{-1};\sigma] \subseteq A\{s^{-1}\}[t,t^{-1};\sigma]$, then from Lemma 1 we know that $GK.\mathrm{dim}A[t,t^{-1};\sigma] \leq GK.\mathrm{dim}A\{s^{-1}\} + 1$. On the other hand $A[t;\sigma^{-1}] \simeq A[t^{-1};\sigma] \subseteq A[t,t^{-1};\sigma]$ implies that $GK.\mathrm{dim}A[t;\sigma^{-1}] \leq GK.\mathrm{dim}A[t,t^{-1};\sigma]$. Now, there is a natural epimorphism from $A[t,\sigma^{-1}]$ onto $A\{s^{-1}\}$ the kernel of which contains the regular element $st - 1$. Thus, by [2], proposition 3.15, $GK.\dim A[t;\sigma^{-1}] \geq GK.\dim A\{s^{-1}\}+1$.

**2.** We have $A[t;\sigma] \subseteq A[s^{-1}t]$ with $s^{-1}t$ being central. Thus, $GK.\dim A[t,\sigma] \leq GK.\dim A[s^{-1}t] = GK.\dim A + 1$, the converse inequality follows from Lemma 1.

Our first result shows that, under some commutativity hypothesis, the Gelfand-Kirillov dimension does not increase under normal localizations.

**Theorem 1.** Let $s \in A$ be a normal element determining an automorphism $\sigma$ of $A$. If there exists a commutative, $\sigma$-invariant subalgebra $B$ of $A$ such that $A$ is a finitely generated right $B[s]$-module, then $GK.\dim A\{s^{-1}\} = GK.\dim A$.

**Proof.** We will first show that $GK.\dim B[s,s^{-1}] = GK.\dim B[s]$. Take $W_V = V + Ks + Ks^{-1}$ where $V$ is a finite dimensional subspace of $B$ with $1 \in V$. Then every element of $W_V^n, n \geq 1$, is a sum of elements of the form $w = s^{i_1}v_1 \dots s^{i_k}v_k s^{\ell}$ where $v_i \in V$ and the integers $0 \leq k \leq n, i_j, \ell \in \mathbb{Z}$ satisfy $\sum_{j=1}^{k} |i_j| + |\ell| + k = n$. It is straightforward to verify that

$$\sigma^{n-1}(w)s^{3n} = s^{n-1}ws^{2n+1} = v_1^{\sigma^{\ell_1}} \dots v_k^{\sigma^{\ell_k}}.s^m \qquad (*)$$

where $\ell_j = i_1 + \dots i_j + n - 1$ for $1 \leq j \leq k$ and $m = 3n + \ell + i_1 + \dots + i_k$. For any $1 \leq j \leq k$ we have

$$\begin{aligned} 0 &\leq i_1 + \dots + i_j + \sum_{r=1}^{k} |i_r| + |\ell| + k - 1 \\ &= \ell_j \leq \sum_{r=1}^{k} |i_r| + n - 1 \\ &= 2n - 1 - (|\ell| + k) < 2n \leq 2n + (n + \ell + \sum_{r=1}^{k} i_r) \\ &= m \end{aligned}$$

Moreover,

$$\begin{aligned} k + m &\leq k + 3n + \ell + i_1 + \dots + i_k \\ &\leq k + 3n + |\ell| + |i_1| + \dots + |i_k| = 4n \end{aligned}$$

hence we get $0 \leq \ell_j \leq m$ for $1 \leq j \leq k$ and $k + m \leq 4n$. Using $\sigma$-invariance and the commutativity of $B$ we may reorder $v_i$'s in formula (*) and additionally assume that $\ell_1 \leq \ell_2 \leq \dots \leq \ell_k$. Then

$$\sigma^{n-1}(w)s^{3n} = s^{\ell_1}v_1 s^{\ell_2-\ell_1}v_2 \dots s^{\ell_k-\ell_{k-1}}s_k s^{m-\ell_k} \in U^{4n}$$

where $U = V + Ks \le B[s]$. This shows that, for any $n \ge 1, \sigma^{n-1}(W_V^n)s^{3n} \subseteq U^{4n}$. Since $s$ is a regular element, this yields $d_{W_v}(n) \le d_U(4n)$. Now, any finite dimensional subspace of $B[s, s^{-1}]$ is contained in a power of a subspace $W_V$ for a suitable $V \subset B$ and the last inequality implies that $GK.\dim B[s, s^{-1}] = GK.\dim B[s]$.

Then, the assertion is a consequence of Proposition 5.5 in [2] and the fact that both $A$ and $A\{s^{-1}\}$ are finitely generated right modules over $B[s]$, $B[s, s^{-1}]$ respectively.

**Corollary 1.** Let $A$ be a finitely module over its centre. Then, for any automorphism $\sigma$ of $A$, $GK.\dim A[t;\sigma] = GK.\dim A[t, t^{-1};\sigma]$.

The above corollary applies if $A$ is either a finitely generated semiprime algebra with $GK.\dim A = 1$, [3], or a simple $PI$-algebra. We do not know whether this may be extended to the class of prime PI-algebras.

Our second positive result is concerned with a special class of automorphisms. We say that an automorphism of $A$ is locally algebraic if for any $a \in A$ the set $\{\sigma^n(a) | n \in N\}$ is contained in a finite dimensional subspace of $A$.

**Proposition 1.** Let $\sigma$ be a locally algebraic automorphism of $A$. Then $GK.\dim A[t;\sigma] = GK.\dim A[t, t^{-1};\sigma] = GK.\dim A[t, \sigma^{-1}] = GK.\dim A + 1$.

**Proof.** In view of lemma 2 we only need to prove that $GK.\dim A[t, t^{-1};\sigma] \le GK.\dim A + 1$. Let $W$ be a finite dimensional subspace of $A[t, t^{-1};\sigma]$. There exists $\ell \in I\!N$ such that $Wt^{\ell} \subseteq A[t,\sigma]$ and there exists a finite dimensional subspace $U$ of $A$ such that $Wt^{\ell} \subseteq U + Ut + \ldots + Ut^{s}$ for some $s \in I\!N$. Since $\sigma$ is locally algebraic, $U$ may be chosen so that $\sigma(U) = U$. Then $W^n \subseteq (Ut^{-\ell} + Ut^{1-\ell} + \ldots + Ut^{s-\ell})^n \le (U + Ut + \ldots + Ut^{s})^n t^{n\ell} \subseteq (U^n + U^n t + \ldots + U^n t^{sn})t^{-n\ell}$ for any $n \in N$. We conclude that $d_W(n) = \dim W^n \le (ns+1)d_U(n)$. Consequently, $GK.\dim A[t, t^{-1};\sigma] \le GK.\dim A + 1$.

**Corollary 2.** Assume that $\sigma$ is an automorphism of $A$ such that $GK.\dim A[t,\sigma] = 1$. Then $GK.\dim A[t, t^{-1};\sigma] = GK.\dim A[t,\sigma] = GK.\dim A + 1 = 1$.

**Proof.** It is enough to show that $\sigma$ is locally algebraic. Suppose otherwise. Then, for some $a \in A$, the elements $\sigma^n(a), n \in I\!N$, are linearly indpendent. Let $W$ be the subspace generated by the set $\{1, t, a\}$. Consider the set $T = \{t^i a t^j | i + j < n\} \subset W^n$. It is clear that the elements $t^i a t^j = \sigma^i(a)t^{i+j}$ are linearly indepent. Thus, $d_W(n) = n + (n-1) + \ldots + 1 = \frac{n(n+1)}{2}$ and $GK.\dim A[t,\sigma] \ge 2$, a contradiction.

It is known that, if $S$ is a commutative, multiplicatively closed set of regular elements of $A$ which is generated by elements $x$ for which the inner derivation determined by $x$ is locally nilpotent, then $S$ is an Ore set and $GK.\mathrm{dim}AS^{-1} = GK.\mathrm{Dim}A$. Our next result is in the same vein. We say that an element $s \in A$ is local normal if $s$ is a normal element determining a locally algebraic automorphism of $A$. Clearly, if $S$ is a commutative multiplicatively closed set generated by local normal elements, then $S$ consists of local normal elements.

**Theorem 2.** Let $S$ be a multiplicatively closed subset of $A$ consisting of local normal elements. Then $S$ is an Ore set and $GK.\mathrm{dim}AS^{-1} = GK.\mathrm{dim}A$.

**Proof.** It is easy to check that $S$ is an Ore set. Let $W$ be a finite dimensional subspace of $AS^{-1}$ let $w$ be a common right denominator of the elements of $W$. Then $Ww \subseteq A$, so $W \subseteq A\{w^{-1}\}$ and we get limsup $\log_n d_W(n) \leq GK.\mathrm{dim}A\{w^{-1}\}$. By Proposition 1, $GK.\mathrm{dim}A[t;\sigma] = GK.\mathrm{dim}A[t,\sigma^{-1}] = GK.\mathrm{dim}A + 1$, where $\sigma$ is the automorphism of $A$ determined by $w$. Now, Lemma 2 implies that $GK.\mathrm{dim}A\{w^{-1}\} = GK.\mathrm{dim}A$. Since $W$ is arbitrary, then $GK.\mathrm{dim}AS^{-1} = GK.\mathrm{dim}A$ follows.

The next proposition will present an important example of a locally algebraic automorphism. For this we need the following definition : a $K$-automorphism $\sigma$ of $K[x_1,\ldots,x_n]$ is triangular if there exist polynomials $p_i \in K[x_{i+1},\ldots,x_n]$ for $1 \leq i < n$ and $p_n \in K$ such that for every $1 \leq i \leq n$ $\sigma(x_i) = \lambda_i x_i + p_i$, where $\lambda_i \in K\backslash\{0\}$.

**Proposition 2.** Let $\sigma$ be an automorphism of $K[x_i,\ldots,x_n]$. If $\sigma$ is triangular, then $\sigma$ is locally algebraic.

**Proof.** Suppose that $\sigma(x_i) = \lambda_i x_i + p_i$ for $1 \leq i \leq n$, where $p_i \in [x_{i+1},\ldots,x_n]$, $\lambda_i \in K$ are as in the above definition of a triangular automorphism. Let $s \geq 1$, be a common upper boundary of degrees of polynomials $p_i$, $1 \leq i \leq n$. First we will show, by induction on $k$, that

$$\deg_{x_i}\sigma^k(x_j) \leq s^{i-j} \text{ for any } k \geq 0 \text{ and } 1 \leq i,j \leq k \qquad (I)$$

If $k = 0$, then the inequality is clear. Suppose $k \geq 1$ and $1 \leq i,j \leq n$. Then

$$\begin{aligned}\sigma^k(x_j) &= \sigma^{k-1}(\lambda_j x_j + p_j(x_{j+1},\ldots,x_n)) \\ &= \lambda_j\sigma^{k-1}(x_j) + p_j(\sigma^{k-1}(x_{j+1}),\ldots,\sigma^{k-1}(x_n))\end{aligned}$$

By the induction hypothesis, we know that $\deg_{x_i} \sigma^{k-1}(x_\ell) \leq s^{i-\ell}$ for any $1 \leq \ell \leq n$. Therefore

$$\begin{aligned}\deg_{x_i} p_j(\sigma^{k-1}(x_{j+1}), \ldots, \sigma^{k-1}(x_n)) &\leq s. \max_{j+1 \leq \ell \leq n} \deg_{x_i} \sigma^{k-1}(x_\ell) \\ &\leq s.s^{i-(j+1)} = s^{i-j}\end{aligned}$$

and

$$\deg_{x_i} \sigma^k(x_j) \leq \max\{\deg_{x_i} \sigma^{k-1}(x_j), \deg_{x_i} p_j(\sigma^{k-1}(x_{j+1}), \ldots \sigma^{k-1}(x_n))\}$$

and thus we have $\deg_{x_i} \sigma^k(x_j) \leq s^{i-j}$.
This proves (I).

Let $V = \operatorname{span}_K\{1, x_1, \ldots, x_n\}$. By making use of (I) one can see see that for any $k \geq 0$, $\sigma^k(V) \subset W = V^{n.s}$. Now, for any $f \in K[x_1, \ldots, x_n]$ there exists $m \geq 1$ such that $f \in V^m$. Hence $\sigma^k(f) \in \sigma^k(V^m) \subset W^m$ for any $K \geq 0$. This shows that $\sigma$ is locally algebraic.

**Corollary 3.** Let $\sigma$ be an automorphism of $A = K[x_1, \ldots, x_n]$ conjugate to a triangular automorphism. Then $GK.\dim A[t, \sigma] = GK.\dim A[t, t^{-1}; \sigma] = n + 1$.

**Proof.** Because of Proposition 2, $\sigma$ is locally algebraic. Now the statement is a consequence of Proposition 1 and the fact that $GK.\dim A = n$.

**Corollary 4.** Let $\sigma$ be an automorphism of $A = K[x, y]$. Then either $GK.\dim A[t, \sigma] = 3$ of $GK.\dim A[t, \sigma] = \infty$.

**Proof.** Using the terminology of [4] we know that either $\sigma$ is conjugate to a triangular automorphism or is a square automorphism. In the latter case, we have $GK.\dim A[t, \sigma] = \infty$ (cf. [4] Corollary 9 and Lemma 10). In the former case $GK.\dim A[t, \sigma] = 3$ by Corollary 3.

We have seen in Lemma 2 and Proposition 1 that under some assumptions on the automorphism $\sigma$ of $A$ the $GK$-dimensions of $A, A[t, \sigma], A[t, \sigma^{-1}], A[t, t^{-1}, \sigma]$ are strongly related. Now we will present some examples indicating that in general this is not the case. We begin with an example showing that the difference $GK.\dim A[t, \sigma] - GK \dim A$ may be an arbitrary natural number even in the case where $A$ is a finitely generated algebra. Notice that this situation is quite different from the case of skew polynomial extensions of derivation type (cf. [2] proposition 3.5).

**Example 1.** For any $\ell \geq 2$ let $G_\ell = H_\ell \rtimes_\sigma, < t >$, where $< t >$ is an infinite cyclic group, $H_\ell$ is a free abelian group on the set $\{x_1, x_2, \ldots, x_{2(\ell-1)}\}$ and $\sigma$ is the automorphism of $H_\ell$ defined by $\sigma(x_i) = x_i x_{i+1}$ if $i$ is odd, $\sigma(x_i) = x_i$ if $i$ is even. Then $G_\ell$ is a nilpotent group of index 2 and the sets $\{x_i | i \text{ is odd } 1 \leq i \leq 2\ell - 3\} \cup \{t\}$ $\{x_i | i \text{ is even } 2 \leq i \leq 2\ell - 2\}$ are contained in the first and the second quotients of the lower central series of $G_\ell$, respectively. Therefore, by making use of Bass formula ([2], Theorem 11.14), we get $GK.\mathrm{dim}K[G_\ell] = \ell + 2(\ell - 1) = 3\ell - 2, GK.\mathrm{dim}K[H_\ell] = 2\ell - 2$. Now, because $K[G_\ell] = K[H_\ell][t, t^{-1}, \sigma]$ and $K[H_\ell]$ is commutative, Theorem 1 implies that

$$GK.\mathrm{dim}K[H_\ell][t, \sigma] - GK.\mathrm{dim}K[H_\ell] = GK.\mathrm{dim}K[G_\ell] - GK.\mathrm{dim}K[H_\ell] = \ell$$

If $GK.\mathrm{dim}A[t, \sigma] = 1$ then by Corollary 2, $GK.\mathrm{dim}A[t, \sigma] = GK.\mathrm{dim}A[t, \sigma^{-1}] = GK.\mathrm{dim}A[t, t^{-1}, \sigma]$. In the following two examples we show that dimensions of the above algebras do not have to be related if $GK.\mathrm{dim}A[t, \sigma] = 2$.

**Example 2.** Let $A = K[\{x_\alpha\}_{\alpha \in \mathbb{Z}} | x_\alpha x_\beta^\delta x_\gamma = 0]$ where $\delta \in \{1, 2\}$ and $\alpha, \beta, \gamma \in \mathbb{Z}$ satisfy one of the following conditions :

(0) $\alpha = \beta = \gamma$;

(i) $\alpha < \beta < \gamma$;

(i') $\alpha > \beta > \gamma$;

(ii) $\alpha \leq \gamma < \beta$;

(ii') $\alpha \geq \gamma > \beta$.

One can check that the rule $\sigma(x_\alpha) = x_{\alpha+1}$ for $\alpha \in \mathbb{Z}$ defines a $K$-automorphism $\sigma$ of $A$. We will show that the algebra $A$ has the following properties :

(1) $GK.\mathrm{dim}A = 0$

(2) $GK.\mathrm{dim}A[t, \sigma] = GK.\mathrm{dim}A[t, \sigma^{-1}] = 2$

(3) $GK.\mathrm{dim}A[t, t^{-1}, \sigma] = \infty$

In order to do this some preparation is needed. First we will show that :

If $x_{\alpha_1} \ldots, x_{\alpha_5} \in A$ is non-zero (II)
then $\min\{\alpha_2, \ldots, \alpha_5\} < \alpha_1 < \max\{\alpha_2, \ldots, \alpha_5\}$.

Suppose that $0 \neq x_{\alpha_1} \ldots x_{\alpha_5} \in A$ and $\alpha_1 = \min\{\alpha_2, \ldots, \alpha_5\}$. If $\alpha_1 < \alpha_2$, then the fact that the relations of types (i), (ii) are satisfied in $A$ implies that $\alpha_3 = \alpha_2$. Then, from the relations (0), (i), (ii), it follows that $x_{\alpha_1}, \ldots, x_{\alpha_4} = 0$. Thus $\alpha_1 = \alpha_2$.

Since $x^3_{\alpha_1} = 0$, then $\alpha_2 < \alpha_3$. Now, we can apply the above reasoning to the monomial $x_{\alpha_2} \dots x_{\alpha_5}$ to get $x_{\alpha_2} \dots x_{\alpha_5} = 0$. This contradiction shows that $\min\{\alpha_2 \dots \alpha_5\} < \alpha_1$.

Since the relations in $A$ are symmetric with respect to the natural order in $\mathbb{Z}$, the element $x_{-\alpha_1} \dots x_{-\alpha_5} \in A$ is also non-zero. Therefore $\min\{-\alpha_2, \dots, -\alpha_5\} < -\alpha_1$. This proves the second inequality in (II).

Using the property (II) one can easily deduce that the following holds :

Let $F, S \subset \mathbb{Z}$ be subsets such that $F$ is finite and $S$ has either an upper or a lower bound. (III)

Then the set of all non-zero monomials from $A$ of the form $x_k W$, where $k \in F$ and $W$ is a word in $\{x_\alpha\}_{\alpha \in S}$, is finite.

**(1)** The property (III) implies that $A$ is locally finite dimensional. Hence $GK.\dim A = 0$.

**(2)** Let $\epsilon \in \{-1, 1\}$. Taking $V = \text{span}_K\{1, x_0, t\} \subset A[t, \sigma^\epsilon]$ it is clear that $GK.\dim A[t, \sigma^\epsilon] \geq 2$.
For any any $\ell \geq 0$, let $V_\ell = \text{span}_K\{1, t, \{x_\alpha\}_{|\alpha| < \ell}\}$.

By (III), the number of non-zero monomials of the form $w = x^{k_1}_{\alpha_1} t^{\ell_1} x^{k_2}_{\alpha_2} t^{\ell_2} \dots t^{\ell_{r-1}} x^{k_r}_{\alpha_r} \in A[t, \sigma^\epsilon]$, where $r \in \mathbb{N}$ and $|\alpha_i| < \ell, k_i \ell_i \geq 0$ for $1 \leq i \leq r$, is finite. Now the fact that every monomial from $V^n_\ell$ can be written as $t^{\ell_0} W t^{\ell_r}$, where $W$ is of the above form and $0 \leq \ell_0,\ \ell_r \leq \ell_0 + \ell_r \leq n$, implies that $d_{V_\ell}(n) \leq Mn^2$ for some constant $M \geq 0$. The last inequality yields $GK.\dim A[t, \sigma^\epsilon] \leq 2$ since for every finite dimensional subspace $V$ of $A[t, \sigma^\epsilon]$ there exists $\ell \geq 0$ such that $V$ is contained in some power of $V_\ell$.

**(3)** Let $V = \text{span}_K\{1, x_0, t, t^{-1}\} \subset A[t, t^{-1}, \sigma]$ then $V$ is a generating space of $A[t, t^{-1}, \sigma]$. For any $m \geq \ell$ let us consider monomials $p(\delta_1, \overline{\delta_1}, \dots, \delta_m, \overline{\delta_m}) = x^{\delta_1}_1 x^{\overline{\delta_1}}_{-1} \dots x^{\delta_m}_m x^{\overline{\delta_m}}_{-m}$ where $\delta_i, \overline{\delta_i} \in \{1, 2\}$ for $1 \leq i \leq m$.

It is straightforward to verify that these monomials are non-zero and hence linearly independent over $K$. Now $d_V(m^3) \geq 2^{2m} = 4^m$ for $m > 3$ follows from the fact that

$$p(\delta_1, \overline{\delta_1}, \dots, \delta_m, \overline{\delta_m}) t^{-m} = t x^{\delta_1}_0 t^{-2} x^{\overline{\delta_1}}_0 t^3 x^{\delta_2}_0 \dots t^{2m-1} x^{\delta_m}_0 t^{-2m} x^{\overline{\delta_m}}_0$$

belongs to $V^{4m(1+2m)m} = V^{m(2m+5)} \subset V^{m^3}$.

The above inequality implies that $GK.\dim A[t,t^{-1},\sigma]=\infty$.

**Example 3.** For any natural $r\geq 2$ take $A_r = K[\{x_\alpha\}_{\alpha\in\mathbb{Z}}|x_\alpha x_\beta = 0$ if $\alpha\leq\beta$; $x_{\alpha_1}\ldots x_{\alpha_r}=0$ for any $\alpha_1,\ldots,\alpha_r\in\mathbb{Z}]$. Let $\sigma$ be a $K$-automorphism of $A_r$ given by $\sigma(x_\alpha)=x_{\alpha+1}$ for $\alpha\in\mathbb{Z}$. Then :

(1) $GK.\dim A_r[t,\sigma]=2$

(2) $GK.\dim A_r[t,\sigma^{-1}] = GK.\dim A_r[t,t^{-1},\sigma]=r$.

(1) Taking $V=\operatorname{span}_K\{1,x_0,t\}\subset A[t,\sigma]$ it is clear that $GK.\dim A[t,\sigma]\geq 2$.
Let $W\subset A_r[t,\sigma]$ be a finite dimensional subspace. There exists $k\geq 0$ such that $W$ is contained in some power of a subspace $V_k$ spanned by $1,t,\{x_\alpha\}_{|\alpha|\leq k}$. Thus, while computing the $GK$.dimension of $A[t,\sigma]$ we may replace $W$ by $V_k$ for some $k\geq 0$. Every word in $\{t,x_{-k},x_{-k+1},\ldots,x_k\}$ can be written in a form $t^iwt^j$ where $0\leq i,j$ and either $w$ is the empty word or $w$ is a word with the first and the last letter from $\{x_{\alpha,|\alpha|\leq k}\}$. Using relations in our algebra (it is enough to consider relations $x_\alpha x_\beta=0$ for $\alpha\leq\beta$) one can check that the number of non-zero elements from $A_r[t,\sigma]$ of the form of $w$ is bounded by a constant $M_k$ depending only on $k$. Therefore $d_{V_k}(n)<M_k(n+1)^2$. This yields the statement (1).

(2) $A_r[t,t^{-1},\sigma]=K[V]$ where $V=\operatorname{span}_k\{1,x_0,t,t^{-1}\}$ it is straightforward to verify that for any $n\geq 1$ elements $t_k$ for $-n\leq k\leq n$ and $t^{\ell_0}x_0\ldots t^{\ell_s-1}x_0t^{\ell_s}$ where $0<s<r$; $\ell_0,\ldots,\ell_s\in\mathbb{Z}$ satisfy $\ell_1,\ell_2,\ldots,\ell_{s-1}<0$ and $\sum_{i=0}^{s}|\ell_i|+s\leq n$ form a basis for $V^n$.

Therefore $d_v(n)\leq 2n+1+\sum_{s=1}^{r-1}(2n+1)^2\binom{n}{s-1} = f(n)$, since $1\leq s\leq r-1, |\ell_0|,|\ell_s|\leq n$ and $\sum_{i=1}^{s-1}|\ell_i|\leq n$. $f(n)$ is a polynomial in $n$ of degree $r$, so $GK.\dim A_r[t,t^{-1},\sigma]\leq r$. Let $W=\operatorname{span}_k\{1,x_0,t^{-1}\}\subset A[t^{-1},\sigma]=A[t,\sigma^{-1}]$. Then for $n\geq r, W^{4n}$ contains the basis element described above for $s=r-1,-n\leq\ell_0,\ell_{r-1}<0\sum_{i=1}^{r-2}|\ell_i|\leq n$.

Thus $d_W(4n)\geq n^2\binom{n}{r-2}$ which is a polynomial in $n$ of degree $r$.
This yields

$$r\geq GK.\dim A[t,t^{-1},\sigma]\geq GK.\dim A[t,\sigma^{-1}]\geq r$$

**References.**


[1] W. Borho, H. Kraft, *Uber die Gelfand-Kirillov Dimension*, Math. Ann. 220 (1976), 1-24.

[2] G. R. Krause, T. H. Lenagan, *Growth of algebras and Gelfand-Kirillov Dimension*, Research Notes in Math., 116, Pitman, London.

[3] L. W. Small, J. T. Stafford, R. B. Warfield, *Affine algebras of Gelfand-Kirillov dimension one are PI*, Math. Proc. Camb. Philos. Soc. 97(1985), 407-414.

[4] M. K. Smith, *Growth of twisted Laurent extensions*, Duke Math. J. 49 (1982), 79-85.

# GOLDIE RANKS OF PRIME POLYCYCLIC CROSSED PRODUCTS

Martin Lorenz
Department of Mathematical Sciences
Northern Illinois University
DeKalb, Illinois 60115 USA

**Abstract.** We use Moody's "Brauer induction theorem" for crossed products $S*\Gamma$ of polycyclic-by-finite groups $\Gamma$ over right Noetherian rings $S$ to determine the Goldie rank of $S*\Gamma$ in certain cases.

**Introduction.** In [6], S. Rosset stated a conjecture on the structure of the Grothendieck group $G_0(k[\Gamma])$ of all finitely generated modules over the group ring $k[\Gamma]$ of a polycyclic-by-finite group $\Gamma$ over a commutative right Noetherian domain $k$. He also showed that this conjecture implies the so- called Goldie rank conjecture for prime polycyclic group algebras. Recently, J. A. Moody [3] has confirmed Rosset's conjecture on $G_0$, even for *crossed products* $S*\Gamma$, where $\Gamma$ is a polycyclic-by-finite group and $S$ is a right Noetherian ring. In this note, we apply Moody's result to obtain estimates for the Goldie rank of $S*\Gamma$ in the case where $S*\Gamma$ is prime. These estimates do in particular yield a slightly generalized version of the Goldie rank conjecture for group rings, as well as a zero divisor theorem for polycyclic crossed products. Our approach to Goldie rank differs from Rosset's in that it is non-homological and is based directly on Goldie's reduced rank function.

**1.**

Let $S*\Gamma$ be a crossed product of the group $\Gamma$ over the ring $S$, and assume that $S*\Gamma$ is right Noetherian. Then, for any finitely generated $S*\Gamma$-module $V$, we can define the *normalized reduced rank* of $V$ by

$$\chi_{S*\Gamma}(V) = \rho(V)/\rho(S*\Gamma) ,$$

where $\rho$ denotes the usual (Goldie-) reduced rank function.

**Lemma.** *Suppose that $S*\Gamma$ is prime, and let $H$ be a subgroup of $\Gamma$ having finite index in $\Gamma$. Then, for any finitely generated $S*\Gamma$-module $V$, we have*

$$\chi_{S*H}(V) = [\Gamma:H]\cdot\chi_{S*\Gamma}(V).$$

**Proof.** Fix a normal subgroup $N$ of $\Gamma$ with $N \subseteq H$ and with $[\Gamma:N]$ finite. Let $C$ denote the set of regular elements of $S*N$. Since $S*\Gamma$ is prime, $S*N$ must be semiprime (in fact, $\Gamma/N$-prime). Thus $C$ is a right Ore set in $S*N$, and also in $S*H$ and $S*\Gamma$ (cf. [4], proof of Lemma 13.3.5(ii) ). Moreover, letting $Q(.)$ denote classical rings of fractions, we have



*F. van Oystaeyen and L. Le Bruyn (eds.), Perspectives in Ring Theory, 215–219.*

$$Q = Q(S^*\Gamma) = (S^*\Gamma)C^{-1} \supseteq T = Q(S^*H) = (S^*H)C^{-1},$$

because $(S^*\Gamma)C^{-1}$ and $(S^*H)C^{-1}$ both are finitely generated modules over the Artinian ring $Q(S^*N) = (S^*N)C^{-1}$, and hence are themselves Artinian. Inasmuch as $S^*\Gamma$ is prime, $Q$ is simple Artinian. So

$$Q \simeq X^{(r)}$$

for some simple right ideal $X$ of $Q$, and $r = \rho(S^*\Gamma)$. Furthermore, by the above explicit form of $Q$ and $T$,

$$Q_T \simeq T_T^{\,|\Gamma:H|}.$$

Therefore, with $l(.)$ denoting composition length, we have

$$l(X_T) = \frac{1}{r}\cdot l(Q_T) = \frac{1}{r}\cdot[\Gamma:H]\cdot l(T_T)$$

$$= \frac{1}{r}\cdot[\Gamma:H]\cdot\rho(S^*H).$$

Similarly, for any finitely generated $S^*\Gamma$-module $V$, we have

$$V \otimes_{S^*\Gamma} Q \simeq X^{(v)} \quad \textit{with} \quad v = \rho(V).$$

Thus

$$\chi_{S^*\Gamma}(V) = \frac{v}{r} = [\Gamma:H]^{-1}\cdot\rho(S^*H)^{-1}\cdot l(X_T)\cdot v$$

$$= [\Gamma:H]^{-1}\cdot\frac{l(V \otimes_{S^*\Gamma} Q\,|_T)}{\rho(S^*H)}.$$

Finally, $V \otimes_{S^*\Gamma} Q\,|_T \simeq V \otimes_{S^*H} T$, since ${}_{S^*\Gamma}Q \simeq S^*\Gamma \otimes_{S^*H} T$, and so $l(V \otimes_{S^*\Gamma} Q\,|_T) = \rho(V_{S^*H})$. This proves the lemma. □

**2.**

The main additional ingredient that we will need in the proof of our main result is Moody's theorem [3] which we state for the convenience of the reader as Theorem 1. Here and in the following

$$\mathbf{F}(\Gamma)$$

denotes the set of *finite* subgroups of the group $\Gamma$.

**Theorem 1 (J. Moody).** *Let $S^*\Gamma$ be a crossed product with $S$ a right Noetherian ring and with $\Gamma$ a polycyclic-by-finite group. Then the induction map*

$$Ind = \bigoplus_{U\in\mathbf{F}(\Gamma)} (.) \otimes_{S^*U} S^*\Gamma : \bigoplus_{U\in\mathbf{F}(\Gamma)} G_0(S^*U) \to G_0(S^*\Gamma)$$

*is onto.*

Here, $G_0(.)$ is of course the usual Grothendieck group of all finitely generated modules over the ring in question. The relevance of $G_0$ in the present context comes from the fact that the normalized reduced rank function defines a homomorphism

$$\chi_{S*\Gamma} : G_0(S^*\Gamma) \to \frac{1}{\rho(S^*\Gamma)} \cdot \mathbf{Z} \subseteq \mathbf{Q} .$$

Instead of letting $U$ run over *all* finite subgroups of $\Gamma$ in the above theorem, it clearly suffices to take a representative set for the conjugacy classes of the maximal finite subgroups of $\Gamma$.

**3.**

From now on, assume that $S$ is right Noetherian and that $\Gamma$ is polycyclic-by-finite. Then $S^*\Gamma$ is also right Noetherian and so the foregoing applies. We put

$$\mathbf{F}_*(\Gamma) = \left\{ U \in \mathbf{F}(\Gamma) \mid \textit{there exists a map } \phi : S^*U \to S \textit{ with } \phi|_S = id_S \right\}.$$

Furthermore, we let

$$f(\Gamma) = l.c.m. \left\{ \#U \mid U \in \mathbf{F}(\Gamma) \right\}, \textit{ and } f_*(\Gamma) = l.c.m. \left\{ \#U \mid U \in \mathbf{F}_*(\Gamma) \right\}.$$

Thus, clearly, $f_*(\Gamma) \mid f(\Gamma) \mid [\Gamma : N]$ holds for any torsion-free normal subgroup $N$ of $\Gamma$ having finite index in $\Gamma$. (Here, $|$ stands for "divides".)

**Theorem 2.** *Let $S^*\Gamma$ be given with $S$ right Noetherian and $\Gamma$ polycyclic-by-finite. Suppose that $S^*\Gamma$ is prime. Then*

$$\rho(S) \cdot f_*(\Gamma) \mid \rho(S^*\Gamma) \mid \rho(S) \cdot f(\Gamma).$$

**Proof.** Let $U$ be a finite subgroup of $\Gamma$, and let $M$ be a finitely generated $S^*U$-module. We claim that

$$\chi_{S*\Gamma}(M \otimes_{S*U} S^*\Gamma) = \frac{1}{\#U} \cdot \chi_S(M).$$

Before proving this, we show how the theorem follows from the claim. Indeed, by Moody's theorem, we have

$$Im\,\chi_{S*\Gamma} = \sum_{U \in \mathbf{F}(\Gamma)} Im\,\chi_{S*\Gamma}(Ind_{S*U}^{S*\Gamma}\ G_0(S^*U)).$$

Our claim further implies that

$$Im\,\chi_{S*\Gamma}(Ind_{S*U}^{S*\Gamma}\ G_0(S^*U)) \subseteq \frac{1}{\#U} \cdot Im\,\chi_S,$$

with equality occuring in case $U \in \mathbf{F}_*(\Gamma)$. Therefore, using the obvious equalities $Im\,\chi_{S*\Gamma} = \rho(S^*\Gamma)^{-1} \cdot \mathbf{Z}$ and $Im\,\chi_S = \rho(S)^{-1} \cdot \mathbf{Z}$, we get

$$\sum_{U\in \mathbf{F}_*(\Gamma)} \frac{1}{\#U}\cdot\frac{1}{\rho(S)}\cdot\mathbf{Z} \subseteq \frac{1}{\rho(S^*\Gamma)}\cdot\mathbf{Z} \subseteq \sum_{U\in \mathbf{F}(\Gamma)} \frac{1}{\#U}\cdot\frac{1}{\rho(S)}\cdot\mathbf{Z}.$$

This implies the formula in the theorem.

In order to establish the claim, choose a normal subgroup $N$ of $\Gamma$ having finite index in $\Gamma$ and such that $N\cap U = \langle 1\rangle$ . (Any torsion-free normal subgroup having finite index in $\Gamma$ will do.) Put $H = \langle N,U\rangle$ and $V = M\otimes_{S^*U} S^*\Gamma$. Then the lemma gives

$$\chi_{S^*\Gamma}(V) = [\Gamma:N]^{-1}\cdot\chi_{S^*N}(V).$$

But $V \simeq W\otimes_{S^*H} S^*\Gamma$, where $W = M\otimes_{S^*U} S^*H$, and so $V|_{S^*N} \simeq \bigoplus_x W^{(x)}|_{S^*N}$. Here $x$ runs through a transversal for $H$ in $\Gamma$ and $W^{(x)} = W\otimes x$ denotes the $x$-conjugate of $W$. Since, clearly, $\chi_{S^*N}(W^{(x)}) = \chi_{S^*N}(W)$, we have

$$\chi_{S^*N}(V) = [\Gamma:H]\cdot\chi_{S^*N}(W).$$

Moreover, $W|_{S^*N} \simeq M\otimes_S S^*N$, since ${}_{S^*U}S^*H \simeq S^*U\otimes_S S^*N$. Hence

$$\chi_{S^*\Gamma}(V) = \frac{[\Gamma:H]}{[\Gamma:N]}\cdot\chi_{S^*N}(W) = \frac{1}{\#U}\cdot\chi_{S^*N}(M\otimes_S S^*N).$$

Thus it suffices to show that

$$\chi_{S^*N}(M\otimes_S S^*N) = \chi_S(M).$$

For this, note that $S^*\Gamma$ prime forces $S$ to be $\Gamma$-prime. Therefore, $Q = Q(S)$ exists and has the form

$$Q \simeq \bigoplus_{i=1}^{q} X_i$$

for simple right ideals $X_i$ of $Q$ such that each $X_i$ is isomorphic to some $\Gamma$-conjugate of $X_1$, and $q = \rho(S)$. Moreover, $S^*N$ extends to $Q^*N$, and

$$\begin{aligned} 1 &= \chi_{Q^*N}(Q^*N) = \chi_{Q^*N}(Q\otimes_Q Q^*N) \\ &= \sum_{i=1}^{q}\chi_{Q^*N}(X_i\otimes_Q Q^*N) = \rho(S)\cdot\chi_{Q^*N}(X_1\otimes_Q Q^*N). \end{aligned}$$

After a suitable renumbering of the $X_i$'s, we have $M\otimes_S Q \simeq \bigoplus_{i=1}^{m} X_i$ with $m = \rho(M_S)$. Therefore,

$$\begin{aligned} \chi_{S^*N}(M\otimes_S S^*N) &= \chi_{Q^*N}(M\otimes_S Q^*N) = \sum_{i=1}^{m}\chi_{Q^*N}(X_i\otimes_Q Q^*N) \\ &= \rho(M_S)\cdot\chi_{Q^*N}(X_1\otimes_Q Q^*N) = \frac{\rho(M_S)}{\rho(S)} = \chi_S(M). \end{aligned}$$

This proves our claim, and hence the theorem. □

**4.**

Necessary and sufficient conditions for $S^*\Gamma$ to be prime can be found in [5]. Here we just note the following two applications of Theorem 2. The first is a slightly generalized version of the affirmative solution of the Goldie rank conjecture [3]. The second corollary falls under the general heading "zero divisor conjecture" and extends the theorem of Farkas, Snider [2] and Cliff [1] to crossed products.

**Corollary 1.** *Let* $S[\Gamma]$ *be the group ring of the polycyclic-by-finite group* $\Gamma$ *over the prime right Noetherian ring* $S$. *Suppose that* $\Gamma$ *has no non-identity finite normal subgroups. Then* $S[\Gamma]$ *is prime right Noetherian, and*

$$\rho(S[\Gamma]) = \rho(S)\cdot f(\Gamma).$$

**Corollary 2.** *Let* $S{*}\Gamma$ *be a crossed product with* $S$ *a prime right Noetherian ring and* $\Gamma$ *a torsion-free polycyclic-by-finite group. Then*

$$\rho(S{*}\Gamma) = \rho(S).$$

*In particular, if* $S$ *is a domain, then so is* $S{*}\Gamma$.

**References.**


1. G. H. Cliff, Zero divisors and idempotents in group rings, Canadian J. Math. **32** (1980),596-602.
2. D. R. Farkas and R. L. Snider, $K_0$ and Noetherian group rings, J. Algebra **42** (1976),192-198.
3. J. A. Moody, Ph.D. thesis, Columbia University, 1986.
4. D. S. Passman, *The Algebraic Structure of Group Rings* , John Wiley & Sons, New York, 1977.
5. D. S. Passman, Infinite crossed products and group-graded rings, preprint (UW-Madison, 1986).
6. S. Rosset, The Goldie rank of virtually polycyclic groups, in: Lect. Notes in Math., Vol. 844, pp. 35-45, Springer, Berlin-Heidelberg-New York, 1981.

# THE GELFAND-KIRILLOV DIMENSION OF POINCARE-BIRKHOFF-WITT EXTENSIONS

J. Matczuk
University of Warsaw
00-901 Warsaw, PKiN

It is well known ([3]), that if d is a derivation of an algebra R, then the Gelfand-Kirillov dimension of the skew polynomial algebra R[x,d] is equal to GK-dim R + 1, provided R is finitely generated. The main objective of this note is to present a generalization of the above mentioned result to the case of PBW-extensions of finitely generated algebras.

Throughout the paper K will denote a fixed field and all considered algebras will be unital algebras over K. For the basic material concerning the Gelfand-Kirillov dimension we refere to [3]. The Gelfand-Kirillov dimension of an algebra R is denoted by GK-dim R.

We say, as in [1], that an overalgebra T of the algebra R is a Poincare-Birkhoff-Witt extension of R (PBW-extension, for short) if there exist elements $x_1,\ldots,x_t \in T$ such that:

(i) the monomials $x_1^{k_1}\ldots x_t^{k_t}$, $k_1,\ldots,k_t \in \mathbb{N}$, form a basis for T as a free left R-module;

(ii) $[x_i,R] \subset R$ for $1 \leqslant i \leqslant t$;

(iii) $[x_i,x_j] \in R+Rx_1+\ldots+Rx_t$ for $1 \leqslant i,j \leqslant t$,

where $[a,b]= ab-ba$ for all $a,b \in T$.

If the above conditions hold, we will write "$T=R[x_1,\ldots,x_t]$ is a PBW-extension of R".



*F. van Oystaeyen and L. Le Bruyn (eds.), Perspectives in Ring Theory, 221–226.*

Suppose that $T=R[x_1,\dots,x_t]$ is a PBW-extension of R satisfying the condition (iii') $[x_i,x_j]\in R+Kx_1+\dots+Kx_t$ for $1\leq i,j\leq t$. Then it can be easily seen that T is a twisted smash product $R\#U(L)$ where L is the Lie algebra $R+Kx_1+\dots+Kx_t/_R$ with the usual Lie bracked operation. Conversly, the Poincare-Birkhoff-Witt theorem implies that any twisted smash product $R\#U(L)$, where L is a finite dimensional Lie algebra acting on R by derivations, is a PBW-extension of R satisfying the condition (iii') with $x_1,\dots,x_t$ - a basis of $L\subset U(L)$.

<u>Theorem A</u>. Suppose that $T = R[x_1,\dots,x_t]$ is a PBW -extension of the algebra R. If R is finitely generated then GK-dim T = GK-dim R +t .

<u>Proof</u>. Suppose that R is finitely generated and let V be a finite dimensional K-vector subspace of R generating R as an algebra. X will stand for a K-linear subspace of T spanned by $1,x_1,\dots,x_t$. For $1\leq i\leq t$ $d_i$ will denote the derivation of R determined by $x_i$ i.e. $d_i(r)=[x_i,r]$ for all $r\in R$.

Since V generates R, there exists $m\geq 1$ such that $d_i(V)\subset V^m$ and $[x_i,x_j]\in V^mX$ for all $1\leq i,j\leq t$. Thus, eventually replacing V by $V^m$, we may additionaly assume that the subspace V satisfies the following conditions:
(1) $1\in V$, (2) $d_i(V)\subset V^2$ for $1\leq i\leq t$, (3) $[x_i,x_j]\in VX$ for $1\leq i,j\leq t$.
The condition (2) yields immediately that
(4) $d_i(V^k)\subset V^{k+1}$ for any $k\in\mathbb{N}$ and $1\leq i\leq t$.
For any $k\geq 1$ let $U_k\subset X^k$ denote the subspace spanned by all monomials of the form $x_1^{k_1}\dots x_t^{k_t}$ where $k_1+\dots+k_t\leq k$, $0\leq k_i$ for $i=1,\dots,t$.

First we will show that between subspaces V, X, $U_k$ the following relations hold:
for any $n\geq 1$

$$X^nV \subset VX^n + V^2X^{n-1} + \dots + V^nX + V^{n+1} \qquad (I_n)$$

$$X^n \subset V^{n-1}U_n \qquad (II_n)$$

The statement $(I_1)$ follows directly from the choice of V. Suppose that $(I_k)$ holds for $1 \leqslant k < n$. Then

$$X^nV \subset X\sum_{i=1}^{n} V^iX^{n-i} \subset \sum_{i=1}^{n}(V^iX + V^{i+1})X^{n-i} \subset$$

$$\subset VX^n + \sum_{i=2}^{n} V^iX^{n+1-i} + \sum_{i=1}^{n-1} V^{i+1}X^{n-i} + V^{n+1} =$$

$$= VX^n + V^2X^{n-1} + \dots + V^nX + V^{n+1} .$$

This shows $(I_n)$.

Since $U_1 = X$, $(II_1)$ is clear. Suppose that $(II_k)$ holds for $1 \leqslant k < n$. Let $y = y_1 \dots y_n \in X^n$, where $y_i \in \{1, x_1, \dots, x_t\}$ for $1 \leqslant i \leqslant n$. If $y_i = 1$ for some i, then $y \in X^{n-1}$. Hence, by the induction hypothesis, $y \in V^{n-2}U_{n-1} \subset \subset V^{n-1}U_n$. Thus suppose that $y_i \neq 1$ for all $1 \leqslant i \leqslant n$. For any permutation $\sigma \in S_n$ we define $y_\sigma = y_{\sigma(1)} \dots y_{\sigma(n)}$. Let $1 \leqslant l < n$ and $\sigma \in S_n$ be a transposition $(l, l+1)$. Then, by making use of the statement $(I_l)$ and the induction hypothesis, we get

$$y \in y_\sigma + X^{l-1}VX^{n-l} \subset y_\sigma + \sum_{i=1}^{l} V^iX^{l-i}X^{n-l} =$$

$$= y_\sigma + \sum_{i=1}^{l} V^iX^{n-i} \subset y_\sigma + \sum_{i=1}^{l} V^iV^{n-i-1}U_{n-i} \subset$$

$$\subset y_\sigma + V^{n-1}U_{n-1} .$$

The above implies that for any $\sigma \in S_n$ $y \in y_\sigma + V^{n-1}U_{n-1}$. This shows that $y \in U_n + V^{n-1}U_{n-1} \subset V^{n-1}U_n$ and gives $(II_n)$

Now we are in a position to finish the proof. The subspace $W = X + V$ generates the algebra T. Moreover $W^n \subset V^nX^n$ for any $n \geqslant 1$. In fact $W = X+V \subset VX$ and if $W^k \subset V^kX^k$ for some $k \geqslant 1$, then inclusions

$VW^k \subset V^{k+1}X^k \subset V^{k+1}X^{k+1}$ and

$XW^k \subset XV^kX^k \subset V^kX^{k+1} + V^{k+1}X^k \subset V^{k+1}X^{k+1}$

show that $W^{k+1} \subset V^{k+1}X^{k+1}$.

Now, using property $(II_n)$, we get

$W^n \subset V^n X^n \subset V^n V^{n-1} U_n \subset V^{2n} U_n$ for any $n \geq 1$. Therefore $d_W(n) \leq d_V(2n)\dim_K U_n = d_V(2n)\cdot f(n)$, where $d_W(n)$ denotes $\dim_K W^n$.

The growth of the function $f(n)$ is polynomial of degree t, since it is equal to the growth of the polynomial algebra in t commuting indeterminates. Thus, the above inequality yields GK-dim T $\leq$ GK-dim R + t .

Because T is a free left R-module with a basis $x_1^{k_1} \ldots x_t^{k_t}$, $k_1, \ldots, k_t \in \mathbb{N}$, $\dim_K V^n U_n = d_V(n) \dim_K U_n$ . Now the inequality GK-dim T $\geq$ GK-dim R + t follows from the inclusion $V^n U_n \subset W^{2n}$ for any $n \geq 1$. This completes the proof of the theorem.

As an application of the above theorem we will show some results concerning incomparability and prime lenght. Henceforth R will be a noetherian, finitely generated algebra of finite Gelfand-Kirillov dimension. $T = R[x_1, \ldots, x_t]$ will be a fixed PBW-extension of R. For any prime ideal P of T, ht(P) will denote the height of P i.e. $\sup \{ m \mid$ there is a chain of prime ideals $P \supsetneq P_1 \supsetneq \ldots \supsetneq P_m \}$ .

<u>Corollary 1</u>. If P is a prime ideal of T such that $P \cap R = 0$, then $ht(P) \leq t$.

<u>Proof</u>. The algebra T poseses a natural filtration $T = \bigcup_{n \geq 0} RU_n$ and it is straightforward to verify that the graded algebra associated to this filtration is isomorphic to the polynomial algebra $R[y_1, \ldots, y_t]$ in commuting indeterminates $y_1, \ldots, y_t$. This implies that T is a noetherian algebra, since R is noetherian. Therefore we may use Corollary 3.16 [3] getting GK-dim T $\geq$ GK-dim $T/P$ + ht(P). Because $P \cap R = 0$, $T/P$ contains an isomorphic copy of R. Hence GK-dim $T/P \geq$ GK-dim R . Now Theorem A and the above

two inequalities lead us to
GK-dim R + $t$ = GK-dim T $\geqslant$ GK-dim R + ht(P).
This shows that ht(P) $\leqslant t$.

Let D denote the set of derivations of R determined by elements $x_1,\dots,x_t$. If I is an ideal of R which is D-invariant / i.e. $d(I) \subset I$ for any $d \in D$ /, then it is straightforward to verify that IT = TI is an ideal of T and $IT \cap R = I$. Thus $R/I$ imbeds in $T/IT$ and it is not hard to see that $T/IT = R/I[\bar{x}_1,\dots,\bar{x}_t]$ is a PBW extension of $R/I$, where $\bar{x}_i$'s denote the natural images of $x_i$'s . Knowing the above we get the following result:

<u>Proposition 2</u>. Suppose that $P_0 \subset P_1 \subset \dots \subset P_{t+1}$ is a chain of prime ideals of T such that $P_0 \cap R = P_{t+1} \cap R = I$. Then $P_i = P_{i+1}$ for some $0 \leqslant i \leqslant t$.

<u>Proof</u>. Since I is a D-invariant ideal of R, $\bar{T} = T/IT = \bar{R}[\bar{x}_1,\dots,\bar{x}_t]$ is a PBW extension of the algebra $\bar{R} = R/I$. In $\bar{T}$ we have a chain of prime ideals $\bar{P}_0 \subset \bar{P}_1 \dots \subset \bar{P}_{t+1}$ such that $\bar{P}_0 \cap \bar{R} = \bar{P}_{t+1} \cap \bar{R} = 0$, where $\bar{P}_i$'s denote the natural images of $P_i$'s in $\bar{T}$. Thus, replacing the extension $R \subset T$ by $\bar{R} \subset \bar{T}$ , we may assume that $I = 0$. Now the thesis is a direct consequence of Corollary 1.

In the following proposition cl-K-dim R denotes the classical Krull dimension of R and D-cl-K-dim R - its D-invariant version, i.e. the maximal lengh of a chain of D-prime ideals of R.

<u>Proposition 3</u>. cl-K-dim T < (t+1)(D-cl-K-dim R + 1). Moreover, if char K = 0, then
cl-K-dim T < (t+1) (cl-K-dim R + 1).

<u>Proof</u>. Let P be a prime ideal of T. Then $P \cap R$ is a D-invariant ideal of R and it easy to check that $P \cap R$

is a D-prime ideal. Therefore any chain of prime ideals of T, while intersecting with R, leads to a chain of D-prime ideals of R of lengh not greater than D-cl-K-dim R + 1. Now, the first statement of the proposition is a consequence of Proposition 2.
Suppose that char K = 0. Then, since R is noetherian, every D-prime ideal of R is prime / cf. Lemma 1.2.[2]/ and the second statement follows.

Observe that in the case of algebras over a field of characteristic 0 the above result was proved in [4].

References:


[1] A.D.Bell, K.R.Goodearl, Uniform rank over differential operator rings and Poincare-Birkhoff-Witt extensions, preprint.

[2] J.Bergen, S.Montgomery, D.S.Passman, Radicals of Lie algebras smash products, preprint.

[3] G.R.Krauze, T.H.Lenagan, Growth of algebras and Gelfand-Kirillov dimension, Research Notes in Math., 116, Pitman, London.

[4] D.S.Passman, Prime ideals in enveloping rings, Trans. Amer. Math. Soc., to appear.

# THE NULSTELLENSATZ AND GENERIC FLATNESS

J.C. McConnell and J.C. Robson
School of Mathematics
University of Leeds
LEEDS LS2 9JT
England

ABSTRACT. An extended version of generic flatness is used to give an elementary proof that certain noncommutative affine algebras over a field "satisfy the Nullstellensatz".

In this note it will be shown that, for certain noncommutative affine algebras R over a field k, the following properties hold:

(i) (Endomorphism property) For each simple right R-module M, End M is algebraic over k.

(ii) (Strong endomorphism property) For each simple right R-module M, End M is finite dimensional over k.

(iii) (Radical property) The Jacobson radical of each factor ring of R is nil.

In the case when R is commutative these are readily converted to standard conclusions of the Nullstellensatz. In the algebras with which we are mainly concerned, (ii) is a consequence of (i), as shown by Small (unpublished; see [6, 9.5.5]). Therefore we concentrate attention on (i) and (iii); saying that R satisfies the Nullstellensatz over k if both hold.

The type of algebra for which these properties will be established includes crossed products of enveloping algebras of Lie algebras, crossed product algebras of polycyclic by finite groups, skew polynomial rings, etc. Such results have already been obtained by Hall [3], Quillen [7], Gabriel (see Dixmier [1, 2.6.9]), Duflo [2], Irving [4] and McConnell [5]. The aim here is to demonstrate a new technique, a development of one used before, which leads easily to these results and also can be used in dealing with algebras over non-Noetherian commutative coefficient rings. It is described in more detail in McConnell-Robson [6, Chapter 9], to which we will refer for some subsidiary results. However there are some additions here.

Our concern throughout will be with algebras over a field k, a



*F. van Oystaeyen and L. Le Bruyn (eds.), Perspectives in Ring Theory, 227–232.*

commutative integral domain $D$ or a commutative ring $K$. We fix $k$, $D$ and $K$ henceforth.

1. DEFINITION A $D$-algebra $R$ is generically flat (sometimes called generically free) over $D$ if for each finitely generated right $R$-module $M$ there exists $0 \neq d \in D$ such that $M_d$ is free over $D_d$. (The suffix here denotes localization with respect to the powers of $d$.) In practice, as one can see, it is sufficient to check this for each cyclic module $M$. There is a well-known connection between this and the Nullstellensatz.

2. LEMMA (Quillen [7]) Let $R$ be a $k$-algebra and $y$ be a central indeterminate. If $R[y]$ is generically flat over $k[y]$ then $R$ has the endomorphism property.

Proof. If $M_R$ is simple and $\operatorname{End} M$ is not algebraic over $k$ then there is an embedding $k[y] \hookrightarrow \operatorname{End} M$. By hypothesis, $M_d$ is free over $k[y]_d$ for some $0 \neq d \in k[y]$. Thus, if $A$ is a proper nonzero ideal of $k[y]_d$ then $AM_d$ is a proper nonzero submodule of $M_d$. However since $k[y] \hookrightarrow \operatorname{End} M$ then $k(y) \hookrightarrow \operatorname{End} M$ and so $M_d = M$. It follows that $k[y]_d$ must be a field; and that is false.

□

Next we note a link with the radical property.

3. LEMMA (Duflo [2]) If $R$ is a $k$-algebra, $x$ a central indeterminate and $R[x]$ has the endomorphism property then $R$ satisfies the Nullstellensatz.

Proof. See [2] or [6, 9.2.6].

□

This helps to explain the concentration, in what follows, on the endomorphism property and on generic flatness. Note that combining Lemmas 2 and 3 gives

4. COROLLARY If $R$ is a $k$-algebra such that $R[x,y]$ is generically flat over $k[y]$ then $R$ satisfies the Nullstellensatz.

□

With this in mind we now strengthen the earlier definition.

5. DEFINITION A $D$-algebra $R$ is $(\mathbb{N},\mathbb{N})$-generically flat over $D$ if $R[x_1,\ldots,x_n,y_1,\ldots,y_m]$ is generically flat over $D[y_1,\ldots,y_m]$ for all $n,m \in \mathbb{N}$.

Next comes a sequence of results which will demonstrate the existence of a large class of such algebras.

6. LEMMA Let $R \subseteq S$ be $D$-algebras and let $R$ be $(\mathbb{N},\mathbb{N})$-generically flat over $D$. Suppose that either

(i) $S$ is a finite extension of $R$ (i.e. $S$ is finitely

generated as a right R-module; or

(ii) S is generated over R by an element z such that $zR = Rz$.

Then S is $(\mathbb{N},\mathbb{N})$-generically flat over D.

Proof. (i) The fact that $S[x_1,\ldots,x_n,y_1,\ldots,y_m]$ is a finitely generated $R[x_1,\ldots,x_n,y_1,\ldots,y_m]$-module shows that it is enough, here, to prove that S is generically flat over D. However any finitely generated right S-module is also a finitely generated right R-module.

(ii) Once again it is enough to show that S is generically flat over D, and it is enough, therefore, to consider a cyclic S-module M, say $M \simeq S/I$ with I a right ideal of S. If one defines, for each n,

$$I_n = \{r \in R \mid rz^n \in z^{n-1}R + \ldots + zR + R + I\}$$

then one obtains a chain of R-modules

$$0 = M_o \subseteq M_1 \subseteq \ldots \subseteq M_n \subseteq \ldots \subseteq M = \bigcup M_n$$

in which $M_n = (I + \sum_{i=0}^{n-1} Rz^i)/I$ and $M_{n+1}/M_n \simeq R/I_n$ as a D-module. Let $N = R[x]/\Sigma x^n I_n$ where x is a central indeterminate. This is a cyclic R[x]-module and so, by hypothesis, $N_d$ is free over $D_d$ for some $0 \neq d \in D$. (It is at this point that $(\mathbb{N},\mathbb{N})$-generic flatness is needed.) It follows that each $D_d$-direct summand $(R/I_n)_d$ of $N_d$ is projective and hence $M_d$ splits; $M_d \simeq \oplus(R/I_n)_d \simeq N_d$. Thus $M_d$ is free. □

7. LEMMA Let S be a filtered D-algebra with $D \subseteq S_o$ and suppose that the associated graded ring gr S is $(\mathbb{N},\mathbb{N})$-generically flat over D. Then so too is S.

Proof. Again it will suffice to show that S is generically flat. If M is a finitely generated right S-module, it can be filtered so that gr M is finitely generated over gr S (see [6, 7.6.11]). Therefore $(\mathrm{gr}\, M)_d$ is free over $D_d$ for some d; and so, arguing as in 6 (ii), $M_d \simeq (\mathrm{gr}\, M)_d$ and thus is free. □

8. DEFINITION Let R,S be K-algebras with $R \subseteq S$. Then S is an almost normalizing extension of R if S is generated over R by a finite set of elements, $z_1,\ldots,z_n$ say, with $z_iR + R = Rz_i + R$ and

$$z_iz_j - z_jz_i \in \sum_{h=1}^{n} Rz_h + R, \text{ for all } i,j.$$

9. LEMMA If R is $(\mathbb{N},\mathbb{N})$-generically flat over D, and S is an almost normalizing extension of R then S is $(\mathbb{N},\mathbb{N})$-generically

flat over D.

Proof. We filter S by "degree" in the $z_i$'s; i.e. we set $S_n = \Sigma Rw$ where w ranges over all words of length at most n in the $z_i$. The associated graded ring gr S is generated by R and the images of the $z_i$ and so is obtainable from R by a finite number of extensions, as covered by Lemma 6 (ii). Therefore gr S, and hence S, is $(\mathbb{N},\mathbb{N})$-generically flat over D. □

The reader may have noticed that, so far, it has not been shown that any examples exist of $(\mathbb{N},\mathbb{N})$-generic flatness. That is remedied next.

10. LEMMA D is $(\mathbb{N},\mathbb{N})$-generically flat over D.

Proof. This is proved in a similar vein to Lemma 6 (ii). However, one can only reduce, initially, to showing that $R = D[x_1,\ldots x_n]$ is generically flat over D. We let W be the semigroup of all words in $x_1,\ldots,x_n$ and order W first by total degree and, subject to that, lexicographically. Suppose $M \simeq R/I$ with I a right ideal of R and, for each $w \in W$, let

$$I(w) = \{d \in D \mid dw \in \Sigma\{Dv \mid v < w\} + I\}.$$

Note that if v divides w in W then $I(v) \subseteq I(w)$.

One can show that any subset S of W has finitely many elements which are not divisible within S. We let $S = \{w \in W \mid I(w) \neq 0\}$ and let $w_1,\ldots,w_t$ be its nondivisible elements. We choose $0 \neq d \in I(w_1) \cap \ldots \cap I(w_t)$. Then $(D/I(w))_d$ is free over $D_d$ for all $w \in W$. One can now argue, as in Lemma 6 (ii), that $M_d \simeq \oplus(D/I(w))_d$. Thus $M_d$ is free. □

11. DEFINITION A K-algebra S is called constructible if it can be obtained from K via a finite number of iterations of finite extensions and almost normalizing extensions.

We note that the classes of k-algebras mentioned in paragraph 3 of the introduction are all constructible k-algebras.

12. THEOREM (i) Any constructible D-algebra is $(\mathbb{N},\mathbb{N})$-generically flat over D.

(ii) Any constructible k-algebra satisfies the Nullstellensatz.

Proof. (i) This combines Lemmas 6, 9 and 10.

(ii) This follows from (i) and Corollary 4. □

There are also consequences for K-algebras

13. COROLLARY Let S be a constructible K-algebra, M a simple

S-module and $P$ a prime ideal of $S$.

(i) If $k$ is the field of fractions of $K/P \cap K$ then $S/P \otimes_K k$ is a constructible $k$-algebra, and $J(S/P \otimes_K k) = 0$.

(ii) If $P = \mathrm{ann}_S M$ then $\mathrm{End}\, M$ is algebraic, indeed finite dimensional, over $k$.

Proof. (i) It is clear that $S/P \otimes_K k$ is constructible over $k$. Moreover it is prime (being a localization with respect to an Ore set) and Noetherian, and so has no nonzero nil ideals.

(ii) The embedding $K/P \cap K \hookrightarrow \mathrm{End}\, M$ extends to an embedding $k \hookrightarrow \mathrm{End}\, M$. It follows that $M$ is also a simple module over $S/P \otimes k$ and so its endomorphism ring over $S/P \otimes k$ is algebraic over $k$. However $\mathrm{End}(M_S) \hookrightarrow \mathrm{End}(M_{S/P \otimes k})$ (in fact they are isomorphic). The finite dimensionality follows, as mentioned before. □

Recall that a Jacobson ring $S$ is one such that $J(S/P) = 0$ for all prime ideals $P$. This implies that the Jacobson radical of each factor ring of $S$ is nil.

14. THEOREM Let $S$ be a constructible $K$-algebra with $K$ being a Jacobson ring, and let $M$ be a simple right $S$-module. Then

(i) $S$ is a Jacobson ring; and

(ii) $K/\mathrm{ann}_K M$ is a field over which $\mathrm{End}\, M_S$ is finite dimensional.

Proof. See [6, 9.4.21]. □

REFERENCES


[1] J. Dixmier, Enveloping Algebras, North-Holland Math. Library 14, Amsterdam, 1977 (translated from Algèbres enveloppantes, Cahiers Scientifiques 37, Gauthier-Villars, Paris, 1974).

[2] M. Duflo, Certaines algèbres de type fini sont des algèbres de Jacobson, J. Algebra 27 (1973) 358-365.

[3] P. Hall, On the finiteness of certain soluble groups, Proc. London Math. Soc. 9 (1959) 595-622.

[4] R.S. Irving, Generic flatness and the Nullstellensatz for Ore extensions, Comm. Algebra 7 (1979) 259-277.

[5] J.C. McConnell, The Nullstellensatz and Jacobson properties for rings of differential operators, J. London Math. Soc. 26 (1982) 37-42.

[6] J.C. McConnell and J.C. Robson, Noncommutative Noetherian Rings, J. Wiley, Chichester-New York, 1987.

[7] D. Quillen, On the endomorphism ring of a simple module over an enveloping algebra, Proc. Amer. Math. Soc. 21 (1969) 171-172.

# GELFAND-KIRILLOV DIMENSION, HILBERT-SAMUEL POLYNOMIALS AND RINGS OF DIFFERENTIAL OPERATORS

J.C. McConnell and J.C. Robson
School of Mathematics
University of Leeds
LEEDS LS2 9JT
England

ABSTRACT. We show that the well known Hilbert-Samuel theory can be extended from the class of universal enveloping algebras of Lie algebras to a wide class of rings of differential operators.

INTRODUCTION. We will work in the class of algebras over a field k which have finite Gelfand-Kirillov dimension. (For background see [6] or [8], Chapter 8.)

There are many algebras of finite GK dimension but the most successful applications of GK dimension are to enveloping algebras of finite dimensional Lie algebras. One reason is that with each finitely generated module M over such an algebra, there is associated a Hilbert-Samuel polynomial whose degree gives the GK dimension of M and whose leading coefficient gives a "multiplicity" to M. In this paper we show that for a large class of algebras (which includes any filtered algebra for which the associated graded algebra is commutative affine), one can develop much of the Hilbert-Samuel machinery (though there are some significant differences) and obtain all the usual consequences. These consequences are outlined below in Section 1. A key step is to show that if R is filtered and gr R is commutative affine then there is an alternative filtration, F say, by finite dimensional subspaces such that $gr_F R$ is also commutative affine. The main applications of these results are to rings of differential operators on smooth affine varieties and on some non-smooth varieties. Most of the results of this paper can be found in [8], Chapter 8, §6, but there are some additions here.

## 1. A SURVEY OF THE ENVELOPING ALGEBRA RESULTS

We summarise, without proof, some results which may be found in [6], Chapter 7, or [8], Chapter 8, §4.

Let R be a homomorphic image of the enveloping algebra $U(\mathfrak{g})$, where $\mathfrak{g}$ is a finite dimensional Lie algebra over k. The usual filtration of $U(\mathfrak{g})$, in which $U(\mathfrak{g})_0 = k$ and $U(\mathfrak{g})_1 = k + \mathfrak{g}$, induces



F. van Oystaeyen and L. Le Bruyn (eds.), Perspectives in Ring Theory, 233–238.

a filtration on $R$ for which $\operatorname{gr} R$ is a commutative affine algebra $k[x_1,\dots,x_n]$, where each $x_i$ has degree one. A good filtration $\{M_n\}$ on a finitely generated R-module $M$ is a filtration for which $\operatorname{gr} M$ is finitely generated over $\operatorname{gr} R$. Given a good filtration $\{M_n\}$ on $M$, there is a polynomial $f \in \mathbb{Q}[x]$ such that $\dim M_n = f(n)$ for $n \gg 0$. This polynomial is called a Hilbert-Samuel polynomial for $M$. There is some uniqueness, viz. if $f(x) = a_d x^d + \dots + a_o$ then $\deg f = d = GK(M)$ and $e := d!a_d \in \mathbb{N}$ and is called the multiplicity. Thus $d$ is independent both of the choice of filtration of $R$ and the good filtration of $M$ while $e$ is independent of the choice of good filtration of $M$. These will be denoted by $f_M$, $d_M$, $e_M$ when necessary. If $0 \to L \to M \to N \to 0$ is a short exact sequence of finitely generated R-modules, then good filtrations can be chosen so that $f_M = f_L + f_N$. As a consequence, one obtains the following properties of $d$ and $e$.

1/ Each finitely generated $M_R$ has associated with it two numbers $d_M, e_M; d_M, e_M \in \mathbb{N}$ and $d_M = GK(M)$.

2/ GK dimension is exact on short exact sequences; i.e. $GK(M) = \max\{GK(L), GK(N)\}$.

3/ $e$ is "additive" over short exact sequences; i.e.

(a) if $d_L = d_M = d_N$ then $e_M = e_L + e_N$

(b) if $d_L < d_M = d_N$ then $e_M = e_N$

(c) if $d_L = d_M > d_N$ then $e_M = e_L$.

4/ GK dimension is finitely partitive; i.e. given a chain of submodules $M = M^{(o)} \supset M^{(1)} \supset \dots \supset M^{(n)}$ with $GK(M^{(i)}/M^{(i+1)}) = GK(M)$ for all $i$, then $n \leq e_M$.

Note that 4/ follows readily from 3/(a). 4/ enables one to relate the Krull dimension and GK-dimension of R-modules as follows. If there exist $a, \alpha \in \mathbb{N}$ such that for all finitely generated $M$,

$$\mathcal{K}(M) = a \text{ implies that } GK(M) \geq \mathcal{K}(M) + \alpha$$

then

$$\mathcal{K}(M) \geq a \text{ implies that } GK(M) \geq \mathcal{K}(M) + \alpha.$$

So, for example, with $a = \alpha = 0$, $\mathcal{K}(M) \leq GK(M)$ for all finitely generated $M$.

One might hope to get results 1/ to 4/ for any algebra of finite GK dimension but this is false. GK dimension need not take integral values and need not be exact on short exact sequences; see [4], 2.11 and [1]. However 1/ to 4/ should hold for the natural examples of algebras of finite GK dimension like group algebras of nilpotent by finite groups and rings of differential operators, and we will indeed prove this for (most of) the latter class.

## 2. SOMEWHAT COMMUTATIVE ALGEBRAS

2.1 Let $R = \bigcup R_n$ be a filtered algebra. Such a filtration is called a standard filtration if $R_n = R_1^n$ for all $n \geq 1$ and is called a finite dimensional filtration if $R_o = k$ and $\dim_k R_n < \infty$ for all $n$.

2.2 These definitions allow the characterization of factors of enveloping algebras.

THEOREM. $R$ is a homomorphic image of the enveloping algebra of a finite dimensional Lie algebra if and only if $R$ has a standard finite dimensional filtration such that $\operatorname{gr} R$ is commutative affine over $k$.

Proof. See [6], Theorem 7.2 or [8], 8.4.3

□

These algebras were considered by Duflo [5] who called them almost commutative algebras.

2.3 The class of algebras considered in this paper is defined next. An algebra $R$ is somewhat commutative if $R$ has a finite dimensional filtration such that $\operatorname{gr} R$ is commutative affine.

The class of somewhat commutative algebras is quite large as will be seen in the next section. Examples of somewhat commutative algebras which are not almost commutative are the Ore extensions $k[y][x;y^n d/dy]$ for $n \geq 2$; see [8], 8.6.10. In the remainder of this section we show how to develop a "Hilbert-Samuel" theory for somewhat commutative algebras.

2.4 Let $R$ be somewhat commutative, $\operatorname{gr} R$ be the corresponding graded ring and choose a standard finite dimensional filtration of $\operatorname{gr} R$. These filtrations of $R$ and $\operatorname{gr} R$ we now fix. Let $M_R$ be finitely generated and choose a good filtration of $M$ and a good filtration of $\operatorname{gr} M$ over $\operatorname{gr} R$. Then, by the results quoted in Section 1, the growth of the filtration of $\operatorname{gr} M$ is given by a Hilbert-Samuel polynomial which has associated integers $d(\operatorname{gr} M)$ and $e(\operatorname{gr} M)$. With this notation we have

PROPOSITION (i) $d(\operatorname{gr} M) = GK(\operatorname{gr} M) = GK(M)$ and so $d(\operatorname{gr} M)$ is independent of the choice of filtrations of $R$, $\operatorname{gr} R$, $M$ and $\operatorname{gr} M$.

(ii) $e(\operatorname{gr} M)$ is independent of the choice of the filtrations of $M$ and $\operatorname{gr} M$.

Proof. Straightforward, see [8], 8.6.4 and 8.6.18.

□

Thus we can define $d_M = d(\operatorname{gr} M)$ and $e_M = e(\operatorname{gr} M)$ respectively. With this notation we have

2.5 THEOREM Let $R$ be somewhat commutative and $M_R$ be finitely generated. Fix a standard finite dimensional filtration of $\operatorname{gr} R$ and define $d_M$, $e_M$ as above. Then properties 1/, 2′/, 3/ and 4/ of Section 1 hold.

Proof. See [8], 8.6.20. □

2.6 Although we have all the consequences of the Hilbert-Samuel theory we do not have an actual Hilbert-Samuel polynomial as is easily seen from the following example. Let $R = k[x]$, which is filtered or graded by letting $x$ have degree 2 and hence $x^n$ has degree $2n$ for all $n$. With this filtration $R$ is somewhat commutative and $\mathrm{gr}R \simeq R$ as graded rings. But, letting $R_n$ be the $n^{th}$ filtration subspace, $R_{2n} = R_{2n+1}$ for all $n$, and so one cannot have a polynomial $g$ such that $\dim R_n = g(n)$ for $n >> 0$. However there is a family of "Hilbert-Samuel polynomials" since results of Lorenz, [7], Chapter III Lemma 1.3, show that if $R$ is somewhat commutative and $M_R$ is finitely generated with a good filtration $\{M_n\}$ then there exists $m \in \mathbb{N}$ and a family of polynomials $g_{\bar{n}} \in \mathbb{Q}[x]$ (one polynomial for each residue $\bar{n} \bmod m$), such that for $n >> 0$, $\dim M_n = g_{\bar{n}}(n)$. Further all the $g_{\bar{n}}$ have the same leading coefficient. This approach leads (via the leading coefficient) to a multiplicity which depends on the choice of filtration of $R$ but, unlike the multiplicity of 2.5, is independent of the choice of a filtration for $\mathrm{gr}\,R$.

## 3. THE CLASS OF SOMEWHAT COMMUTATIVE ALGEBRAS

We now consider some classes of algebras which we would like to include in our theory.

3.1 Let $A$ be a commutative affine algebra over a field $k$ of characteristic zero and $\Delta(A)$ be the subalgebra of $\mathrm{End}_k A$ generated by $A$ and $\mathrm{Der}\,A$, the k-algebra derivations of $A$. Then $\Delta(A)$ has the standard filtration in which $(\Delta(A))_o = A$ and $(\Delta(A))_1 = A + \mathrm{Der}\,A$ and $\mathrm{gr}\Delta(A)$ is commutative affine. (See for example [8], 15.1.20.) For $A$ regular, $\Delta(A)$ coincides with the ring of differential operators $\mathcal{D}(A)$ on $A$; see [8], 15.5.6 for example. If $A$ is not regular, it can happen that $\mathcal{D}(A) \supset \Delta(A)$ but there are interesting situations where one still has that $\mathrm{gr}\,\mathcal{D}(A)$ is commutative affine; see, for example, [11], Theorem 3.12 and [10], Theorem 4.1.

3.2 Let $R$ be somewhat commutative, $\mathbf{g}$ a finite dimensional Lie algebra and $S = R*U(\mathbf{g})$ a crossed product of $R$ by $U(\mathbf{g})$. Then $S$ has a standard filtration with $S_o = R$, $S_1 = R + R\mathbf{g}$ such that $\mathrm{gr}\,S = R[X_1,\dots,X_n]$, a polynomial algebra over $R$ on the central indeterminates $X_1,\dots,X_n$.

3.3 With these two sets of examples in mind we have

THEOREM Let $S$ be a filtered ring with $S_o = R$ such that

(i) $R$ is somewhat commutative,

(ii) $\operatorname{gr} S$ is generated over $R$ by finitely many (homogeneous) central elements.

Then $S$ is somewhat commutative.

Proof. Rather than giving the full proof we indicate how the proof goes in a special case which is sufficient to display the main ideas. (The full proof will appear in [9].) Let $R$ be commutative affine, $R = k[y_1,\dots,y_c]$ say, and $S = R[x_1,\dots,x_b]$ have a standard filtration, $S_o = R$, $S_1 = R + \Sigma_p Rx_p$ with $\operatorname{gr} S$ being commutative affine. Then for all $i,j$,

$$x_i y_j - y_j x_i = f_{ij} \in R$$

and

$$x_i x_j - x_j x_i = \Sigma_p g_{ijp} x_p + h_{ij} \in \Sigma_p Rx_p + R.$$

Now $R$ has a finite dimensional filtration in which $R_n$ is spanned by the monomials in the y's of degree at most $n$. Choose $m$ so that all the $f_{ij}$, $g_{ijp}$ and $h_{ij}$ lie in $R_m$.

Now define a filtration on $S$ by setting $S_n = R_n$ for $0 \leq n \leq m$, $S_{m+1} = R_{m+1} + \Sigma_p kx_p$ and then letting the filtration 'grow' for $n > m + 1$. This gives a finite dimensional filtration of $S$ such that $\operatorname{gr} S$ is affine by construction and commutative by the choice of $m$. Thus $S$ is somewhat commutative. □

3.4 Let $S$ be a filtered algebra such that $\operatorname{gr} S$ is commutative affine (as in the examples in 3.1) and let $M_S$ be finitely generated. The reader might ask why one does not work directly with $\operatorname{gr} M$ as a $\operatorname{gr} S$-module, since over $\operatorname{gr} S$ one has the Hilbert-Samuel machinery. The main difficulty is whether, given a good filtration on $M$, the equality $\operatorname{GK}(\operatorname{gr} M) = \operatorname{GK}(M)$ holds. This has only been known in the special case of the Weyl algebra $A_n$ (with respect to the filtration coming from regarding $A_n$ as a ring of differential operators); see [2], Theorem 3.1 or [3], Chapter 3, Theorem A.2.4. However this result is now known, under the sole condition that $\operatorname{gr} S$ is commutative affine over $k$. There is also an example which shows that one can have $\operatorname{GK}(\operatorname{gr} M) \neq \operatorname{GK}(M)$ if the condition that $\operatorname{gr} S$ is commutative is deleted. These results will appear in [9].

REFERENCES


[1] G.M. Bergman, Gelfand-Kirillov dimension can go up in extension modules, Comm. Algebra 9 (1981) 1567-70.

[2] I.N. Bernstein, Modules over a ring of differential operators. Study of the fundamental solutions of equations with constant coefficients, Funct. Anal. Appl. 5 (1971) 89-101.

[3] J.E. Bjork, Rings of Differential Operators, North-Holland Mathematics Library 21, North-Holland, Amsterdam, 1979.

[4] W. Borho and H. Kraft, Über die Gelfand-Kirillov Dimension, Math. Ann. 220 (1976) 1-24.

[5] M. Duflo, Certaines algèbres de type fini sont des algèbres de Jacobson, J. Algebra 27 (1973) 358-365.

[6] G. Krause and T.H. Lenagan, Growth of Algebras and Gelfand-Kirillov Dimension, Research Notes in Mathematics 116, Pitman, London, 1985.

[7] M. Lorenz, Gelfand-Kirillov dimension and Poincaré series, Proceedings of the 1986 Granada Conference, to appear.

[8] J.C. McConnell and J.C. Robson, Noncommutative Noetherian Rings, J. Wiley, Chichester-New York, 1987.

[9] J.C. McConnell and J.T. Stafford, Gelfand-Kirillov dimension and associated graded modules, in preparation.

[10] I.M. Musson, Rings of differential operators on invariant rings of tori, Trans. Amer. Math. Soc. 303 (1987) 805-827.

[11] S.P. Smith and J.T. Stafford, Differential operators on an affine curve, Proc. London Math. Soc., to appear.

# GABRIEL - POPESCU TYPE THEOREMS AND GRADED MODULES

C. Menini
Dipartimento di Matematica
Università degli Studi de L'Aquila, via Roma
I - 67100 L'AQUILA (ITALY)

ABSTRACT. Using a system of generators of a Grothendieck category, we prove general versions of Gabriel - Popescu Theorem. From these results we easily deduce some theorems of equivalence involving the category of (graded) modules over a group graded ring recently obtained by M. Beattie and T. Albu e C. Năstăsescu and we solve a question concerning such equivalences posed by the two last authors.

## INTRODUCTION

Throughout this paper the word "rng" will mean an associative ring while we will refer to a rng with identity $1 \neq 0$ as a "ring".

Let G be a multiplicative group, R a G-graded ring, $U = \bigoplus_{x \in G} R(x)$ the canonical generator of the category R-gr of graded left R-modules. In a recent paper (see [B] ), M. Beattie proved that there exists a rng S such that the category R-gr is equivalent to that one of unital left S-modules. Still in a recent paper (see [AN ] ) T. Albu and C. Năstăsescu showed that if B is any subring of $End_{R\text{-}gr}(U)$ containing the smash product $R \# G$, then the category R-gr is equivalent to a quotient category of B-mod. They also proved, using different techniques, that R-mod is equivalent to a quotient category of $END_R(U)$-mod and asked if-more generally-this holds whenever one replaces the ring $END_R(U)$ by an arbitrary subring C of $End_R(U)$ containing $END_R(U)$.



*F. van Oystaeyen and L. Le Bruyn (eds.), Perspectives in Ring Theory, 239–251.*

The aim of this paper is to prove some general versions of Gabriel - Popescu Theorem from which all the above-mentioned results can be easily derived sometime in a more general form. This enables us to give also a positive answer to the quoted question by Albu and Nastasescu.

Namely, let $C$ be a Grothendieck category, $\{U_i\}_{i\in I}$ a system of generators of $C$, $U = \coprod_{i\in I} U_i$, $\varepsilon_i : U_i \longrightarrow U$, $i\in I$, the canonical injection, $A = \mathrm{End}_C(U)$.

For every $M\in \mathrm{Ob}(C)$ let $\overline{\mathrm{Hom}}_C(U,M) = \{f\in \mathrm{Hom}_C(U,M) \mid f\circ\varepsilon_i = 0$ for almost every $i\in I\}$, $\overline{A} = \overline{\mathrm{Hom}}_C(U,U)$. We prove that the functor $S = \mathrm{Hom}_C(U,\text{—})$ from $C$ to the category mod-A of unital right modules over the rng $\overline{A}$ is full and faithful. When all the $U_i$'s, i I, are small we also prove that S has a left adjoint T so that T is exact and the adjunction $TS \longrightarrow 1_C$ is an isomorphism. Therefore S induces an equivalence between $C$ and the quotient category of $\overline{A}$-mod by Ker(T). From this, one easily gets, in the case when the $U_i$'s, i I, are small and projective, that S is an equivalence between $C$ and $\overline{A}$-mod. This last equivalence was first noted by M. Harada in [ H ] as a consequence of Freyd's Theorem (see [ P ] Corollary 6.4 page 103) and a Gabriel's results (see [G ] Proposition 2 page 347).

Let now B be any subring of $\mathrm{End}_C(U)$ containing $\overline{A}$. Then we prove that the functor $\mathrm{Hom}_C(U,\text{—})$ is an equivalence between $C$ and a quotient category of B-mod.

In section 3 we apply this results to the category R-gr and R-mod, R being a G-graded ring. Then when $C$ = R-gr, $U = \bigoplus_{x\in G} R(x)$ Freyd-Gabriel-Harada Theorem is exactly Beattie's result while our last Theorem becomes a slight generalization of the first one of the quoted results by Albu & Nastasescu. Finally by applying the same theorem to the category R-mod one gets an affermative answer to their above-mentioned question.

Further applications of these theorems will appear elsewhere.

## 1. NOTATIONS AND PRELIMINARIES

As stated in the Introduction, by a rng we will mean an associative ring, while a ring will be a rng with identity $1\neq 0$. If $\overline{A}$ is a rng, a left $\bar{A}$-module M is called unital if $\bar{A}M = M$. We will denote by A-mod (resp. by $\bar{A}$-mod) the category of left unital modules over the ring A (resp. over the rng $\bar{A}$). Moreover if A is a ring every module over A is intended to be unital. The notation ${}_AM$ will be used to emphasize that M is a left A-module. Analogous notations will hold on the right. If A and B are two rings we will write ${}_AM_B$ to mean that M is an A-B-bimodule (left A-module and right B-module). Maps between modules will be written on the opposite side to that of the scalars.

Following Fuller [F] , we say that a rng $\bar{A}$ has <u>enough idempotents</u> in case there exist ortogonal idempotents $(\eta_i)_{i\in I}$ in $\bar{A}$ (called a complete set of idempotents for $\bar{A}$) such that $\bar{A} = \bigoplus_{i\in I} \bar{A}\eta_i = \bigoplus_{i\in I} \eta_i \bar{A}$.

Let G be a multiplicative group with identity element e.Let $R = \bigoplus_{x\in G} R_x$ be a graded ring of type G. We denote by R-gr (gr-R) the category of graded left (right) R-modules. If $M = \bigoplus_{x\in G} M_x$ and $N = \bigoplus_{x\in G} N_x$ are two graded left R-modules, for every $x\in G$ we set

$$HOM_R(M,N)_x = \{f \in Hom_R(M,N) \mid (M_1)f \leq N_{1x} \text{ for all } y \in G\}$$

In particular $HOM_R(M,N)_e = Hom_{R\text{-}gr}(M,N)$ is the set of morphism in the category R-gr from M to N. Clearly, for every $x\in G$, $HOM_R(M,N)_x$ is an additive subgroup of $Hom_R(M,N)$ so that

$$HOM_R(M,N) = \bigoplus_{x\in G} HOM_R(M,N)_x$$

is a graded abelian group of type G. If M=N then $END_R(M)=HOM_R(M,M)$ is a subring of $End_R(M)$ and is a graded ring of type G.

Let $M = \bigoplus_{x\in G} M_x \in$ R-gr and let $x\in G$. The <u>x-suspension</u> M(x) of M is the graded module obtained from M by setting $M(x)_y = M_{yx}$ for all $y\in G$. If $z\in M$, $x\in G$, $z_x \in M_x$ will denote the x-th component of z.

For all the basic facts concerning graded modules the reader is referred to [NV].

## 2. <u>GABRIEL - POPESCU TYPE THEOREMS</u>

<u>2.1</u>. Throughout this section $\mathcal{C}$ denotes a fixed, but arbitrary,

Grothendieck category, $\{U_i\}_{i\in I}$ a family of objects of $C$, $U=\coprod_{i\in I} U_i$ their coproduct. For every $i\in I$, $\varepsilon_i : U_i \longrightarrow U$ and $\pi_i : U \longrightarrow U_i$ are the canonical injection and projection respectively and $\eta_i = \varepsilon_i \circ \pi_i$. For every finite subset F of I we set $\eta_F = \sum_{i\in F} \eta_i$. $A = \text{End}_C(U)$ is a ring with respect to the multiplication $fg=f\circ g$, $f,g\in A$ and for every $X\in \text{Ob}(C)$, $\text{Hom}_C(U,X)$ has a natural structure of right A-module.

For every $X\in \text{Ob}(C)$ and for every $f\in \text{Hom}_C(U,X)$ let $\text{Supp}(f)=\{i\in I \mid f\eta_i \neq 0\}$ and set

$$\overline{\text{Hom}}_C(U,X) = \{f\in \text{Hom}_C(U,X) \mid \text{Supp}(f) \text{ is finite}\},$$

$$\overline{A} = \overline{\text{Hom}}_C(U,U).$$

In the following proposition we list some basic facts whose proof is straightforward.

2.2 Proposition. 1) For every $X\in \text{Ob}(C)$, if $f\in \text{Hom}_C(U,X)$ and G is a finite subset of I containing supp(f), then $f\eta_G = f$

2) $\overline{A}$ is a left idempotent ideal of A and $\overline{A} = \bigoplus_{i\in I} A\eta_i$

3) For every A-submodule L of a right A-module M it is

$$M\overline{A} \cap L = L\overline{A}$$

4) For every exact sequence

$$0 \longrightarrow M_1 \longrightarrow M \longrightarrow M_2 \longrightarrow 0 \quad \text{in A-mod}$$

the sequence

$$0 \longrightarrow M_1\overline{A} \longrightarrow M\overline{A} \longrightarrow M_2\overline{A} \longrightarrow 0 \quad \text{in } \overline{A}\text{-mod is exact}$$

5) For every $X\in \text{Ob}(C)$, $\overline{\text{Hom}}_C(U,X) = \text{Hom}_C(U,X)\overline{A}$ and the assignement $X \longmapsto \overline{\text{Hom}}_C(U,X)$ defines a covariant left exact functor $\overline{\text{Hom}}_C(U,—)$ from $C$ to $\overline{A}$-mod.

6) If each $U_i$, $i\in I$, is small then the functor $\text{Hom}_C(U,—)$ commutes with coproducts and $\overline{A}$ is a rng with enough idempotents. In fact $(\eta_i)_{i\in I}$ is a complete set of idempotents for $\overline{A}$.

7) If each $U_i$, $i\in I$, is projective then the functor $\overline{\text{Hom}}_C(U,—)$ is exact.

2.3 Lemma. If U is a generator of $C$, then the functor

$\overline{\mathrm{Hom}}_{\mathcal{C}}(U,—): \mathcal{C} \longrightarrow \text{mod-}\bar{A}$ <u>is faithful</u>.

<u>Proof</u>. Let $X,Y \in Ob(\mathcal{C})$, $o \neq \alpha \in \mathrm{Hom}_{\mathcal{C}}(X,Y)$. As U is a generator of $\mathcal{C}$, there is a $\beta \in \mathrm{Hom}_{\mathcal{C}}(U,X)$ such that $\alpha\beta \neq o$. Then $\alpha \circ \beta \circ \eta_i \neq 0$ for some $i \in I$ and $\beta \circ \eta_i \in \overline{\mathrm{Hom}}_{\mathcal{C}}(U,X)$.

<u>2.4</u> Let $X \in Ob(\mathcal{C})$ and let M be an $\bar{A}$-submodule of $\mathrm{Hom}_{\mathcal{C}}(U,X)$ (we do not necessarily assume that M is unital), $i: M \longrightarrow \mathrm{Hom}_{\mathcal{C}}(U,X)$ the canonical injection. Set $H = \sum_{f \in M} \mathrm{Im}(f)$ and let $\chi: H \longrightarrow X$ be the canonical injection. Then every $f \in M$ factors through H so that there is an ambedding $j: M \longrightarrow \mathrm{Hom}_{\mathcal{C}}(U,H)$ such that $i = \mathrm{Hom}_{\mathcal{C}}(U,\chi) \circ j$.

<u>2.5 Lemma</u>. <u>If U is a generator of</u> $\mathcal{C}$ <u>then, within the notations of 2.4, for</u> every $Z \in Ob(\mathcal{C})$ <u>and for</u> every $\phi \in \mathrm{Hom}_{\bar{A}}(M, \mathrm{Hom}_{\mathcal{C}}(U,Z))$ <u>there is an</u> $\alpha \in \mathrm{Hom}_{\mathcal{C}}(H,Z)$ <u>such that for every</u> $f \in M$, $\phi(f) = \alpha \circ j(f)$.

<u>Proof</u>. This proof is modelled on part of the proof of "(1) $\Longrightarrow$ (2)" of Gabriel - Popescu Theorem in [ P ] (see [ P ] Theorem 7.9 page 111). Thus we will only sketch it outlining the differences but referring to [ P] for common details.

For each $f \in M$ let $i_f: U \longrightarrow U^{(M)}$ denote the corresponding injection. Let $\lambda: U^{(M)} \longrightarrow H$ be the unique morphism such that $\lambda \cdot i_f = j(f)$ for every $f \in M$. Analogously, let $\mu: U^{(M)} \longrightarrow Z$ such that $\mu \circ i_f = \phi(f)$ for each $f \in M$. Then, by the definition of H, $\lambda$ is an epimorphism and it is enough to show that $\mathrm{Ker}(\mu)$ contains $\mathrm{Ker}(\lambda)$. Let F be a finite subset of M, $i_F: U^{(F)} \longrightarrow U^{(M)}$ the canonical injection and $K_F = \mathrm{Im}(i_F) \cap \mathrm{Ker}(\lambda)$. Since $\mathcal{C}$ is an Ab-5 category it is enough to show that $K_F$ is contained in $\mathrm{Ker}(\mu)$. Let $K'_F = i_F^{-1}(K_F)$ an let $\kappa: K'_F \longrightarrow U^{(F)}$ be the canonical inclusion. Also let $\pi'_f: U^{(F)} \longrightarrow U$ denote, for each $f \in F$, the corresponding canonical projection. Then, for every $g \in \overline{\mathrm{Hom}}_{\mathcal{C}}(U,K'_F)$, $\pi'_f \kappa g \in \overline{\mathrm{Hom}}_{\mathcal{C}}(U,U) = \bar{A}$ for every $f \in F$. Thus $\phi(f)\pi'_f \kappa g = \phi(f\pi'_f \kappa g)$ and hence $\mu i_F \kappa g = \sum_{f \in F} \phi(f)\pi'_f \kappa g = \phi(\lambda i_F \kappa g) = 0$ as $\lambda i_F \kappa = 0$. Honce $\mu i_F \kappa g = 0$ for every $g \in \overline{\mathrm{Hom}}_{\mathcal{C}}(U,\kappa'_F)$ and, by

Lemma 2.3, $\mu i_F \kappa = 0$.

2.6 Theorem. Let $\{U_i\}_{i\in I}$ be a system of generators of a Grothendieck category $C$, $U = \coprod_{i\in I} U_i$. Then the functor $\overline{\mathrm{Hom}}_C(U,—): C \longrightarrow \text{mod-}\bar{A}$ is full and faithful.

Proof. Let X,Y be two objects of $C$ and let

$$\psi_{X,Y}: \mathrm{Hom}_C(X,Y) \longrightarrow \mathrm{Hom}_{\bar{A}}(\overline{\mathrm{Hom}}_C(U,X), \overline{\mathrm{Hom}}_C(U,Y))$$

be the canonical map. By Lemma 2.3, $\psi_{X,Y}$ is injective.

Let $M = \overline{\mathrm{Hom}}_C(U,X)$, $H = \sum_{f\in M} \mathrm{Im}(f)$. Then H = X. Indeed, let $\chi: H \longrightarrow X$ be the canonical injection, $\eta: X \longrightarrow \mathrm{Coker}(\chi)$ the Cokernel of $\chi$. If $\nu \neq 0$ then, by Lemma 2.3, it would be $\nu \cdot f \neq 0$ for some $f \in M$. But $\mathrm{Im}(f) \hookrightarrow H$. Contradiction. Hence H=X and Lemma 2.5 implies that $\psi_{X,Y}$ is surjective.

2.7 Proposition. Let $\{U_i\}_{i\in I}$ be a system of small generators of a Grothendieck category $C$, $U = \coprod_{i\in I} U_i$. Then the functor $S = \overline{\mathrm{Hom}}_C(U,—): C \longrightarrow \text{mod-}\bar{A}$ has a left adjoint T such that $T(\eta_i \bar{A}) = U_i$, for every $i \in I$.

Proof. Since the $U_i$'s are small, $\bar{A}$ is a rng with enough idempotents (see Prop. 2.2) : $\bar{A} = \bigoplus_{i\in I} \bar{A}\,\eta_i = \bigoplus_{i\in I} \eta_i \bar{A}$. Is easily follows that, in this case, the $\eta_i \bar{A}$'s, $i \in I$ form a system of projective small generators of mod-$\bar{A}$. Thus the assignement $T(\eta_i \bar{A}) = U_i$, $i \in I$, can be uniquely extended to a functor $T: \text{mod-}\bar{A} \longrightarrow C$ which commutes with inductive limites (see [P] Theorem 6.2 page 99). As $\mathrm{Hom}_C(U_i Z) = \mathrm{Hom}_{\bar{A}}(\eta_i \bar{A}, \mathrm{Hom}_C(U,Z)) = \mathrm{Hom}_{\bar{A}}(\eta_i \bar{A}, S(Z))$ naturally for each $i \in I$, $Z \in C$, then it is straightforward to prove that T is a left adjoint of S (see e.g. the proof of Theorem 6.3 page 101 in [ P ] ).

2.8 In the hypothesis of Proposition 2.7, we denote by

$$\psi_{-,-}: \mathrm{Hom}_{\bar{A}}(—, \overline{\mathrm{Hom}}_C(U,—)) \longrightarrow \mathrm{Hom}_C(T(—),—)$$

the natural equivalence of the adjointess. Note that for every $X,Y \in Ob(\mathcal{C})$, $L \in$ mod-A, if $v = s(\tau)\cdot u$ for $\tau \in Hom_{\mathcal{C}}(X,Y)$,

$$u \in Hom_{\bar{A}}(M,S(X)) \quad v \in Hom_{\bar{A}}(M,S(Y)) \text{ then } \psi_{M,Y}(v) = \tau \cdot \psi_{M,X}(u)$$

In the following we adapt to our case the ideas of M. Takeuchi in [T] .

Let $X \in Ob(\mathcal{C})$ and $M \overset{i}{\hookrightarrow} S(X)$ a monomorphism in mod-$\bar{A}$ and set $H = \sum_{f \in Im(i)} Im(f)$. As we remarked in 2.4 there is an embedding $j:M \longrightarrow S(H)$ such that $i = s(\chi)\cdot j$ where $\chi : H \longrightarrow X$ is the canonical injection. Then $\psi_{M,X}(i) = \chi \cdot \psi_{M,H}(j)$.

2.9 Lemma. In the hypothesis of Proposition 2.7 and within the above notations $\psi_{M,H}(j):T(M) \longrightarrow H$ is an isomorphism. Thus $\psi_{M,X}(i):T(M) \longrightarrow X$ is a monomorphism.

Proof. Let $Z \in Ob(\mathcal{C})$, $\beta \in Hom_{\mathcal{C}}(T(M),Z)$. Then $\beta = \psi_{M,Z}(\phi)$ for an unique $\phi \in Hom_{\bar{A}}(M,\overline{Hom}_{\mathcal{C}}(U,Z))$. In view of Lemma 2.5 there is an $\alpha \in Hom_{\mathcal{C}}(H,Z)$ such that, for every $f \in M$, $\phi(f) = \alpha \circ j(f)$. If follows that $\beta = \alpha \circ \psi_{M,H}(j)$ moreover, $\alpha$ is unique. In fact if $\beta = \alpha' \circ \psi_{M,H}(j)$ with $\alpha' \in Hom_{\mathcal{C}}(H,Z)$ then $0 = \gamma \circ \psi_{M,H}(j)$, for $\gamma = \alpha - \alpha'$. Then $0 = \gamma \circ j(f)$ for every $f \in M$ and hence, by definition of H, $\gamma = 0$ i.e. $\alpha = \alpha'$. Therefore $\psi_{M,H}(j)$ is an isomorphism.

2.10 Lemma. In the hypothesis of Proposition 2.7 if $i:M \hookrightarrow F$ is an embedding in mod-$\bar{A}$ and F is a direct sum of copies of the $\eta_i A$'s, i I, then $T(i):T(M) \longrightarrow T(F)$ is a monomorphism in $\mathcal{C}$ .

Proof. Since $\eta_i A = S(U_i)$, $i \in I$, the proof follows as in Lemma 2 of [T ] using our Lemma 2.9 and observing that for each finite coproduct L of copies of the $U_i$'s, $TS(L) = L$.

2.11 Theorem let $\{U_i\}_{i \in I}$ be a system of small generators of a Grothendieck category $\mathcal{C}$, $U = \coprod_{i \in I} U_i$. Then the functor $S = \overline{Hom}_{\mathcal{C}}(U,—): \mathcal{C} \longrightarrow$ mod-$\bar{A}$ has a left adjoint T.

Moreover the adjunction $TS \longrightarrow 1_{\mathcal{C}}$ is an isomorphism and T is exact. Thus S induces an equivalence between $\mathcal{C}$ and the quotient category of $\bar{A}$-mod by Ker(T).

Proof. Let $X \in Ob(\mathcal{C})$. Then, as we noted in the proof of Theorem 2.6, $X = \sum_{f \in M} Im(f)$ where $M = S(X)$. Thus, by Lemma 2.9, $\zeta_{S(X),X}(i): T(S(X)) \longrightarrow X$, where $i: S(X) \longleftrightarrow S(X)$ is the identity on S(X), is an isomorphism. As the $\eta_i A$'s, $i \in I$, are a system of generators of mod-$\bar{A}$ and T is right exact, using Lemma 2.10 and an easy diagram chasing one prove that T is also left exact (see [T] Lemma 3 and corollary). The last assertion now follows by [G] Proposition 5 page 374.

2.12 Corollary (freyd-Gabriel-Harada). Let $\{U_i\}_{i \in I}$ be a system of small projective generators of a Grothendieck category $\mathcal{C}$, $U = \coprod_{i \in I} U_i$. Then the functor $\overline{Hom}_{\mathcal{C}}(U, -)$ is an equivalence between $\mathcal{C}$ and mod-$\bar{A}$.

Proof. Let $M \in$ mod-$\bar{A}$. As $\bar{A}$ is a projective generator of mod-$\bar{A}$, we have an exact sequence in mod-$\bar{A}$ of the form

$$\bar{A}^{(J)} \xrightarrow{\alpha} \bar{A}^{(L)} \longrightarrow M \longrightarrow 0$$

Since $\bar{A} = \overline{Hom}_{\mathcal{C}}(U, U)$, $\alpha$ induces in a natural way a morphism $\beta: U^{(J)} \longrightarrow U^{(L)}$ in $\mathcal{C}$ such that $\overline{Hom}_{\mathcal{C}}(U, \beta) = \alpha$. Let $X = CoKer(\beta)$. As the functor $\overline{Hom}_{\mathcal{C}}(U, -)$ is exact and commutes with coproducts, one gets $\overline{Hom}_{\mathcal{C}}(U, X) = CoKer(\alpha) = M$.

2.13 Let now B be a subring (with 1!) of $A = Hom_{\mathcal{C}}(U, U)$ containing $\bar{A} = Hom_{\mathcal{C}}(U, U)$. Then for each $X \in \mathcal{C}$, $Hom_{\mathcal{C}}(U, X)$ is in a natural way a right B-module so that assignement $X \longmapsto Hom_{\mathcal{C}}(U, X)$ defines a covariant functor $S_B: \mathcal{C} \longrightarrow$ mod-B. $S_B$ has a left adjionit $T_B$: mod-B $\longrightarrow \mathcal{C}$ such that $T_B(B) = U$ (see [P] Theorem 7.2 page 108).

2.14 Theorem. If $\{U_i\}_{i \in I}$ is a system of generators of the Grothendieck category $\mathcal{C}$ then the functor $T_B$ is exact and the adjunction $T_B \circ S \longrightarrow 1_{\mathcal{C}}$ is an isomorphism. Thus $S_B$ induces an equivalence between $\mathcal{C}$ and the quotient category of B-mod by Ker(T).

Proof. Let X, $Z \in Ob(\mathcal{C})$, M a B-submodule of $Hom_{\mathcal{C}}(U,X)$. Then, as $B \leq \bar{A}$, $Hom_B(M, Hom_{\mathcal{C}}(U,Z)) \leq Hom_{\bar{A}}(M, Hom_{\mathcal{C}}(U,Z))$. As M in Lemma 2.5 was not assumed to be a unital $\bar{A}$-module, an analogous of Lemmata 2.5 and 2.9, with $\bar{A}$ replaced by B, and T and S replaced by $T_B$ and $S_B$, hold. Thus, as in Theorem 2.11 one gets that $T_B \cdot S_B \longrightarrow 1$ is an isomorphism.

Let now L be a B-submodule of $B^n$, $n \in N$. Then $L \overset{i}{\hookrightarrow} B^n \hookrightarrow A^n = S(U^n)$ As $T_B(B^n) = U^n = T_B S_B(U^n)$ using the analogous of Lemma 2.9 one gets that $T_B(L) \longrightarrow T_B(B^n)$ is an isomorphism.

Using this fact, one can prove, as in [ T ], that $T_B$ is left exact.

2.15 Remark. If I is finite then Theorem 2.14 is exactly Gabriel-Popescu Theorem.

## 3. Applications

3.1 Let G be group, $R = \bigoplus_{x \in G} R_x$ be a graded ring of type G. Then $\{R(x)\}_{x \in G}$ is a system of small projective generators of the Grothendieck category R-gr (see [NV] ).

Let $U = \bigoplus_{x \in G} R(x)$ and for every $x \in G$ let $1_{(x)}$ be the identity element of the ring R considered as an homogeneous element of degree x of $R(x^{-1})$. Thus $R(x^{-1}) = R\, 1_{(x)}$.

For every $N \in$ R-gr, the assignement $f \longmapsto ((1_{(x)})f)_{x \in G}$ allow us to identify $Hom_{R\text{-}gr}(U,N)$ with $\prod_{x \in G} N_x$.

In particular, if N=U, $A = Hom_{R\text{-}gr}(U,U)$ can be identified with $\prod_{x \in G} U_x = \prod_{x \in G} (\bigoplus_{y \in G} R_{xy^{-1}} 1_{(y)})$. Let $\alpha \in A$. Then for every x, $y \in G$, there is a unique $(x,y)\alpha \in R_{xy^{-1}}$ such that

$$(1_{(x)})\alpha = \sum_{y \in G} (x,y)\alpha\, 1_{(y)}$$

Thus, by means of our identification, α will coincide with

$(\sum_{y\in G}(x,y)\alpha\, 1_{(y)})_{x\in G}$ and this matches with the matricial form $R\nabla G$ of A given in [AN] .

Moreover for every $N\in R\text{-gr}$, $f=((1_{(x)}f)_{x\in G}\in \mathrm{Hom}_{R\text{-gr}}(U,N)$ and $\alpha\in A$ we have

$$\alpha f=((1_{(x)})(\alpha\cdot f))_{x\in G}$$

and since $(1_{(x)})\alpha=\sum_{y\in G}(x,y)\alpha\, 1_{(y)}$ for every $x\in G$, then

$$(1_{(x)})(\alpha\cdot f)=\sum_{y\in G}(x,y)\alpha\,(1_{(y)})f$$

Let $\bar{A}=\mathrm{Hom}_{R\text{-gr}}(U,U)$. Then, by our identification, for every $x\in G$, the element $\eta_{x^{-1}}\in\bar{A}$ coincides with the element $p_x=(p_{xy})_{y\in G}$ where $p_{xy}=\delta_{xy}1_{(y)}$ for all $y\in G$, and $\delta_{xy}$ is the Kronecker's symbol. Thus, in view of proposition 2.2, $A=\bigoplus_{x\in G}p_x\bar{A}$ and for every subring B of A containing $\bar{A}$ we have $\bar{A}=\bar{A}B=\bigoplus_{x\in G}p_xB$. Note now that for every $N\in R\text{-gr}$, the assignement $f=((1_{(x)})f)_{x\in G}\longmapsto\sum_{x\in G}(1_{(x)})f$ yields an abelian group isomorphism $\zeta_N$ between $\overline{\mathrm{Hom}}_{R\text{-gr}}(U,N)$ and N. In particular $\zeta_U$ is an isomorphism between $\bar{A}$ and U and $(p_x)\zeta_U=1_{(x)}$ for every $x\in G$. Clearly, via $\zeta_U$, we can define a left R-module structure on $\bar{A}$ by setting, forevery $r\in R$, $\alpha=(\sum_{y\in G}(x,y)\alpha\,1_{(y)})_{x\in G}\in\bar{A}$

$$r:\alpha=(r\cdot(\alpha)\zeta_U)\zeta_U^{-1}=(r\cdot\sum_{x\in G}\sum_{y\in G}(x,y)\alpha\,1_{(y)})\zeta_U^{-1}=$$

$$=(\sum_{x\in G}\sum_{t\in G}\sum_{y\in G}r_{xt^{-1}}(t,y)\alpha\,1_{(y)})\zeta_U^{-1}=(\sum_{y\in G}\sum_{t\in G}r_{xt^{-1}}(t,y)\alpha\,1_{(y)})_{x\in G}$$

In particular $rp_x=(r_{yx^{-1}}1_{(x)})_{y\in G}$ $\quad r\in R,\ x\in G$

In this way $\zeta_U$ becomes an isomorphism of left R-modules and as ${}_RU$ is a free left R-module on the generators $1_{(x)}$, $x\in G$, we get that ${}_R\bar{A}=\bigoplus_{x\in G}Rp_x$ is a free left R-module on the generators $p_x$, $x\in G$.

Elements $rp_x$ and $sp_z$, for r,s in R, x and z in G, multiply according to the following rule:

$$(rp_x)\cdot(sp_z)=((1_{(y)})(rp_x)\cdot(sp_z))_{y\in G}=((r_{yx^{-1}}1_{(x)})(sp_z))_{y\in G}=$$

$$=(r_{yx^{-1}}\cdot s_{xz^{-1}}1_{(z)})_{y\in G}=((rs_{xz^{-1}})_{yx^{-1}}1_{(z)})_{y\in G}=(rs_{xz^{-1}})p_z$$

Thus, if $r \in R_x$, we have

$$(*) \quad r_x p_z = p_{xz} r p_z \in p_{xz} \bar{A} \quad \text{for every } z \in G.$$

Hence $\bar{A}$ coincides with the rng $R\#G^*$ defined in [B] .
Let now $N \in R$-gr. Then the left $\bar{A}$-module structure of $\overline{\mathrm{Hom}}_{R\text{-gr}}(U,N)$ agrees with the rule

$$(rp_x)(y)\zeta_N^{-1} = (ry_x)\zeta_N^{-1} \quad \text{for every } r \in R,\ x \in G,\ y \in N \text{ and}$$

hence $\overline{\mathrm{Hom}}_{R\text{-gr}}(U,N)$ coincides with N" in Beattie's notations.
Conversely let $M \in \bar{A}$-mod. Then $M = \bar{A}M = \bigoplus_{x \in G} p_x M$ and in view of (*) $\bar{A}$ induces on M a graded R-module structure defined by setting

$$r\cdot(p_x y) = (rp_x)y \quad r \in R, \quad x \in G, \quad y \in M.$$

Note that, in this way, for every $N \in R$-gr, $\zeta_N$ becomes an isomorphism of graded left R-modules.
In particular ${}_R\bar{A} \cong {}_R U$ in R-gr
and, for every $x \in G$ ${}_R(\bar{A}p_x) = R(x^{-1})$
Therefore, in view of a well Known Mitchell's result (see [P] Theorem 6.5 page 103) the left adjoint functor T of the functor $S = \overline{\mathrm{Hom}}_{\mathcal{C}}(U,—)$ (see Proposition 2.7) coincides with the functor $(—)'$ defined by Bettie and by Corollary 2.12 we get

<u>3.2</u> <u>Theorem</u> (Beattie) <u>The functor</u> $\overline{\mathrm{Hom}}_{R\text{-gr}}(U,—)$ <u>is an equivalence between R-gr and R#G*-mod</u>.

<u>3.3</u> Let now B be any subring of A containing $\bar{A}$. Then in view of 3.1 the functor $S_B$ defined in 2.13 coincides, when $\mathcal{C}$ = R-gr and $U = \bigoplus_{x \in G} R(x)$, with that one defined in [AN] , § 2 (see also [AN] Remark 2.4,1).
For every $M \in B$-mod, $\bar{A}M \in \bar{A}$-mod so that, by 3.1 we can consider a graded left R-module structure on $\bar{A}M$ and the assignement $M \longmapsto \bar{A}M$ gives rise to a functor $T_B$: B-mod $\longrightarrow$ R-gr. Since $T_B(B) = {}_R(\bar{A}B) = {}_R\bar{A} \cong {}_R U$ by the quoted Mitchell's result we get that $T_B$ is exactly the left adjoint functor of $S_B$ (See 2.13).
By Theorem 2.14 we get the following slight improvement of Theorem 2.1 in [AN] .

3.4 Theorem. Let B be a subring of $\mathrm{End}_{R\text{-gr}}(U)$ containing $\bar{A}=R\# G^*$. Then the functors.

$$R\text{-gr} \underset{S_B}{\overset{T_B}{\rightleftarrows}} B\text{-mod}$$

defined in 3.3 have the following properties:

1) $S_B$ is a right adjoint of $T_B$

2) $T_B$ and $S_B$ are exact functors

3) $T_B \circ S_B = 1_{R\text{-gr}}$

Thus R-gr is equivalent to the quotient category of B-mod by the class $\mathrm{Ker}(T_B) = \{ M \in B\text{-mod} \mid \bar{A}M=0\}$

3.5 Consider now the rng $\bar{D}=\overline{\mathrm{Hom}}_{R\text{-mod}}(U,U)$ and let B be any subring of $D=\mathrm{End}_R(U)$ containing $\bar{D}$. Then U is in a natural way a right B-module and the functor $T'_B=U_B \otimes_B -$ : B-mod $\longrightarrow$ R-mod is the left adjoint of the functor $S'_B=\mathrm{Hom}_R({}_R U_B, -)$ : R-mod $\longrightarrow$ B-mod. Thus if we apply our Theorem 2.14 to this situation we get the following generalization of theorem 5.8 in [AN] (see Remark 3.7 below).

3.6 Theorem. Within the notations introduced in 3.5, the functors $T'_B$ and $S'_B$ have the following properties:

1) $T'_B$ and $S'_B$ are exact functors

2) $T'_B \circ S'_B = 1_{R\text{-mod}}$

Thus R-mod is equivalent to the quotient category of B-mod by $\mathrm{Ker}(T'_B)$.

3.7 Remark. Let $x \in G$. Then $\mathrm{Hom}_R(R(x),U)=\overline{\mathrm{Hom}}_R(R(x),U)$ (see [NV] Corollary I.2.11) and thus $\bar{D} \leq \mathrm{END}_R(U)$. Therefore our theorem 3.6 gives an answer to question (2) in Remark 5.9 of [AN] .

ACKNOWLEDGEMENT

This paper was written while the author was a member of the G.N.S.A.G.A. of the "Consiglio Nazionale delle Ricerche", with a partial financial support from Ministero della Pubblica Istruzione.

REFERENCES

[AN] T.ALBU - C.NĂSTĂSESCU: "Infinite group graded ring, rings of endomorphisms, and localization", preprint.

[B] M. BEATTIE: "A Generalization of the Smash Product of a Graded Ring", J. Pure Appl. Alg., to appear.

[G] P.GABRIEL: "Des catégories abéliennes", Bull. Soc. Math. France, 90 (1962), 323 - 448.

[F] K. R. FULLER: "On rings whose left modules are direct sums of finitely generated modules", Proc. Amer. Soc., 54 (1976), 39-44.

[H] M. HARADA: "Perfect categories I" and "Perfect categories II", Osaka J. Math. 10 (1973), 329-341 and 343-355.

[NV] C. NĂSTĂSESCU-F. VAN OYSTAEYEN: "Graded Ring Theory", North-Holland Mathematical Library, Volume 28, North-Holland Publishing Company, Amsterdam-New York-Oxford, 1982.

[P] N. POPESCU: "Abelian Categories with Applications to Rings and Modules", Academic Press, London & New York, 1973.

[T] M. TAKEUCHI: "A Simple Proof of Gabriel and Popesco's Theorem", J. Algebra, 18 (1971), 112-113.

# CROSSED PRODUCTS OF HOPF ALGEBRAS AND ENVELOPING ALGEBRAS

S. Montgomery*
University of Southern California
Los Angeles, CA 90089

ABSTRACT. The notion of crossed products for Hopf algebras acting (weakly) on non-commutative algebras is discussed together with the "splitting theorem" of Blattner-Cohen-Montgomery that a Hopf algebra H may be written as a crossed product over a homomorphic image $\bar{H}$ provided the map $H \to \bar{H}$ is split as a coalgebra map.

Some new examples and applications are then given, concerning the existence of coalgebra splittings, the relationship to various definitions in the literature of crossed products of enveloping algebras, and a method of explicitly extending a Lie cocycle on a Lie algebra L to a Hopf cocycle on U(L).

## INTRODUCTION

For A a k-algebra, k a field, and H a Hopf algebra, crossed products $A \#_\sigma H$ were introducted by Sweedler [S68] in the case H was cocommutative, A was an H-module, and A was commutative (more generally, the cocycle $\sigma$ could take values in the center of A). Since then, various crossed product constructions have also been defined for group algebras H = kG, where the group G acts as automorphisms, and for enveloping algebras H = U(L), where the underlying Lie algebra L acts as derivations (see for example [L], [Mc], [Ch]). Recently Blattner, Cohen, and Montgomery [BCM] have given a general construction of a crossed product $A \#_\sigma H$, for arbitrary A and H, in which A is not necessarily an H-module and $\sigma$ does not necessarily have central values.

In §1 of this paper we discuss this crossed product construction, as well as a theorem of [BCM] which gives a criterion for H to be a crossed product: namely, if which A is not $H \xrightarrow{\pi} \bar{H} \to 0$ is an exact sequence of Hopf algebras, and A denotes the (left) Hopf kernel of $\pi$, then $H \cong A \#_\sigma \bar{H}$ provided that the sequence is split as a coalgebra sequence. This generalizes the known fact for groups, that if $G \to G/N \to 0$, then $kG \cong kN \#_\sigma k[G/N]$, since choosing coset representatives for G/N in G induces a coalgebra splitting. An example is given to show that not every $H \to \bar{H} \to 0$ can be split as a coalgebra sequence.

*Work partially supported by NSF Grant DMS 87-00641.



*F. van Oystaeyen and L. Le Bruyn (eds.), Perspectives in Ring Theory, 253–268.*

In §2 we turn to the case $H = U(L)$, the enveloping algebra of the Lie algebra L. First, as in [BCM], we show that for Lie algebras, if $L \to L/N \to 0$, then $U(L) \cong U(N) \#_\sigma U(L/N)$. As an application, we prove that if H is cocommutative and k is algebraically closed, then every Hopf algebra sequence $H \to \overline{H} \to 0$ is split as a coalgebra sequence. We then prove that the other definitions of "crossed products" $R * U(L)$, appearing in [Mc], [Ch], [McR], are in fact equivalent to one of our crossed products $R \#_\sigma U(L)$ (Theorem 2.8).

In §3 we consider the relationship between our Hopf cocycles $\sigma : U(L) \times U(L) \to R$ and Lie cocycles $\tau : L \times L \to R^-$. Although a lot is known about the relationship between the cohomology of Lie algebras and enveloping algebras, the general theory does not appear to give an explicit method for "extending" $\tau$ to $\sigma$. We are able to do this, in the following sense (Theorem 3.6): let R be an algebra on which the Lie algebra L acts as derivations, and let $\mathfrak{L} = R^- \times_\tau L$ be a Lie extension of L with kernel $R^-$ and Lie cocycle $\tau$. This Lie extension determines a crossed product $R \#_\sigma U(L)$ with Hopf cocycle $\sigma : U(L) \times U(L) \to R$ (as in 2.8) and the values of $\sigma$ can be computed inductively from the values of $\tau$ on L and the action of L on R. The proof of this result depends ultimately on [BCM 4.14] discussed in §1 above. An example is given to illustrate the computation of $\sigma$ in terms of $\tau$.

Throughout, A and R denote algebras with 1 over a field k, and H denotes a Hopf algebra with comultiplication $\Delta$, counit $\varepsilon$, and antipode S. We also use the summation notation of [S69].

The author would like to thank R. J. Blattner for helpful discussions on the topic of this paper, in particular regarding Theorem 2.8.

## §1 CROSSED PRODUCTS OF HOPF ALGEBRAS

We motivate our study of crossed products with the example of group algebras. Let G be a group and N any normal subgroup of G. We would like to formulate things in a general enough way so that we can write kG in the form $kN \#_\sigma k[G/N]$. Here $k[G/N]$ is our Hopf algebra, kN is the algebra on which it acts, and $\sigma$ is a suitable cocycle. In fact such a crossed product (for groups) has been studied for almost ten years [L, MP] and is constructed as follows:

For each $\overline{x} \in \overline{G} = G/N$, choose a coset representative $\gamma(\overline{x}) \in \overline{x}$, and assume that $\gamma(\overline{1}) = 1$. Extending $\gamma$ linearly to $k\overline{G}$, we see that $\gamma$ linearly splits the exact sequence $kG \xrightarrow{\pi} k\overline{G} \to 0$. Although $\gamma$ is not an algebra morphism, it is a coalgebra morphism, since $\Delta g = g \otimes g$, all $g \in G$.

Clearly $kG = kN\, \gamma(\overline{G})$. Writing the elements of G in the form $n\, \gamma(\overline{x})$, $n \in N$, $\overline{x} \in \overline{G}$, we see that they multiply as in a classical crossed product. That is,

$$(1.1) \qquad (n\, \gamma(\overline{x}))\, (m\, \gamma(\overline{y})) = n(\overline{x} \cdot m)\, \sigma(\overline{x}, \overline{y})\, \gamma(\overline{x}\, \overline{y})$$

where

$$(1.2) \qquad \overline{x} \cdot m = (\mathrm{ad}\, \gamma(\overline{x}))\, m = \gamma(\overline{x})\, m\, \gamma(\overline{x})^{-1}$$

and

(1.3) $\sigma(\bar{x}, \bar{y}) = \gamma(\bar{x})\,\gamma(\bar{y})\ \ \gamma(\bar{x}\,\bar{y})^{-1} \in N.$

Since N is normal in G, kN is stable under this action of $\bar{G}$ and $\sigma$ has values in N. One may check that $\sigma$ satisfies

(1.4) $[\bar{x} \cdot \sigma(\bar{y}, \bar{z})]\,\sigma(\bar{x}, \bar{y}\,\bar{z}) = \sigma(\bar{x}, \bar{y})\,\sigma(\bar{x}\,\bar{y}, \bar{z}),$

the usual cocycle condition.

However, unlike the classical situation, kN is not a $\bar{G}$-module in general, kN is not commutative, and the values of $\sigma$ do not necessarily lie in the center of N. Instead, $\bar{G}$ is mapped to Aut (kN) non-homomorphically (we say $\bar{G}$ acts "weakly" on kN) and this action satisfies a "twisted module" condition:

(1.5) $[\bar{x} \cdot (\bar{y} \cdot n)]\,\sigma(\bar{x}, \bar{y}) = \sigma(\bar{x}, \bar{y})\ \ (\bar{x}\,\bar{y} \cdot n).$

Consequently kN is a $\bar{G}$-module if and only if $\sigma$ has values central in N.

Thus our general set-up must involve "weak" actions of a Hopf algebra H on a non-commutative algebra A, and cocycles $\sigma$ which take arbitrary values. In addition we allow H to be non-cocommutative, in order that the theory will apply to group-graded rings. We formalize this as follows:

1.6 DEFINITION. Let H be a Hopf algebra and A an algebra. By a weak action of H on A we mean a bilinear map $(h,a) \to h \cdot a$ of $H \times A \to A$ such that, for $h \in H$, $a, b \in A$,
(1) $h \cdot ab = \Sigma_{(h)}\,(h_{(1)} \cdot a)\ \ (h_{(2)} \cdot b)$
(2) $h \cdot 1 = \varepsilon(h)1$
(3) $1 \cdot a = a$
Conditions (1) and (2) say that H measures A. An action of H on A is a weak action which makes A into an H-module.

1.7 DEFINITION. Let H be a Hopf algebra with a weak action on the algebra A, and let $\sigma : H \times H \to A$ be a k-bilinear map. Let $A\,\#_\sigma H$ be the (possibly nonassociative) algebra with underlying vector space $A \otimes_k H$ and with multiplication

(1.8) $(a \otimes h)\,(b \otimes \ell) = \Sigma_{(h)(\ell)} a(h_{(1)} \cdot b)\,\sigma(h_{(2)}, \ell_{(1)}) \otimes h_{(3)}\ell_{(2)}$

for all $a, b \in A$, $h, \ell \in H$. The algebra $A\,\#_\sigma H$ is called a crossed product if it is associative with identity element $1 \otimes 1$.

The element $a \otimes h$ of $A\,\#_\sigma H$ will be written $a \# h$.

The cocycle $\sigma$ will be called normal if $\sigma(h, 1) = \sigma(1, h) = \varepsilon(h)1$, all $h \in H$. $A\,\#_\sigma H$ has identity element $1\#1$ if and only if $\sigma$ is normal [BCM 4.4].

1.9 LEMMA [BCM 4.5] $A\,\#_\sigma H$ is a crossed product if and only if $\sigma$ is normal and the following two conditions hold, for all $h, \ell, m \in H$, $a \in A$:

(1) (cocycle condition)

$$\Sigma_{(h)(\ell)(m)}[h_{(1)}\cdot\sigma(\ell_{(1)}, m_{(1)})]\,\sigma(h_{(2)}, \ell_{(2)}m_{(2)}) = \Sigma_{(h)(\ell)}\sigma(h_{(1)}, \ell_{(1)})\,\sigma(h_{(2)}\ell_{(2)}, m)$$

(2) (twisted module condition)

$$\Sigma_{(h)(\ell)}(h_{(1)}\cdot(\ell_{(1)}\cdot a))\,\sigma(h_{(2)}, \ell_{(2)}) = \Sigma_{(h)(\ell)}\sigma(h_{(1)}, \ell_{(1)})\,(h_{(2)}\ell_{(2)}\cdot a)$$

This construction of $A \#_\sigma H$ generalizes work of Sweedler [S68], who considered the case when H was cocommutative, A was an H-module, and A was commutative (or more generally, $\sigma$ took values in the center Z(A)); in his situation, condition 1.9(2) was not required. In our situation, if we assume H is cocommutative and $\sigma$ is invertible under convolution, then in fact A is an H-module if and only if $\sigma$ has values in Z(A) [BCM 4.7, 4.8].

Condition (1) and (2) in 1.9 clearly generalize (1.4) and (1.5), and thus kG = kN $\#_\sigma$k[G/N] in the terminology of 1.7. Recall that in this situation, we began with $kG \xrightarrow{\pi} k[G/N]$, a surjection of Hopf algebras. More generally, if $H \xrightarrow{\pi} \overline{H} \to 0$ is a sequence of Hopf algebras, when can we write $H = A \#_\sigma \overline{H}$, for suitable A and $\sigma$?

To answer this, we require another definition.

1.10 DEFINITION. If $\pi : H \to K$ is a Hopf algebra morphism, the left Hopf kernel of $\pi$ is the set

$$\text{LH Ker}\,\pi = \{h \in H : (\mathrm{id} \otimes \pi)\Delta h = h \otimes \overline{1}\}$$

As an example, if $kG \xrightarrow{\pi} k[G/N]$, it is easy to see that LH Ker $\pi$ = kN.

We also recall the definition of the adjoint action of a Hopf algebra H on itself: if h, k $\in$ H, then

$$(\mathrm{ad}\,h)\,k = \Sigma_{(h)}\, h_{(1)}k\,(S\,h_{(2)})$$

This clearly extends the usual notions for H = kG, in which $(\mathrm{ad}x)y = xyx^{-1}$ for x,y $\in$ G, and for H = U(L), in which (ad x)h = xh - hx for x $\in$ L, h $\in$ H.

We come to the main theorem.

1.11 THEOREM [BCM 4.14]. Let $H \xrightarrow{\pi} \overline{H} \to 0$ be an exact sequence of Hopf algebras which is split as a coalgebra sequence; that is, there exists a coalgebra map $\gamma : \overline{H} \to H$ such that $\pi \circ \gamma = \mathrm{id}$. Suppose also $\gamma(\overline{1}) = 1$, and let A = LH Ker $\pi$. Then $H \cong A \#_\sigma \overline{H}$ as algebras, where the weak action of $\overline{H}$ on A is given by

(1.12) $\quad \overline{h}\cdot a = (\mathrm{ad}\,\gamma(\overline{h}))a$ for $h \in \overline{H}$, $a \in A$;

and where $\sigma : \overline{H} \times \overline{H} \to A$ is given by

(1.13) $\sigma(\bar{h}, \bar{\ell}) = \Sigma_{(\bar{h})(\bar{\ell})} \gamma(\bar{h}_{(1)}) \gamma(\bar{\ell}_{(1)})[S\gamma(\bar{h}_{(2)}\bar{\ell}_{(2)})]$ .

We give an outline of the proof. A number of fairly straightforward things must be checked: A is an adH-stable subalgebra of H, the action (1.12) is a weak action in the sense of 1.6, and $\sigma(\bar{H} \times \bar{H}) \subseteq A$. Thus the algebra $A \#_\sigma \bar{H}$ can be constructed. To show it is a crossed product (and so satisfies 1.9 (1) (2)), we show it is algebra isomorphic to H, certainly an associative algebra, via a map $\phi : A \#_\sigma \bar{H} \to H$ such that $\phi(1\#1) = 1$.

Define $\phi$ by $\phi(a \# h) = a\gamma(\bar{h})$ for $a \in A$, $\bar{h} \in \bar{H}$; it is not difficult to check that $\phi$ is a homomorphism. The difficulty comes in showing that $\phi$ is bijective. To see this, we construct an inverse $\psi : H \to A \#_\sigma \bar{H}$, as follows. First define $Q : H \to H$ by $Q(h) = \Sigma_{(h)} h_{(1)}(S\gamma\pi h_{(2)})$; one may check that $Q(H) \subseteq \text{LH Ker } \pi$. Then define $\psi$ by $\psi h = \Sigma_{(h)} Qh_{(1)} \# \pi h_{(2)}$. It is straightforward that $\phi\psi = \text{id}$; with a bit more work it follows that $\phi\psi = \text{id}$. Thus $A \#_\sigma \bar{H} \cong H$.

We record a fact we will need later. Although $\psi$ looks complicated in general, it takes a very simple form on A. For if $a \in A$, $\Sigma_{(a)} a_{(1)} \otimes \pi a_{(2)} = a \otimes \bar{1}$. Thus $\psi a = \Sigma_{(a)} Qa_{(1)} \# \pi a_{(2)} = Qa \# 1 = \Sigma_{(a)} a_{(1)} (S\gamma \pi a_{(2)}) \# 1 = a \# 1$. That is :

(1.14) $\psi(a) = a \# 1$, all $a \in A = \text{LH Ker } \pi$.

The theorem gives rise to several questions.

1.15. Since H is a Hopf algebra, $A \#_\sigma \bar{H}$ also has a Hopf algebra structure obtained by pulling back $\Delta_H$ and $S_H$ via $\phi$. The resulting formulas are too complicated to be useful, except in the case that A itself is a Hopf subalgebra of H [BCM 4.17]. This raises the question: when is A a Hopf subalgebra?

Sweedler [S69] proves that this is always true when H is cocommutative; however cocommutativity is certainly not necessary. The key to answering this question is to define the <u>right Hopf kernel</u> of a Hopf morphism $\pi : H \to K$. Thus, we set

$$A' = \text{RH Ker } \pi = \{h \in H : (\pi \otimes \text{id})\ \Delta h = \bar{1} \otimes h\} .$$

If the antipode S of H is bijective, then A is a Hopf subalgebra of H if and only if A = A' [BCM 4.19].

One might hope that A = A' in general; however this is false. [BCM 4.21] gives an example in which $\pi : H \to \bar{H}$ is split as a coalgebra map (and thus $H \cong A \#_\sigma \bar{H}$) but such that $A \neq A'$ and A is not a Hopf subalgebra of H.

1.16. A more interesting question may be: which exact sequences $H \xrightarrow{\pi} \bar{H} \to 0$ of Hopf algebras are split as coalgebra sequences?

This does not always happen, as the following example shows. Let G be a group with subgroup K, and let $H = (kG)^*$, $\overline{H} = (kK)^*$. The restriction map $\pi : (kG)^* \to (kK)^*$ is a Hopf algebra morphism, since $\pi = i^*$ where $i : kK \to kG$ is the injection. Now, assume for the moment that $\pi$ is split as a coalgebra map; that is, there exists a coalgebra morphism $\gamma : (kK)^* \to (kG)^*$ satisfying $\pi\gamma = \text{id}$. But then $\gamma^* : kG \to kK$ would be an algebra morphism, surjective since $\gamma$ is injective. All that remains is to produce k, G, and K such that kK is not a homomorphic image of kG. In particular, we may choose $k = \mathbb{C}$, $G = S_3$, and $K = \langle (123) \rangle$.

Note that in this example, $H = (kG)^*$ is not cocommutative. We ask:

1.17. QUESTION: If H is cocommutative, is every exact sequence $H \to \overline{H} \to 0$ of Hopf algebra morphisms split as a coalgebra sequence ?

The answer to the question is yes provided k is algebraically closed of characteristic 0. This will be shown in the next section.

## §2. CROSSED PRODUCTS OF ENVELOPING ALGEBRAS

From now on, L will denote a Lie algebra over k and U(L) its universal enveloping algebra. We first consider what happens given an exact sequence of Lie algebras $L \to \overline{L} \to 0$.

2.1 [BCM 4.20]. Let N be a Lie ideal of L, set $\overline{L} = L/N$, and consider $L \xrightarrow{\pi} \overline{L} \to 0$, where $\pi$ is the canonical projection. $\pi$ extends to U(L), so that $U(L) \xrightarrow{\pi} U(\overline{L}) \to 0$ is an exact sequence of Hopf algebras. Then $U(L) \cong U(N) \#_\sigma U(\overline{L})$ as Hopf algebras.

The proof of this fact is an application of 1.11. There are several things to check. First we define $\gamma$. Choose an ordered basis $\{\overline{x}_\alpha\}$ of $\overline{L}$, and let $\gamma : \overline{L} \to L$ be a vector space embedding such that $\pi\gamma = \text{id}$. For any ordered monomial $\overline{x}_{\alpha_1} \cdots \overline{x}_{\alpha_n}$ of $U(\overline{L})$ with $\alpha_1 \leq \cdots \leq \alpha_n$, define $\gamma(\overline{x}_{\alpha_1} \cdots \overline{x}_{\alpha_n}) = \gamma(\overline{x}_{\alpha_1}) \cdots \gamma(\overline{x}_{\alpha_n})$; extend $\gamma$ linearly to a map $U(\overline{L}) \to U(L)$. Since $\Delta x = x \otimes 1 + 1 \otimes x$, all $x \in L$, it is easy to see that $\gamma$ is a coalgebra splitting of $\pi$. Second, we claim that $A = LH \operatorname{Ker} \pi = U(N)$. One sees directly that $U(N) \subseteq LH \operatorname{Ker} \pi$; equality follows using the PBW theorem and the fact that $U(L) \cong A \otimes U(\overline{L})$ as vector spaces.

2.1 has several interesting consequences. The first concerns the question raised in §1.

2.2 PROPOSITION. Let H be cocommutative and k be algebraically closed of characteristic 0. Then any exact sequence $H \to \overline{H} \to 0$ of Hopf algebras is split as a coalgebra sequence.

proof: This requires a fundamental result of Kostant: such an H is of the form $U(L) \# kG = U(L)G$, a skew group ring of a group G acting as automorphisms on an enveloping algebra U(L) (see [S69, 8.1.5 and 13.0.1]). Then $\overline{H} = \pi(H) = \pi(U(L))\ \pi(G)$. Certainly $\pi(G) = \overline{G}$ is a group, and it is also true that $\pi(U(L)) = U(\overline{L})$, the enveloping algebra of the Lie algebra $\overline{L} = \pi(L)$. For, U(L) is a pointed irreducible coalgebra so [S69, 8.0.9] implies that $\pi(U(L))$ is also pointed and irreducible. Thus since k has characteristic 0, $\pi(U(L)) = U(P)$, the enveloping algebra of the Lie algebra P of primitive elements of $\pi(U(L))$, by [S69, 13.0.1]. Since $\pi$ preserves primitive elements, $\pi(L) = \overline{L}$ is a Lie subalgebra of P, and so $U(\overline{L}) \subseteq U(P)$. But $\pi(U(L))$ is generated by $\pi(L)$, so in fact $\pi(U(L)) = U(\overline{L})$. Thus $\overline{H} = U(\overline{L})\overline{G}$.

We now define $\gamma: \overline{H} \to H$ by combining the map $\gamma: \overline{G} \to G$ from §1 together with the $\gamma: U(\overline{L}) \to U(L)$ in 2.1 : for any ordered monomial $\overline{m} \in U(\overline{L})$ and $\overline{g} \in \overline{G}$, let $\gamma(\overline{m}\,\overline{g}) = \gamma(\overline{m})\,\gamma(\overline{g})$. This $\gamma$ will be a coalgebra splitting of $\pi$.

The above proof does not work in characteristic $p \neq 0$, because of the presence of divided power algebras. In particular a Hopf homomorphic image of U(L) is not necessarily another enveloping algebra.

The second application of 2.1 will be to show that several other "enveloping algebra crossed products", of [BGR, Ch, Mc, McR], are in fact Hopf algebra crossed products of the type discussed here. We first review these other constructions.

2.3 DEFINITION [Ch, McR]. Let R be a k-algebra and L a Lie algebra over k, and fix a k-basis $\{x_\alpha\}$ for L. A k-algebra S containing R is called a crossed product of R by U(L), written $S = R*U(L)$, provided there is a vector space embedding of L into S, $x \to \tilde{x}$, such that for all $x, y \in L$, $r \in R$,

(i) $\tilde{x}r - r\tilde{x} = \delta_x(r) \in R$, where $\delta_x \in \mathrm{Der}_k(R)$

(ii) $\tilde{x}\tilde{y} - \tilde{y}\tilde{x} = \widetilde{[x, y]} + \tau(x,y)$, where $\tau : L \times L \to R$

(iii) S is a free right (and left) R-module with the standard monomials in $\{\tilde{x}_\alpha\}$ as basis.

Before considering the second definition, we recall some well-known facts about arbitrary Lie algebra extensions. Given an exact sequence of Lie algebras

$0 \to N \to \mathcal{L} \xrightarrow{\pi} L \to 0$, so that $\mathcal{L}$ is an extension of L with kernel N, let $\gamma : L \to \mathcal{L}$ be a linear splitting of $\pi$. Define $\tau: L \times L \to N$ by $\tau(x, y) = [\gamma(x), \gamma(y)] - \gamma[x, y]$, all $x, y \in L$, and note $\tau(x, y) \in \operatorname{Ker} \pi = N$. Define $\delta_x(n) = (\operatorname{ad} \gamma(x))n$, all $x \in L$, $n \in N$.

Then it is not difficult to verify that $\tau$ and $\delta_x$ satisfy

(2.4) $\tau[[x,y],z] + \tau[[z,x],y] + \tau[[y,z],x] = \delta_x(\tau(y,z)) + \delta_z(\tau(x,y)) + \delta_y(\tau(z,x))$

(2.5) $\delta_x(\delta_y(n)) - \delta_y(\delta_x(n)) = \delta_{[x,y]}(n)) + [\tau(x,y), n]$

for all $x, y, z \in L$, $n \in N$.

Conversely, given Lie algebras L and N, a map $L \to \mathrm{Der}_k(N)$ given by $x \to \delta_x$, and an alternating bilinear map $\tau : L \times L \to N$ satisfying (2.4) and (2.5), then the vector space $N \oplus L$ can be made into a Lie algebra $\mathfrak{L}$, written $\mathfrak{L} = N \times_\tau L$, via the Lie multiplication

(2.6) $[n + x, m + y] = [n, m] + \delta_x(m) - \delta_y(n) + \tau(x,y) + [x,y]$.

One may compare (2.4), (2.5), and (2.6) with 1.9(1), (2), and (1.8) respectively.

We remark that non-abelian extensions of Lie algebras were classified in [H] in terms of a third cohomology group, analogously to the Eilenberg-Maclane result for group extensions.

The second definition of crossed product extends work of [BGR, Theorem 4.2], in which no cocycle was present, and work of [Mc, §2.4], in which a cocycle appeared but R was a commutative U(L)-module. The fact that their definition should extend to the general situation was suggested by W. Chin.

2.7 DEFINITION. Let R be a k-algebra and L a Lie algebra over k. Assume there exists a linear map $L \to \mathrm{Der}_k R$, given by $x \to \delta_x$, and (considering R as $R^-$, a Lie algebra under [ ]) a Lie cocycle $\tau : L \times L \to R$ satisfying (2.4) and (2.5). Then form the Lie algebra $\mathfrak{L} = R^- \times_\tau L$ as above, and its enveloping algebra $U(\mathfrak{L})$. In $U(\mathfrak{L})$, let I be the ideal generated by the set $\mathcal{S} = \{1_{U(R)} - 1_R,\ a.b - ab : a,b \in R\}$, where ab denotes the usual multiplication in R and a.b denotes multiplication in $U(R^-)$. Now define

$$R \times_\tau U(L) = U(\mathfrak{L}) / I$$

2.8 THEOREM. Let R be an algebra and L a Lie algebra over k. Then the following are equivalent, for a k-algebra S:

(i) $S \cong R * U(L)$ as in Definition 2.3

(ii) $S \cong R \times_\tau U(L)$ as in Definition 2.7, where $\tau : L \times L \to R$ is a Lie cocycle

(iii) $S \cong R \#_\sigma U(L)$ as in Definition 1.7, where $\sigma : U(L) \times U(L) \to R$ is a Hopf cocycle.

proof: Following [BGR, Mc], we first make an observation concerning the ideal I in 2.7. Let $I_0$ be the ideal of $U(R^-)$ generated by the set $\mathcal{S}$; then $U(R)/ I_0 \cong R$. We claim that $I = I_0 U(\mathcal{L})$, the right ideal of $U(\mathcal{L})$ generated by $I_0$. For, choose $x \in L$, $a,b \in R$. Then

$$x(a.b - ab) = (a.b - ab)\, x + \delta_x(a).b - \delta_x(a)b + a.\, \delta_x(b) - a\, \delta_x(b)$$
$$\in (a.b - ab)\, x + \mathcal{S}$$

and

$$x.(1_{U(R)} - 1_R) = (1_{U(R)} - 1_R)\, x$$

Thus $L\, \mathcal{S} \subseteq \mathcal{S} L + \mathcal{S}$, and so $U(\mathcal{L})I_0 \subseteq I_0\; U(\mathcal{L})$, proving the claim.

(i) $\Rightarrow$ (ii). Assume $S = R * U(L)$ as in 2.3, and let $\tilde{L}$ denote the image of L in S. By 2.3, (iii) $R \cap \tilde{L} = (0)$ ; by (i) and (ii) $\mathcal{L} = R \oplus \tilde{L}$ is a Lie algebra. It follows that $\tau : L \times L \to R$ is a Lie cocycle satisfying (2.4) and (2.5). Moreover $\mathcal{L}$ generates S. Thus by the universal property of the enveloping algebra U(L), there exists a homomorphism $\psi : U(\mathcal{L}) \to S$ which is surjective and such that $\psi|_{\mathcal{L}} = \mathrm{id}$. Let I be the ideal of $U(\mathcal{L})$ as constructed in 2.7; it will suffice to prove that $I = \mathrm{Ker}\, \psi$. Since $\psi(U(R)) = R$, clearly $I \subseteq \mathrm{Ker}\, \psi$ and $I_0 = \mathrm{Ker}\, \psi \cap U(R)$. However, PBW gives that $U(\mathcal{L})$ is free over U(R) with basis the standard monomials in a basis for $\tilde{L}$. Since by hypothesis S is free over R with the same basis (or the image of the same basis), it follows that $\mathrm{Ker}\, \psi$ is generated by $\mathrm{Ker}\, \psi \cap U(R) = I_0$. But then $\mathrm{Ker}\, \psi = I$.

(ii) $\Rightarrow$ (iii). Assume $S = R \times_\tau U(L)$ as in 2.7. We then have an exact sequence of Lie algebras $R^- \times_\tau L \xrightarrow{\pi} L \to 0$ with kernel $R^-$. It follows from 2.1, since the exact sequence of Hopf algebras $U(R^- \times_\tau L) \xrightarrow{\pi} U(L) \to 0$ has a coalgebra splitting, that we may write

$$U(R^- \times_\tau L) \cong U(R) \#_\sigma U(L)\,,$$

where $\sigma : U(L) \times U(L) \to U(R)$ is a Hopf cocycle. The isomorphism $\psi : U(R^- \times_\tau L) \to U(R) \#_\sigma U(L)$ is described in the proof of 1.11 [BCM 4.14] ; note that by (1.14), $\psi(a) = a \# 1$ for all $a \in \mathrm{LH\,Ker}\, \pi = U(R)$.

From the claim at the beginning of the proof, $I = I_0\, U(R^- \times_\tau L)$ ; thus $\psi(I) = \psi(I_0)\, \psi(\, U(R^- \times_\tau L)\,) = I_0 \#_\sigma U(L)$. It follows that

$$R \times_\tau U(L) = U(R^- \times_\tau L)/ I \cong U(R) \#_\sigma U(L) / I_0 \#_\sigma U(L)$$
$$\cong (U(R) / I_0) \#_{\bar{\sigma}} U(L) \cong R \#_{\bar{\sigma}} U(L)\,.$$

Here $\bar{\sigma} : U(L) \times U(L) \to R$ denotes $\sigma$ under the homomorphism. It is clear that properties 1.9(1) and (2) are preserved in this homomorphism, and thus $R \#_{\bar{\sigma}} U(L)$ is a Hopf algebra crossed product with cocycle $\bar{\sigma}$.

(iii) ⇒ (i) Assume that $S = R \#_\sigma U(L)$. For each $x \in L$, write $\tilde{x} = 1 \# x$, so that $x \to \tilde{x}$ is a vector space embedding of L into S. We first check (i) and (ii) of 2.3, using the definition of multiplication in (1.8) and the fact that $\Delta x = x \otimes 1 + 1 \otimes x$. For (i),

$$\begin{aligned}\tilde{x}r - r\tilde{x} &= (1 \# x)(r \# 1) - (r \# 1)(1 \# x)\\ &= 1\,(1\cdot r)\# x + 1\,(x\cdot r)\# 1 - r \# x\\ &= r \# x + \delta_x(r) \# 1 - r \# x\\ &= \delta_x(r),\end{aligned}$$

since any $x \in L$ must act as a derivation of R.

For (ii),

$$\begin{aligned}\tilde{x}\,\tilde{y} &= (1 \# x)(1 \# y) = \sigma(x, y) \# 1 + \sigma(x, 1)\# y + \sigma(1, y)\# x + \sigma(1, 1)\# xy\\ &= 1 \# xy + \sigma(x, y) \# 1.\end{aligned}$$

Thus

$$\begin{aligned}[\tilde{x}, \tilde{y}] &= 1 \# (xy - yx) + [\sigma(x, y) - \sigma(y, x)] \# 1\\ &= \widetilde{[x, y]} + \tau(x, y),\end{aligned}$$

where $\tau(x, y) = \sigma(x, y) - \sigma(y, x) \in R$.

Finally, for (iii), since $S \cong R \otimes U(L)$ as vector spaces, S is free over $R = R \otimes 1$ with basis any basis for $1 \otimes U(L)$. In particular, the set of all $1 \# m$, where m is a standard monomial in some fixed basis $\{x_\alpha\}$ of L, form such a basis. Using the relation $1 \# xy = \tilde{x}\,\tilde{y} - \sigma(x,y)$ from above, it is not difficult to see (by induction) that the standard monomials in $\{\tilde{x}_\alpha\}$ form a free basis for S over R. Thus $S = R * U(L)$ in the sense of 2.3.

## §3. CONNECTIONS BETWEEN LIE COCYCLES AND HOPF COCYCLES

We have seen in §2, 2.8 that a Lie extension $R^- \times_\tau L$ with Lie cocycle $\tau$ will give rise to a U(L) crossed product $R \#_\sigma U(L)$ with Hopf cocycle $\sigma$. In this section we consider the relationship between $\sigma$ and $\tau$. In one direction this is easy: the "symmetrization" of $\sigma$ gives a Lie cocycle $\tau$. The more interesting direction is the following: given $\tau$ on L, can we explicitly compute $\sigma$ on U(L)? We will see that this is indeed possible.

We remark that if R is commutative and an L-module, the relationship between L-extensions of $R^-$ and U(L) Hopf extensions of R has been known for a long time. Sweedler defined a cohomology theory for algebras over Hopf algebras and proved [S68, Theorem 4.3]

$$H^n(L, R^-) \cong H^n(U(L), R),$$

all $n \geq 1$. A similar statement, with the right hand side denoting Hochschild cohomology, was known much earlier [CE, XIII, 5.1]; and, as mentioned previously, non-abelian extensions of L were studied in [H]. However in all of these results the passage from L-cocycles to U(L)-cocycles was done via existence proofs. An explicit way of obtaining $\sigma$ from $\tau$, as in our 3.6, does not seem to appear in the literature.

Our first lemma is straightforward and presumably well-known.

3.1 LEMMA. Let $R \#_\sigma U(L)$ be a crossed product. Then $R \otimes 1 + 1 \otimes L$ is a Lie extension of L with cocycle $\tau : L \times L \to R$ given by, for all x, y ∈ L,

$$\tau(x, y) = \sigma(x, y) - \sigma(y, x)$$

proof: For simplicity, we identify R with $R \otimes 1$. As before, if x ∈ L, denote $1 \# x \in 1 \otimes L$ by $\tilde{x}$, and $1 \otimes L$ by $\tilde{L}$. Thus we must show that $R + \tilde{L}$ is a Lie algebra. At the end of the proof of 2.8, it was shown that, for all x, y ∈ L, r ∈ R,

$$\tilde{x}r - r\tilde{x} = x \cdot r = \delta_x(r) \in R \text{ and } [\tilde{x},\tilde{y}] = \widetilde{[x,y]} + \tau(x,y).$$

We can now see directly that $R + \tilde{L}$ is a Lie algebra, extending L with cocycle τ. For, if x, y ∈ L, r, s ∈ R,

$$\begin{aligned}[r + \tilde{x}, s + \tilde{y}] &= [r,s] + [\tilde{x},s] + [r,\tilde{y}] + [\tilde{x},\tilde{y}] \\ &= [r,s] + \delta_x(s) - \delta_y(r) + \tau(x,y) + \widetilde{[x,y]}.\end{aligned}$$

This is precisely (2.6); of course the Jacobi identity is satisfied since $R \#_\sigma U(L)$ is a Lie algebra under [ , ]. Since $R + \tilde{L}$ is a Lie algebra, (2.4) and (2.5) must hold.

Alternatively, (2.4) and (2.5) could have been shown directly. (2.5) follows from $[\tilde{x}, \tilde{y}] = \widetilde{[x, y]} + \tau(x, y)$ via the adjoint action, and (2.4) follows from 1.9(1): first substitute x, y, z ∈ L for h, ℓ, m, obtaining

$$\sigma(xy, z) = x \cdot \sigma(y, z) + \sigma(x, yz).$$

Summing this relation over the cyclic permutations of (x, y, z) gives exactly (2.4).

Going from τ to σ is considerably more difficult. We first make a definition.

3.2 DEFINITION. Fix an ordered basis $\{x_\alpha\}$ for L, so that the standard monomials in $\{x_\alpha\}$ form a basis for U(L). A bilinear map $\sigma : U(L) \times U(L) \to R$ is called order preserving with respect to the basis $\{x_\alpha\}$ if

$$\sigma(x_{\alpha_1} \cdots x_{\alpha_n}, x_{\beta_1} \cdots x_{\beta_m}) = 0$$

whenever $x_{\alpha_1} \leq \cdots \leq x_{\alpha_n} \leq x_{\beta_1} \leq \cdots \leq x_{\beta_m}$.

We use the notation $\hat{x}_i$ to mean that the term $x_i$ is omitted from a monomial. Thus if $x_1, \ldots, x_m \in L$,

$$\Delta(x_1 \cdots x_m) = \sum_{1 \leq i_1 < \cdots < i_r \leq m} (\cdots \hat{x}_{i_1} \cdots \hat{x}_{i_r} \cdots) \otimes x_{i_1} \cdots x_{i_r} \tag{3.3}$$

3.4 LEMMA. Let $\sigma : U(L) \times U(L) \to R$ be order-preserving with respect to the ordered basis $\{x_\alpha\}$ of L, and assume that σ satisfies the cocycle condition 1.9(1).

<u>Choose</u> $x_i, y_i, y$ <u>in the basis of</u> L <u>so that</u> $x_1 \le \cdots \le x_m, y_1 \le \cdots \le y_n$, and $y \le x_m$. <u>Then</u>

(1) $\sigma(x_1 \cdots x_m, y) =$

$$\sigma(x_1 \cdots x_{m-1}, [x_m, y]) + \sigma(x_1 \cdots x_{m-1} y, x_m) + (x_1 \cdots x_{m-1}) \cdot \sigma(x_m, y)$$

(2) $\sigma(x_1 \cdots x_m, y_1 \cdots y_n) =$

$$\sum_{\substack{1 \le i_1 < \cdots < i_r \le m \\ 1 \le j_1 < \cdots < j_s \le n-1}} (\cdots \hat{x}_{i_1} \cdots \hat{x}_{i_r} \cdots, \cdots \hat{y}_{j_1} \cdots \hat{y}_{j_s} \cdots)\, \sigma(x_{i_1} \cdots x_{i_r} y_{j_1} \cdots y_{j_s}, y_n)$$

<u>proof</u> We first show (2), as it is easier; it follows directly from the cocycle condition by setting $h = x_1 \cdots x_m$, $\ell = y_1 \cdots y_n$, and $m = y_n$. The right hand side is immediate, using 3.3. For the left hand side, since $\sigma$ is order-preserving, $\sigma(\ell_{(1)}, m_{(1)}) = 0$ unless $\ell_{(1)} = m_{(1)} = 1$. In that case $h_{(1)} \cdot \sigma(1, 1) = h_{(1)} \cdot 1 = 0$ unless also $h_{(1)} = 1$. Thus there is only one surviving term, namely

$$[1 \cdot \sigma(1, 1)]\, \sigma(h, \ell m) = \sigma(x_1 \cdots x_m, y_1 \cdots y_n).$$

To prove (1), we will first prove

(*) $\sigma(x_1 \cdots x_m, y) = \sigma(x_1 \cdots x_{m-1}, x_m y) + (x_1 \cdots x_{m-1}) \cdot \sigma(x_m, y)$.

Interchange the left and right hand sides of the cocycle condition and set $h = x_1 \cdots x_{m-1}, \ell = x_m$, and $m = y$. Since $\sigma$ is order-preserving and $x_1 \le \cdots \le x_m$, $\sigma(h_{(1)}, \ell_{(1)}) = 0$ unless $h_{(1)} = \ell_{(1)} = 1$. Thus

$$\Sigma_{(h)(\ell)}\, \sigma(h_{(1)}, \ell_{(1)})\, \sigma(h_{(2)} \ell_{(2)}, m) = \sigma(1, 1)\sigma(h\ell, m) = \sigma(x_1 \cdots x_m, y)$$

On the other side, using $\sigma(x_m, 1) = 0 = \sigma(1, x_m)$ and $h_{(1)} \cdot \sigma(1, 1) = 0$ unless $h_{(1)} = 1$, and $\sigma(h_{(2)}, 1) = 0$ unless $h_{(2)} = 1$, we obtain

$$\Sigma_{(h)(\ell)(m)} [h_{(1)} \cdot \sigma(\ell_{(1)}, m_{(1)})]\, \sigma(h_{(2)}, \ell_{(2)} m_{(2)})$$

$$= \sum_{(h)} [h_{(1)} \cdot \sigma(x_m, y)]\, \sigma(h_{(2)}, 1) + 1 \cdot \sigma(1, 1)\, \sigma(h, x_m y)$$

$$= (x_1 \cdots x_{m-1}) \cdot \sigma(x_m, y) + \sigma(x_1 \cdots x_{m-1}, x_m y)$$

proving (*).

Since $x_m y = [x_m, y] + y x_m$, to finish (1) it suffices to prove that $\sigma(x_1 \cdots x_{m-1}, y x_m) = \sigma(x_1 \cdots x_{m-1} y, x_m)$. This again follows from the cocycle condition, by setting $h = x_1 \cdots x_{m-1}$, $\ell = y$, and $m = x_m$. Since $y \le x_m$, $\sigma(y, x_m) = 0$, and thus

$$\Sigma_{(h)(\ell)(m)}[h_{(1)} \cdot \sigma(\ell_{(1)}, m_{(1)})]\, \sigma(h_{(2)}, \ell_{(2)} m_{(2)}) = \sigma(h, yx_m)\,.$$

If $h_{(1)} \ne 1$, then $\sigma(h_{(1)}, \ell_{(1)}) = 0$ unless $\ell_{(1)} = y$. But then $\ell_{(2)} = 1$, and so $\sigma(h_{(2)}\ell_{(2)}, m) = \sigma(h_{(2)}, x_m) = 0$ since $\sigma$ is order-preserving. Thus $\sigma(h_{(1)}, \ell_{(1)})\, \sigma(h_{(2)}\ell_{(2)}, m) = 0$ unless $h_{(1)} = 1$. But then $\sigma(1, \ell_{(1)}) = 0$ unless $\ell_{(1)} = 1$. It follows that

$$\Sigma_{(h)(\ell)}\, \sigma(h_{(1)}, \ell_{(1)})\, \sigma(h_{(2)}\ell_{(2)}, m) = \sigma(1, 1)\, \sigma(h\ell, m) = \sigma(x_1 \cdots x_{m-1}y, x_m).$$

This proves (1) .

3.5 PROPOSITION. Let $R \#_\sigma U(L)$ be a crossed product constructed from a Lie extension $R^{-}\times_\tau L$ of L with Lie cocycle $\tau$, as in 2.7 and 2.8. Let $\{x_\alpha\}$ be a fixed ordered basis for L. Assume

(1) $\sigma$ is order-preserving with respect to the given basis

(2) if $x_1, \ldots, x_m$ are basis elements with $x_1 \le \cdots \le x_m$, then for all $r \in R$,

$$(x_1 \cdots x_m) \cdot r = x_1 \cdot (x_2 \cdots \cdot (x_m \cdot r) \cdots)$$

(3) $\sigma(x, y) - \sigma(y, x) = \tau(x, y)$, for all $x,y \in R$.

Then for all $a, b \in U(L)$, $\sigma(a, b)$ can be computed inductively from the values of $\tau$ on L and the action of L on R.

proof: It suffices to assume that a and b are standard monomials in the $\{x_\alpha\}$. Consider basis elements $x_1,\ldots,x_m$ with $x_1 \le \cdots \le x_m$ , and $y, y_1,\ldots,y_n$ with $y_1 \le \cdots \le y_n$.

We first show that $\sigma(x_1 \cdots x_m, y)$ can be computed, by induction on m. If $m = 1$, then $\sigma(x_1, y) = 0$ if $x_1 \le y$, and $\sigma(x_1, y) = \tau(y, x_1)$ if $x_1 > y$, using (1) and (3). For $m > 1$, we use 3.4(1). Since $[x_m, y] \in L$, $\sigma(x_1 \cdots x_{m-1}, [x_m, y])$ can be computed by induction on m. Similarly, using (2) and the case $m = 1$, $(x_1 \cdots x_{m-1}) \cdot \sigma(x_m, y)$ can be found from the action of L. Only the middle term remains. By (1) we may assume $y < x_m$ . But then $x_1 \cdots x_{m-1}y = w + u$, where $w = x_1 \cdots y \cdots x_{m-1}$ is a standard monomial and u is a sum of monomials of degree $< m$. Thus $\sigma(x_1 \cdots x_{m-1}y, x_m) = \sigma(w, x_m) + \sigma(u, x_m) = \sigma(u, x_m)$ , which can be computed by induction on m.

We have shown the n = 1 step, for all m, of finding $\sigma(x_1 \cdots x_m, y_1 \cdots y_n)$ by induction on n. For n > 1 we use 3.4(2). The rightmost factor can be done by the n = 1 case above, and the left factor (on the right hand side) has right entry $\cdots \hat{y}_{j1} \cdots \hat{y}_{js} \cdots$ of degree $\leq n - 1$, which is then determined by induction on n.

3.6 THEOREM. Let R be an algebra on which the Lie algebra L acts (weakly) as derivations, and let $\mathcal{L} = R^{-}\times_{\tau}L$ be a Lie extension of L with kernel R and Lie cocycle $\tau$. Let $R \times_{\tau} U(L)$ be the crossed product constructed from $R^{-}\times_{\tau}L$ as in 2.7, and let $\bar{\sigma}$ denote the Hopf cocycle in the isomorphism $R \times_{\tau} U(L) \cong R \#_{\bar{\sigma}} U(L)$ of 2.8. Then $\bar{\sigma} : U(L) \times U(L) \to R$ can be computed inductively from the values of $\tau$ on L and the action of L on R.

proof: It suffices to show that $\bar{\sigma}$ satisfies the three hypotheses of 3.5. First, since $R \#_{\bar{\sigma}} U(L)$ was obtained as a homomorphic image of $U(R) \#_{\sigma} U(L)$, and the three hypotheses are preserved under this homomorphism, it suffices to prove them for the preimage $\sigma$ of $\bar{\sigma}$ and the algebra $U(R) \#_{\sigma} U(L)$. We recall how $\sigma$ was constructed in 2.1: choose an ordered basis $\{x_\alpha\}$ for L, define $\gamma : L \to R^{-}\times_{\tau}L$ to be a vector space embedding such that $\pi\gamma = \mathrm{id}$, where $\pi : R^{-}\times_{\tau}L \to L$ is the projection, and then define $\gamma$ on any standard monomial in the $\{x_\alpha\}$ by $\gamma(x_{\alpha_1} \cdots x_{\alpha_n}) = \gamma(x_{\alpha_1}) \cdots \gamma(x_{\alpha_n})$. In particular $\gamma$ is multiplicative on the factors of a standard monomial. Finally $\sigma$ was defined via (1.13):

$$\sigma(w, z) = \Sigma_{(w)(z)}\, \gamma(w_{(1)})\, \gamma(z_{(1)})\ \ S\gamma(w_{(2)}\, z_{(2)})\ .$$

We can now prove (1), that is, $\sigma$ is order-preserving. For if $w = x_{\alpha_1} \cdots x_{\alpha_n}$ and $z = x_{\beta_1} \cdots x_{\beta_m}$ are standard monomials with $\alpha_n \leq \beta_1$, then by the above $\gamma(w_{(1)})\, \gamma(z_{(1)}) = \gamma(w_{(1)}\, z_{(1)})$. But then $\sigma(w, z) = \gamma * S\gamma(wz)$, where * denotes convolution. Since S is the inverse of the identity under * , $\gamma$ is the inverse of $S\gamma$ under *, so $\gamma * S\gamma = \varepsilon$. Thus $\sigma(w, z) = \varepsilon(wz) = 0$.

For (2) of 3.5, assume that $x_1,\ldots,x_m$ are basis elements with $x_1 \leq \cdots \leq x_m$. By (1.12), the action of U(L) on U(R) is given by $w \cdot a = (\mathrm{ad}\, \gamma(w))a$, for $w \in U(L)$, $a \in U(R)$. Thus $(x_1 \cdots x_m) \cdot a = (\mathrm{ad}\, \gamma(x_1 \cdots x_m))\, a = (\mathrm{ad}\, \gamma(x_1) \cdots \gamma(x_m))a = x_1 \cdot (x_2 \cdots \cdot (x_m \cdot a) \cdots)$.

Finally we show (3) of 3.5. By correct choice of $\gamma : L \to R\times_{\tau}L$ , we recover the cocycle $\tau$ via

$$\tau(x, y) = [\gamma(x), \gamma(y)] - \gamma[x, y]$$

as in the discussion preceeding (2.4). Since $\gamma[x,y] \in \mathfrak{L} = R\times_\tau L$, $S\gamma[x,y] = -\gamma[x, y]$. Thus $\tau(x, y) = [\gamma(x), \gamma(y)] + S\gamma[x, y]$.

We now use (1.13) to evaluate $\sigma$ in terms of $\gamma$:

$$\begin{aligned}\sigma(x, y) &= \gamma(x)\,\gamma(y)\,S\gamma(1) + \gamma(1)\,\gamma(y)\,S\gamma(x) + \gamma(x)\,\gamma(1)\,S\gamma(y) + \gamma(1)\,\gamma(1)\,S\gamma(xy)\\ &= \gamma(x)\,\gamma(y) - [\gamma(y), \gamma(x)] + \gamma(y)\,S\gamma(x) + \gamma(x)\,S\gamma(y) + S\gamma(yx) - S\gamma[y,x]\\ &= \sigma(y, x) + [\gamma(x), \gamma(y)] + S\gamma[x, y]\\ &= \sigma(y, x) + \tau(x, y)\,.\end{aligned}$$

This proves the theorem.

In fact the methods used in proving the theorem - in particular Lemma 3.4 and the proof of 3.5 - give an algorithm for computing $\sigma$. We illustrate this with some computations for the Heisenberg algebra.

3.7 EXAMPLE. Let L be the Heisenberg algebra, with basis $\{p,q,z\}$ satisfying $[p, q] = z$, $[p, z] = [q, z] = 0$. Let L act on an algebra R as derivations so that $\mathfrak{L} = R^{-}\times_\tau L$ is a Lie extension with given Lie cocycle $\tau$. Order the basis of L by declaring $p < q < z$. In $R\,\#_\sigma U(L)$ we will compute $\sigma(pz, q)$, $\sigma(q^2, p)$, $\sigma(pq^2, p)$, and $\sigma(pq, pq)$.

First observe that since we may assume $\sigma$ is order-preserving, $0 = \sigma(p, q) = \sigma(pq, z) = \sigma(p^2q, q)$; also by property (3), $\sigma(q, p) = \tau(q, p) + \sigma(p, q) = \tau(q, p)$. Now

1) $\sigma(pz, q) = \sigma(p, [z, q]) + \sigma(pq, z) + p\cdot\sigma(z, q)$ using 3.4(1)
$\quad = -p\cdot\tau(q, z)$

2) $\sigma(q^2, p) = \sigma(q, [q, p]) + \sigma(qp, q) + q\cdot\sigma(q, p)$
$\quad = \sigma(q, -z) + \sigma(-z + pq, q) + q\cdot\tau(q, p)$
$\quad = -\sigma(z, q) + q\cdot\tau(q, p)$
$\quad = \tau(q, z) + q\cdot\tau(q, p)$

3) $\sigma(pq^2, p) = \sigma(pq, [q, p]) + \sigma(pqp, q) + p\cdot(q\cdot\sigma(q, p))$
$\quad = \sigma(pq, -z) + \sigma(p[q, p] + p^2q, q) + p\cdot(q\cdot\sigma(q, p))$
$\quad = 0 - \sigma(pz, q) + 0 - p\cdot(q\cdot\tau(p, q))$
$\quad = p\cdot\tau(q, z) - p\cdot(q\cdot\tau(p, q))$ using 1) .

4) An example using 3.4(2) is

$$\begin{aligned}\sigma(pq, pq) = {} & \sigma(p, p)\,\sigma(q, q) + \sigma(p, 1)\,\sigma(qp, q) + \sigma(q, p)\,\sigma(p, q) + \sigma(q, 1)\,\sigma(p^2, q)\\ & + \sigma(pq, 1)\,\sigma(p, q) + \sigma(pq, p)\,\sigma(1, q) + \sigma(1, p)\,\sigma(pq, q)\end{aligned}$$

$+ \sigma(1,1)\, \sigma(pqp, q)$

$= \sigma(pqp, q) = p \cdot \tau(q,z)$ as in 3)

A specific example of a Lie cocycle $\tau$ for an action of L can be given as follows. Consider the Lie algebra $\mathfrak{L}$ with basis {x,y,u,v} and relations [x, y] = u, [y, u] = v, [x, u] = 0, and v central. Then $\mathfrak{L}$ / <v> $\cong$ L, a non-split extension, where $\pi : \mathfrak{L} \to L$ is given by $\pi(x) = p$, $\pi(y) = q$, $\pi(u) = z$. L acts trivially on R = k[v], polynomials in v, but the cocycle $\tau$ is non-trivial. For, let $\gamma : L \to \mathfrak{L}$ be given by $\gamma(p) = x$, $\gamma(q) = y$, $\gamma(z) = u$. It follows that $\tau(q, z) = [\gamma q, \gamma z] - \gamma[q, z] = [y, u] - 0 = v$, although $\tau(p, q) = \tau(p, z) = 0$. In the illustration 2) above, it follows that $\sigma(q^2, p) = \tau(q, z) + 0 = v$.

## REFERENCES


[BCM] R.J. Blattner, M. Cohen, and S. Montgomery, "Crossed products and inner actions of Hopf algebras", Trans. Amer. Math. Soc. 298(1986), 671-711.

[BGR] W. Borho, P. Gabriel, and R. Rentschler, Primideale in Einhüllenden auflösbarer Lie-Algebren, Lecture Notes in Math. vol 357, Springer-Verlag, Berlin, 1974.

[CE] H. Cartan and S. Eilenberg, Homological Algebra, Princetorn University Press, Princeton, 1956.

[Ch] W. Chin, "Prime ideals in differential operator rings and crossed products of infinite groups," J. Algebra 106 (1987), 78-104.

[H] G. Hochschild, "Lie algebra kernels and cohomology", American J. Math. 76 (1954), 698-716.

[L] M. Lorenz, "Primitive ideals in crossed products and rings with finite group actions", Math. Z. 158 (1978), 285-294.

[Mc] J.C. McConnell, "Representations of solvable Lie algebras and the Gelfand-Kirillov conjecture," Proc. London Math. Soc. 29 (1974), 453-484.

[McR] J.C. McConnell and J.C. Robson, Noncommutative Noetherian Rings, Wiley, New York, 1987.

[MP] S. Montgomery and D.S. Passman, "Crossed products over prime rings," Israel J. Math. 31 (1978), 224-256.

[S68] M.E. Sweedler, "Cohomology of algebras over Hopf algebras", Trans. Amer. Math. Soc. 133 (1968), 205-239.

[S69] M.E. Sweedler, Hopf algebras, Benjamin, New York, 1969.

# PROGRESS ON SOME PROBLEMS ABOUT GROUP ACTIONS

S. Montgomery
University of Southern California
Los Angeles, CA 90089-1113

At the 1983 Antwerp conference, I raised a number of open questions concerning group actions on rings; these problems were discussed in detail in [M84]. In the intervening four years, progress has been made on a number of these questions, and that is what I wish to report on now. I also hope that this report will stimulate interest in the remaining unsolved problems.

Usually R denotes an algebra with 1 over a commutative ring k and G denotes a finite group of order $|G|$ acting as k-automorphisms of R. Spec R denotes the prime spectrum of R, and $\mathrm{Spec}_m$ R denotes the maximal spectrum.

Problem 1: If $|G|^{-1} \in k$, is R (Schelter) integral over $R^G$?

No progress. Recall it is known to be true if either R satisfies a polynomial identity or if G is abelian.

Problem 2: If P, Q $\in$ Spec R such that $P \cap R^G = Q \cap R^G$, does $Q = P^g$ for some $g \in G$ if either (a) R is semiprime PI or (b) R is prime and G is X-outer?

Note that the "orbit problem", as it was called, was known to be true if R was commutative (or Azumaya), if $|G|^{-1} \in R$, or if R was a PI algebra and either P or Q was identity faithful, although it was known to be false if R was PI but not semiprime.

The two specific cases have been answered in the negative by D.S. Passman, with the same example: let $A = k[x_1,\ldots,x_p]$ be the commutative polynomial ring in p variables over a field k of characteristic $p \neq 0$, and let $M = (x_1,\ldots,x_p)$ be the maximal ideal. Let $\sigma \in \mathrm{Aut}_k A$ permute the $x_i$ cyclically; so $\sigma$ has order p.

Now let $R = \begin{pmatrix} A & A \\ M & A \end{pmatrix}$. $\sigma$ extends to R by letting it act on each entry.

Let $\tau \in \mathrm{Aut}(R)$ be conjugation by $\begin{pmatrix} 1 & 1 \\ 0 & 1 \end{pmatrix}$. Then $\sigma\tau = \tau\sigma$ is an automorphism of R of order p; it is X-outer since it moves the center of R. Let $G = \langle \sigma\tau \rangle$.

Let $P = \begin{pmatrix} M & A \\ M & A \end{pmatrix}$ and let $Q = \begin{pmatrix} A & A \\ M & M \end{pmatrix}$. Then P and Q are G-stable primes of R



*F. van Oystaeyen and L. Le Bruyn (eds.), Perspectives in Ring Theory, 269–274.*

with $P \cap R^G = Q \cap R^G$.

Moreover both R. Guralnick and C.L. Huang have pointed out that Passman's example can be "lifted" to a ring R of characteristic 0. Details of these examples can be found in the expository paper [M 87].

Problem 3: Study properties of primes to determine which are or are not "equivalence invariants" in Spec $R^G$.

For this problem, we are assuming $|G|^{-1} \in R$, and say that p, q $\in$ Spec $R^G$ are equivalent if there exists $P \in$ Spec R such that both p and q are minimal over $P \cap R^G$. This is a well-defined equivalence relation, and an "equivalence invariant" is simply a property common to an equivalence class. Of course when R is commutative each class consists of a single prime $p = P \cap R^G$; in general a class may contain up to $|G|$ primes.

Previously known equivalence invariants were height of p, primitivity of $R^G/p$, GK dimension of $R^G/p$ (provided R is PI), and classical Krull dimension of $R^G/p$ (provided R is affine PI or Noetherian PI). In general classical Krull dimension is not an equivalence invariant.

Recently D.S. Passman and the author [MP 87] have shown that for the same R and G, whether or not a property is an equivalence invariant depends on the action of G. In particular if $R = k<x_1,...,x_d>$ is the free algebra and V the vector space with k-basis $\{x_1,...,x_d\}$, we may consider $G \subset GL(V)$ as automorphisms of R. If $|G|^{-1} \in$ k and G acts as scalars on V, then $R^G/p$ being finite dimensional or PI is an equivalence invariant for Spec $R^G$. This is false whenever G does not act as scalars on V.

Problem 4: If $|G|^{-1} \in R$, is $f_m$ a closed map?

Here $f_m : \mathrm{Spec}_m R \rightarrow \mathrm{Spec}_m R^G$ is the map on the maximal spectra given by $f_m(M) = \{m \in \mathrm{Spec}_m \mid m \supseteq M \cap R^G\}$. It was known that the analogous maps on the prime or primitive spectra were always closed, and thus that $f_m$ was closed if R was a PI algebra.

The answer to Problem 4 is no, in general [MP87]. In fact let $R = k<x_1, x_2, x_3>$ be the free algebra on three generators over a field k of characteristic 0, and let $G = <g> \cong \mathbb{Z}_2$ act as scalars on R by $x_i^g = -x_i$, all i. Then $f_m$ is not closed.

One might still ask whether Problem 4 has a positive answer provided R satisfies some other finiteness condition, such as (left) Noetherian or finite Gelfand-Kirillov dimension.

Problem 5: What is the connection between Problems 1 and 4?

Since in the counterexample for Problem 4, R is integral over $R^G$, it is certainly false that integrality implies the closed map property. The converse is open, but since integrality itself is open this may not be a useful question.

As in 4, one might re-ask the question under suitable finiteness conditions; Problem 5 was motivated originally by work of Artin and Schelter showing that for extensions of affine PI algebras, integrality is closely related to the map on maximal spectra being closed.

Problems 6 and 7: Let R be a Noetherian domain which is affine over a field k. Is $R^G$ Noetherian? Is $R^G$ affine over k?

No progress. Recall it is true if $|G|^{-1} \in k$.

Problem 8: If R is simple Noetherian and G is outer, is $R^G$ Noetherian?

The general problem remains open. However it has now been shown by T. Hodges and J. Osterburg to be true for the Zaleskii- Neroslavskii example [HO]. Their arguments have been extended to more general situations in [LoP].

Problem 9: Let R be semiprime Goldie with $|G|^{-1} \in R$, and let $M_R$ be a right R-module of Goldie rank n. Does $M_{R^G}$ have Goldie rank $\leq n|G|$?

Recall that the answer was yes if M = R or if R was semi-simple Artinian. However, recently Lance Small has given a counterexample to the general case.

Let S = k<x,y>, the free algebra on two variables over a field of char. $\neq 2$, and let D be a division ring containing S. Let t be an indeterminate over D and consider S + tD[t], which is a left and right Ore domain. Form

$$R = \begin{pmatrix} S + t\,D[t] & D[t] \\ t\,D[t] & D[t] \end{pmatrix}$$

R is prime Goldie on the right and left, as it contains $M_2(t\,D[t])$. Let G = <g>, where g denotes conjugation by $\begin{pmatrix} 1 & 0 \\ 0 & -1 \end{pmatrix}$. Then $R^G = \begin{pmatrix} S + t\,D[t] & 0 \\ 0 & D[t] \end{pmatrix}$. Define the right R-module by $M_R = R/\ M_2(t\,D[t]) \cong \begin{pmatrix} S & D \\ 0 & D \end{pmatrix}$.

$M_R$ has Goldie rank n = 2, but $M_{R^G}$ has infinite Goldie rank since S = k<x,y> has infinite Goldie rank.

The question remains open for the case of R an Ore domain. Also, the question may still be true if the module M is torsion-free.

Problem 10: Let $U = k\{X_1,\ldots,X_d\}$ be the ring of d generic $m \times m$ matrices, $d \geq 2$. Consider $G \subset GL(V)$ acting as automorphisms of U by extending the action from the vector space V with k-basis $\{X_1,\ldots,X_2\}$. Assume that $|G|^{-1} \in k$ and that G does not consist of scalar matrices.

Prove that for $m \geq 3$, $U^G$ is not finitely-generated.

No progress on this problem. It was already known that if G was cyclic, with $|G| = n$, then $U^G$ was not finitely generated for $m \geq n - [\sqrt{n}] + 1$.

The motivation for Problem 10 was that for large enough matrices compared to |G|, U should behave like the free algebra.

There has been progress on the case m = 2, however. When m = d = 2, E. Formanek and A. Schofield [FS] have proved that $U^G$ is finitely-generated if $G \subset SL(V)$; in particular this is true for G = <g>, $g = \begin{pmatrix} i & 0 \\ 0 & -i \end{pmatrix}$, a case which was mentioned in [M84] as being open.

More recently C. Huh has answered the question completely when $m = d = 2$ and G is abelian: in that case $U^G$ is finitely-generated if and only if G contains no pseudo-reflections [H].

Since $SL_2$ contains no pseudo-reflections, it is likely that this criterion is the right one for any G in the case $m = d = 2$. There are no known finitely-generated examples if either $m > 2$ or $d > 2$.

Problem 11: Determine $Aut_k U$, for the generic matrix ring $U = k\{X,Y\}$.

Although the problem remains unsolved, some new facts are known. Recall that Bergman had given an example of a non-tame automorphism of U; recently Alev and LeBruyn have shown that Bergman's automorphism (as well as some others) becomes tame on the center of the trace ring of U, when $m = 2$[AL].

Problem 12: Find an internal characterization of the subrings S of a generic matrix ring U which are anti-ideals of U.

No progress. Recall that S is an anti-ideal of R if whenever $a,b \in R$ with $ab \in S$ and $0 \neq b \in S$, then $a \in S$. Such subrings are precisely those which arise as fixed rings $R^G$, when R is a domain and G is X-outer. Thus the problem concerns trying to describe possible fixed rings $U^G$.

Problem 13: Compute the generalized "characteristic invariant" for some specific group actions.

No progress. This invariant is an analog for prime rings and X-inner automorphisms of one of the invariants used by V. Jones in classifying all finite group actions on the hyperfinite $II_1$ factor.

Problems 14 and 15: For a given group action $\alpha : G \to Aut_k(A)$ on a k-algebra A, compute Spec $(\alpha)$. More generally, find an algebraic analog of the Connes spectrum.

As in the previous problem, these ideas come from operator algebras, where they have been very useful. First consider the case where G is finite abelian, $|G|^{-1} \in k$, and k is a field containing all $n^{th}$ roots of unity, where n is the period of G. For each $\lambda \in \hat{G}$, the dual group of G, let $A_\lambda = \{a \in A \mid a^g = \lambda(g)a, \text{ for all } g \in G\}$ Then we may write $A = \sum_{\lambda \in \hat{G}} \oplus A_\lambda$. The spectrum of $\alpha$ is defined by $\text{Spec}(\alpha) = \{\lambda \in \hat{G} \mid A_\lambda \neq 0\}$.

For a $C^*$-algebra, the Connes spectrum is defined by $\Gamma(\alpha) = \bigcap_B \text{Spec}(\alpha|_B)$,

where the intersection runs over all G-stable "hereditary" C*-subalgebras B of A. The term hereditary here is not the usual algebraic notion but is defined analytically in terms of positive elements. Forunately it can be shown that B is hereditary if and only if B is *-stable and $B = R \cap L$, where R (resp. L) is a norm closed right (left) ideal of A (we note that in a von Neumann algebra M, W*-closure is required and thus hereditary subalgebras are replaced by the "corners" $B = eMe$, where $e^* = e$). According to a theorem of Olesen and Pedersen, the skew group ring A*G is prime if and only if A is G-prime and $\Gamma(\alpha) = \text{Spec}(\alpha)$.

Recently D.S. Passman and the author [MP88] have found an algebraic analog of $\Gamma(\alpha)$ and of the Olesen-Pedersen theorem. We define a subalgebra B of A to be hereditary if B = RL ≠ 0, where R (resp. L) is a right (left) ideal of A; then $\Gamma(\alpha)$ is defined as above. We prove that if A is prime, then A*G is prime if and only if $\Gamma(\alpha) = \hat{G} =$ Spec $(\alpha)$.

If G is not abelian, our result generalizes to algbras graded by G, rather than to an action of G as automorphisms. For an action $\alpha : G \to \mathrm{Aut}_k(A)$, a different approach will be needed. One possibility is the following: when G is abelian, $|G|^{-1} \in k$, and k is algebraically closed, each $\lambda \in \hat{G}$ can be identified with an irreducible representation of G, and conversely.

This suggests the following definition. Assume $|G|^{-1} \in k$, and let Irr(G) denote the set of irreducible representations of G. Since $\alpha : G \to \mathrm{Aut}\, A$ is a representation of G, A is a completely reducible G-module, so we may write $A = \Sigma \oplus A_\rho$, where $A_\rho$ is the direct sum of all the irreducible G-subspaces corresponding to $\rho \in \mathrm{Irr}(G)$. Now define Spec $(\alpha) = \{ \rho \in \mathrm{Irr}(G) \mid A_\rho \neq (0)\}$. Does an analog of the Olesen-Pedersen theorem hold for G non-abelian using this definition of Spec $(\alpha)$?

## References


[AL] J. Alev and L. Lebruyn, Automorphisms of generic 2 × 2 matrices, preliminary version.

[FS] E. Formanek and A.H. Schofield, Groups acting on the ring of two 2 × 2 generic matrices and a coproduct decomposition of its trace ring, Proc. AMS 95(1985), 179-183.

[HO] T.J. Hodges and J. Osterburg, A rank two indecomposable projective module over a Noetherian domain of Krull dimension one, Bull. LMS 19(1987), 139-144.

[Hu] C. Huh, Invariants of finite abelien groups acting on the algebra of two 2 × 2 generic matrices, Proc. AMS, to appear.

[LoP] M. Lorenz and D.S. Passman, The structure of $G_0$ for certain polycyclic group algebras and related algebras, to appear.

[M84] S. Montgomery, Group actions on rings: some classical problems, in Methods in Ring Theory (ed. F. van Oystaeyen), Reidel, 1984, 327-346.

[M87] ____________, Prime ideals and group actions in noncommutative algebras, Proceedings of the Denton conference on invariant theory, AMS Contemporary Math, to appear.

[MP87] S. Montgomery and D.S. Passman, Prime ideals in fixed rings of free algebras, Comm. in Alg., to appear.

[MP88] ______________, Algebraic analogs of the Connes spectrum, J. Algebra, to appear.

# A Note on the *PI*-Property of Semigroup Algebras

Jan Okninski
University of Warsaw
00-901 Warsaw, Poland

Let $S$ be a semigroup. We say that $S$ has the property $\mathcal{P}_n, n \geq 2$, if for any $a_1, \ldots, a_n \in S$ there exist $\sigma \neq 1$ in the symmetric group $S_n$ such that $a_1 a_2 \ldots a_n = a_{\sigma(1)} a_{\sigma(2)} \ldots a_{\sigma(n)}$. Further, $S$ is said to have the permutational property $\mathcal{P}$ if $S$ has $\mathcal{P}_n$ for some $n \geq 2$. Semigroups of this type were first studied in [10], ($S$-periodic), [2], (groups), implicitly in [4] ($S - O$-simple) and recently in [5].

It is straightforward that if the semigroup algebra $K[S]$ over a field $K$ satisfies a polynomial identity, then $S$ has the permutational property. The question to what extent the converse holds was suggested in [10]. While the results of [2] show that this may not be the case for the class of (not finitely generated) groups, a positive answer has been obtained for a number of classes of finitely generated semigroups, [10], [7], [8], [9]. The aim of this note is to construct a finitely generated semigroup $S$ with the property $\mathcal{P}$ such that $K[S]$ is not a $PI$-algebra for any field $K$.

**Proposition.** Let $T$ be the completely $O$-simple semigroup with the Rees presentation $M^0(\{1\}^o, Z, Z; P)$ where $Z$ is the set of integers and the sandwich matrix $P = (p_{ij})$ is given by $p_{ii} = 0$, $p_{ij} = 1$ for $i, j \in Z$, $i \neq j$, (cf. [1], §3.2). Then $T$ has the property $\mathcal{P}$.

**Proof.** Observe that $T$ may be identified with the semigroup $\{a_{ij} | i, j \in Z\} \cup \{0\}$ with the multiplication rules

$$a_{ij} a_{kl} = \begin{cases} a_{il} & \text{if } \quad j \neq k \\ 0 & \text{if } \quad j = k \end{cases}, \qquad a_{ij} 0 = 0 a_{ij} = 0$$



*F. van Oystaeyen and L. Le Bruyn (eds.), Perspectives in Ring Theory, 275–278.*

We will check that $T$ has the property $\mathcal{P}_{10}$. (Some more computations show that $T$ has $\mathcal{P}_6$ but not $\mathcal{P}_5$). Assume that some elements $x_1, \ldots, x_{10} \in T$ are given.

The following obvious properties of $T$ will be used :

i. If $y_1 \ldots y_k = 0$ for some $y_1, \ldots, y_k \in T$, then there exists $i \in \{1 \ldots, k-1\}$ with $y_i y_{i+1} = 0$.

ii. $xy = 0$, $x, y \in T$, if and only if there exists $n \in Z$ such that $x$ lies in the $n$-th column and $y$ lies in the $n$-th row of $T$.

iii. If $y_1 \ldots y_k \neq 0$. $z_1 \ldots z_r \neq 0$ for some $y_i, z_j \in T$, such that $y_1, z_1$ lie in the same row and $y_k, z_r$ lie in the same column of $T$, then $y_1 \ldots y_k = z_1 \ldots z_k$.

Here, by the $n$-th row (column) of $T$ we mean the set $\{a_{ni} | i \in Z\}(\{a_{in} | i \in Z\}$ repectively).

If $x_1 \ldots x_{10} = 0$, then **i.** allows to find a nontrivial permutation of $x$'s with the zero product. Thus, suppose that $x_1 \ldots x_{10} \neq 0$ and $x_1 \ldots x_{10} \neq x_{\sigma(1)} \ldots x_{\sigma(10)}$ for any $1 \neq \sigma \in S_{10}$. Assume first that for some $n \in Z$ there exist three integers $r, s, t \in \{1, \ldots, 10\}$, $r < s < t$, such that the elements $x_r, x_s, x_t$ lie in the same row of $T$. Write $x_1 \ldots x_{10} = x_1 \ldots x_{r-1} y_1 y_2 y_3$ where $y_1 = x_r \ldots x_{s-1}$, $y_2 = x_s \ldots x_{t-1}, y_3 = x_t \ldots x_{10}$. Then $y_1, y_2, y_3$ lie in the same row and consequently $(x_1 \ldots x_{r-1}) y_2 \neq 0, y_2 y_1 \neq 0$, $y_1 y_3 \neq 0$. Thus $x_1 \ldots x_{10} = x_1 \ldots x_{r-1} y_2 y_1 y_3$ contradicting our supposition. Hence, for any $n \in Z$, there are at most 2 $x$'s lying in the $n$-th row of $T$. A similar argument shows that there are at most 2 $x$'s in the $n$-th column of $T$.

Let $j < 10$ be maximal with respect to the property $x_j x_2 \neq 0$. From the foregoing it follows that $j \geq 7$ since at most 2 $x$'s may annihilate $x_2$ on the left. Let us consider all elements of the form

$$(*) \qquad x_1 x_i x_{i+1} \ldots x_j x_2 x_3 \ldots x_{i-1} x_{j+1} \ldots x_{10}$$

where $i \in \{2, 3, \ldots, j\}$. We always have $x_i x_{i+1} \ldots x_j x_2 x_3 \ldots x_{i-1} \neq 0$. Further, $x_1 x_i = 0$ for at most 2 integers $i \in \{2, 3, \ldots, j\}$ and $x_{i-1} x_{j+1} = 0$ for at most 2 integers $i-1, i \in \{2, 3, \ldots, j\}$. Since $j \geq 7$, it then follows that at least two elements of the form (*) are non-zero. Now, $x_1 \ldots x_{10}$ is one of them and the property (iii) implies that this product is "permutable". This contradiction completes the proof.

Let $R = M^o(G^o, I, M, P)$ be an arbitrary completely $O$-simple semigroup. One could expect that there is a finiteness condition $\mathcal{F}$ in the class of sandwich matrices such that $R$ has the property $\mathcal{P}$ if and only if $G$ has $\mathcal{P}$ and $\mathcal{F}$ holds for $P$. Finiteness of some suitable notions of rank plays such a role for the $PI$-property of $K[R]$, ([3], cf. [6]). Our example shows that the situation differs with respect to the property $\mathcal{P}$. In fact, let $P_n$ be the submatrix of $P$ defined by $P_n = (p_{ij})_{i,j\in\{1,\dots,n\}}$. Since $P_n$ is invertible as a matrix in $M_n(K)$ if $n-1 \neq 0$ in $K$, then it follows that the rank of $P$ is infinite (regardless the characteristic of $K$). Consequently, $K[T]$ is not a $PI$-algebra, [3], [6].

**Lemma.** (c.f. [5]). let $I$ be an ideal of a semigroup $S$. Then $S$ has the property $\mathcal{P}$ if and only if $I, S/I$ have the property $\mathcal{P}$.

**Proof.** The necessity is clear. Assume that $I$ has $\mathcal{P}_r$ and $S/I$ has $\mathcal{P}_m$ for some $r, m > 1$. Put $n = rm$ and let $x_1, \dots, x_n$ be any elements of $S$. If there exists $i \in \{0, \dots, n-m\}$ such that $x_{i+1} \dots x_{i+m} \in I$, then for some $1 \neq \sigma \in S_m, x_{i+1} \dots x_{i+m} = x_{i+\sigma(1)} \dots x_{i+\sigma(m)}$. Consequently, $x_1 \dots x_n = x_1 \dots x_i x_{i+\sigma(1)} \dots x_{i+\sigma(m)} x_{i+m+1} \dots x_n$ and we are done. On the other hand, if any product of the subsequent $m$ elements of $\{x_1, \dots, x_n\}$ lies in $I$, then putting $y_i = x_{(i-1)m+1} \dots x_{mi}$, $i = 1, 2, \dots, r$ we may use the property $\mathcal{P}_r$ in $I$. Then $x_1 \dots x_n = y_1 \dots y_r = y_{\tau(1)} \dots t_{\tau(r)}$ for some $1 \neq \tau \in S_r$, yielding the assertion in this case.

Now, we are ready to complete our construction.

**Example.** Let $S$ be the monoid generated by $\{T, y, z\}$ subject to the relations :

$$ya_{ij} = a_{i+1,j}, za_{ij} = a_{i-1,j},$$
$$a_{ij}y = a_{i,j-1}, a_{ij}z = a_{i,j+1}, zy = 1 = yz \text{ for } i, j \in Z$$

It is clear that $S$ is finitely generated, for example $T = < a_{00}, y, z >$. Moreover, $T$ is an ideal of $S$ with the quotient $S/T$ isomorphic to the infinite cyclic group with zero adjoined. Hence $S/T$ has $\mathcal{P}_2$ and so, by the lemma, $S$ has the permutational property. Clearly, $K[S]$ is not a $PI$-algebra for any field $K$ since $K[T]$ is not a $PI$-algebra.

**References.**


[1] Clifford, A. H., Preston, G. B., *The Algebraic Theory of Semigroups*, Vol. 1, Providence, 1964.

[2] Curzio, M., Longobardi, P., May, M., Robinson D.J.S., *A permutational property of groups*, Arch. Math. 44(1985), 385-389.

[3] Domanov, O. I., *On identities of semigroup rings of completely O-simple semigroups*, Mat. Zametki 18 (1975), 203-212.

[4] Domanov, O. I. , *Identities of semigroup rings of O-simple semigroups*, Sib. Math. J. 18(1976), 1406-1407.

[5] Justin, J., Pirillo, G.,*Comments on the permutation property for semigroups*, preprint.

[6] Okninski, J., *On semigroup algebras satisfying polynomial identities*, Semigroup Forum 33 (1986), 87-102.

[7] Okninski J., *On monomial algebras*, Arch. Math., to appear.

[8] Okninski, J., *On cancellative semigroup rings*, Comm. Algebra, 15 (1987), 1667-1677.

[9] Okninski, J., *Noetherian property for semigroup rings*, Proceedings of the 1986 Granada Conference on Ring theory, to appear.

[10] Restivo, A., Reutenauer C., *On the Burnside problem for semigroups*, J. Algebra 89(1984), 102-104.

# Invariant fields of linear groups and division algebras

DAVID J. SALTMAN*
Department of Mathematics
University of Texas at Austin
Austin, Texas 78727

**Abstract.** In this paper we show how the invariant fields of reductive groups can be described as multiplicative invariant fields of their Weyl groups. Using this, we show how some such invariant fields are described by Brauer Severi varieties. For example, if $SL_n$ is the special linear group and $G_r$ is the quotient of $SL_n$ by the central cyclic subgroup of order $r$, an invariant field of $G_r$ is stably isomorphic to the center of the "generic division algebra of degree $n$ and exponent $r$."

## Introduction

Let $F$ be an algebraically closed field of characteristic 0 and $G$ a reductive linear algebraic group defined over $F$. Assume $V$ is a finite dimensional $G$ representation over $F$, which we further assume has an element with trivial $G$ stabilizer. Let $F[V]$ be the symmetric (polynomial) algebra on $V$ and $F(V)$ its field of fractions. Then the action of $G$ on $V$ extends naturally to an action on $F(V)$. We denote by $F(V)^G$ the field of $G$ invariant elements. Note immediately that by a result of Bogomolov (see 2.1) $F(V)^G$ is, up to stable isomorphism, independent of the choice of $V$. This allows us to talk about the invariant field of $G$ without specifying $V$. The fields $F(V)^G$ are the subject of this paper.

In this paper we will detail some connections between invariant fields of reductive linear groups and division algebras. To motivate this connection, let $UD(F,n,r)$ be the generic division algebra over $F$ of degree $n$ in $r$ variables (where $r \geq 2$). Set $Z(F,n,r)$ to be the center of $UD(F,n,r)$. Procesi has shown that $Z(F,n,r)$ can be described both as an invariant field of the projective linear group $PGL_n(F)$ and a "multiplicative" invariant field of the symmetric group $S_n$. The goal of this paper is to demonstrate how this connection between invariant field, division algebras, and multiplicative invariant fields extend to other linear groups besides $PGL_n$. We describe this connection using these three concepts plus the notion of Brauer Severi varieties.

The noteworthy results of this paper come in sections 2 and 3, after some remarks about "twisted" multiplicative invariant fields in section 1. In section 2, invariant fields of reductive groups are shown to be, up to stable isomorphism, "twisted" multiplicative invariant fields of the Weyl group (Corollary 2.7). Though the twisting can often be dispensed with, the spin groups are a prominent example where (so far) this is impossible. Also in section 2 is a general

* The author is grateful for support under NSF grant DMS-8601279.



*F. van Oystaeyen and L. Le Bruyn (eds.), Perspectives in Ring Theory, 279–297.*

relationship between invariant fields of isogeneous groups and Brauer Severi varieties (Theorem 2.11). In section 3, several infinite classes of semisimple groups are considered and a specific description (in terms of Brauer Severi varieties) of their invariant fields is given. Perhaps the most noteworthy examples concern the images of the special linear group $SL_n$. To be precise, let $C_n \subseteq SL_n$ be the center of $SL_n$. Of course, $C_n$ is cyclic of order $n$. Let $r$ be a positive integer dividing $n$, and let $C_r \subseteq C_n$ be the subgroup of order $r$. Set $G_r = SL_n/C_r$, so $G_n = PGL_n$. The invariant field of $G_r$ is stably isomorphic to the function field of the Brauer Severi variety of $A/Z$, where $Z = Z(F, n, r)$ and $A$ is a power of $UD(F, n, r)$ (Theorem 3.2). It follows that the invariant field of $G_r$ is stably isomorphic to the center of the "generic division algebra of degree $n$ and exponent $r$". It also follows that if $n$ is even, the invariant fields of $G_2$, $PO_n$, and $PSp_n$ are all stably isomorphic.

Before we begin in earnest, let us recall some definitions and specify some notation. $F$ will always be an algebraically closed field of characteristic 0. All rings and fields will be $F$ algebras and all morphism between such will be $F$ morphisms. If $W$ is a group acting on an $F$ algebra $R$, this action will be assumed to be an $F$ algebra action. In particular, $W$ always acts trivially on $F$. If $R$ is an $F$ algebra, $R^*$ will denote the group of units of $R$. If $R$ is a domain, $q(R)$ will denote the field of fractions of $R$. A field extension $K \subseteq L$ is called *rational* if $L$ has a transcendence basis $\{x_1, \cdots, x_n\}$ over $K$ such that $L = K(x_1, \cdots, x_n)$. Two fields $K$, $L$ are called stably isomorphic if they have extension fields $K' \supseteq K$, $L' \supseteq L$ such that $K'/K$ and $L'/L$ are rational and $K' \cong L'$.

If $W$ is a group, a $W$ module is any module over the integral group ring $Z[W]$. A $W$ representation over a field $L$ is a module over the group ring $L[W]$. If $G$ is an algebraic group, all finite dimensional $G$ representations will have algebraic actions by $G$. Now let $W$ be a finite group. A $W$ lattice is a $W$ module which is finitely generated free as a $Z$ module. A permutation $W$ lattice is a $W$ lattice $P$ with a $Z$ basis $\{x_i\}$ such that $\{x_i\}$ is closed under the $W$ action. If $H \subseteq W$ is a subgroup, then $Z[W/H]$ is the permutation lattice with $Z$ basis isomorphic, as a $W$ set, to the set of left cosets of $H$ in $W$. The element of $Z[W/H]$ corresponding to $H$ itself is called the *canonical generator.* If $M$, $N$ are two $W$ modules, then $\mathrm{Ext}_W(M, N)$ is just the first Ext group $\mathrm{Ext}^1_{Z[W]}(M, N)$.

If $L$ is a field and $V$ a finite dimensional $L$ vector space, we set $L[V]$ to be the symmetric algebra and $L(V) = q(L[V])$. If $M$ is a lattice (over $W$ say) we set $L[M]$ to be the group algebra and $L(M) = q(L[M])$. This makes sense because $L[M]$ has the form of a localized polynomial ring $L[x_1, \cdots, x_n](1/x_1 \cdots x_n)$. Note that no logical confusion between $L(M)$ and $L(V)$ is possible, because a lattice is never a vector space and vice versa. On the other hand, $L(M)$ and $L(V)$ are less similar than the notation suggests; $M$ embeds as a multiplicative subgroup of $L(M)$ but $V$ embeds in $L(V)$ additively.

Finally, if $A$ is a central simple algebra finite dimensional over its center $K$ we will say that $A/K$ is central simple. The degree of $A$ or $A/K$ is the square root of the dimension of $A$ over $K$. We denote by $[A]$ the equivalence class $A$ in the Brauer group $Br(K)$. The exponent of $A$ is the order of $[A]$. If $A/K$ is central simple of degree 2, then $A$ is generated by some $\alpha$, $\beta \in A$ such that $\alpha\beta = -\beta\alpha$. If $a = \alpha^2$, $b = \beta^2$ then $0 \neq a$, $b \in K^*$ and $a$, $b$ determine $A$. We write $A = (a, b)_K$ or just $(a, b)$. And really finally, if $A/K$ is central simple, we denote by $K(A)$ the function field of the Brauer Severi variety defined by $A$ (*e.g.*

[Ar]) or equivalently, the generic splitting field of $A$ ([A]).

## Section One: Multiplicative Invariants

In section two the invariants of reductive groups will be shown to be stably isomorphic to certain multiplicative invariants of the Weyl groups. As compared to the multiplicative field invariants defined and studied in [S1], we will have to consider certain "twisted" multiplicative actions. The purpose of this section is to define and study these multiplicative invariant fields.

In what follows, $W$ will always be a finite group. Let $L$ be field with a $W$ action and $M$ a $W$ lattice. Assume $\alpha$ is an extension $0 \to L^* \to M_\alpha \to M \to 0$ of $W$ modules. Recall the definitions of $L(M)$ and $L[M]$ from the introduction. Using $\alpha$ we define an *$\alpha$-twisted* action of $W$ on $L(M)$ as follows. As abelian groups, $M_\alpha \cong L^* \oplus M$. Fix an embedding $M \subseteq M_\alpha$. Using this embedding, we can define an isomorphism $\exp : M_\alpha \cong (L[M])^* = L^*M$. In $M$ and $M_\alpha$ we will write the operation additively, while the operation in $(L[M])^*$ is written multiplicatively. Thus exp is sort of a "formal" exponential map. For each $\sigma \in W$, $\sigma$ induces $\phi'_\sigma : M \to M_\alpha$ by setting $\phi'_\sigma(m) = \sigma(m)$. By the universal property of group algebras, $\phi_\sigma = (\exp)\phi'_\sigma$ extends to an algebra homomorphism $\phi_\sigma : L[M] \to L[M]$. It is easy to check that $\phi_\sigma\phi_\tau = \phi_{\phi\tau}$ so the $\phi$'s define a $W$ action on $L[M]$, which induces a $W$ action on $L(M)$. This is the "$\alpha$-twisted" action. For simplicity we write $\phi_\sigma(f) = \sigma(f)$. Note that if $m \in M$, $\sigma(\exp(m)) = \exp(m)f$ for some $f \in L^*$.

Another choice of embedding $M \subseteq M_\alpha$ yields an isomorphic action of $W$. Thus the $\alpha$ twisted action is well defined up to isomorphism. We define $L_\alpha(M)^W$ to be the invariant field with respect to this action.

Assume now that $L/K$ is $W$ Galois. Let $U$ be a finite dimensional $L$ vector space with a so called semilinear action of $W$. That is, if $u \in L$, $\sigma \in W$, and $f \in L$, then $\sigma(fu) = \sigma(f)\sigma(u)$. Form $L[U]$ and $L(U)$ as in the introduction. It was observed by Endo and Miyata ([EM] p.16) that:

**Lemma 1.1.** *$L(U)^W$ is rational over $K$.*

As a consequence of the above, we derive a result about the fields $L_\alpha(M)^W$. Let $0 \to L^* \to M_\alpha \to M \to 0$ and $0 \to L^* \to N_\beta \to N \to 0$ be two extensions (called $\alpha$ and $\beta$) of $W$ where $M, N$ are $W$ lattices. Assume there is an embedding $\phi : M_\alpha \to N_\beta$ such that $\phi$ is the identity on $L^*$. Denote the cokernel of $\phi$ by $P$. Use $\phi$ to embed $L[M]$ in $L[N]$ and so $L(M)$ in $L(N)$. Clearly the $\beta$-twisted action of $W$ on $L(N)$ is an extension of the $\alpha$ twisted action on $L(M)$. Thus we can identify $L_\alpha(M)^W$ with a subfield of $L_\beta(N)^W$.

**Corollary 1.2.** *If $P$ is a permutation lattice, and the $\alpha$ twisted action of $W$ on $L(M)$ is faithful, then $L_\beta(N)^W$ is rational over $L_\alpha(M)^W$.*

*Proof.* Set $L' = L(M)$ viewed as a subfield of $L(N)$. Let $x_1, \cdots, x_n \in P$ be a $Z$ basis permuted by $W$. Choose preimages $x'_1, \cdots, x'_n \in N_\beta$ of the $x_i$'s. Set $y_i = \exp(x'_i) \in L(N)$. It is easy to see that the $y_i$ form a transcendence basis for $L(N)$ over $L'$. Furthermore, if $\sigma \in W$, $\sigma(y_i) = fy_i$ for some $f \in L'$. It follows that if $V = L'y_1 + \cdots + L'y_n$, then $V$ is closed under the ($\beta$-twisted) action of $W$ on $L(N)$, and this action is semilinear over $L'$. As $L(N) = L'(V)$ the result follows from 1.1. Q.E.D.

If we specialize to $L = F$, $W$ has the trivial action on $F^*$. However, there are still nontrivial extensions of $F^*$ by $M$ and we can consider $F_\alpha(M)^W$. The multiplicative invariant fields that arise in section 2 have the form $F_\alpha(M)^W$ for $W$ the Weyl group.

The next few results will be devoted to showing how division algebras and Brauer Severi varieties arise in studying the fields $L_\alpha(M)^W$. As above, let $H \subseteq W$ be a subgroup. Fix a positive integer $n$. Define $W$ lattices $I[W/H]$, and $I_n[W/H]$ by the following diagram:

$$
\begin{array}{ccccccccc}
0 & \longrightarrow & I[W/H] & \longrightarrow & Z[W/H] & \stackrel{\varepsilon}{\longrightarrow} & Z & \longrightarrow & 0 \\
 & & \downarrow & & \downarrow{\scriptstyle \mathrm{Id}} & & \downarrow{\scriptstyle \pi} & & \\
0 & \longrightarrow & I_n[W/H] & \longrightarrow & Z[W/H] & \longrightarrow & Z/nZ & \longrightarrow & 0
\end{array}
\tag{7}
$$

where all rows are exact, $Z$ is the integers with the trivial $W$ action, $\varepsilon(\alpha) = 1$ for $\alpha$ a canonical generator of $Z[W/H]$, and $\pi$ is the natural surjection. Let $L/K$ be any $W$ Galois extension of fields.

**Lemma 1.3.**

1) $\mathrm{Ext}_W(I[W/H], L^*) \cong \ker(H^2(W, L^*) \to H^2(H, L^*))$
2) *Suppose* $\alpha \in \mathrm{Ext}(I[W/H], L^*)$ *maps to* $\beta \in H^2(W, L^*)$, $[A] \in Br(K)$ *corresponds to* $\beta$, *and* $A$ *has* $L^H$ *as a maximal subfield. Then* $L_\alpha(I[W/H], W)$ *is isomorphic to the generic splitting field* $K(A)$.
3) $\mathrm{Ext}_W(I_n[W/H], L^*)$ *is isomorphic to the subgroup of* $\mathrm{Ext}_W(I[W/H], L^*)$ *of elements of order dividing* $n$.
4) *Suppose* $\alpha \in \mathrm{Ext}_W(I_n[W/H], L^*)$ *corresponds to* $\beta \in H^2(W, L^*)$, *and* $A$ *is as in 2). Then* $L_\alpha(I_n[W/H], W)$ *is isomorphic to a purely transcendental extension of degree 1 of* $K(A)$.

*Proof.* By [B] p.61, $\mathrm{Ext}^i(Z[W/H], L^*) = H^i(H, L^*)$ and so $\mathrm{Ext}(Z[W/H], L^*) = (0)$. Applying the long exact sequence of Ext to (7) we have:

$$
\begin{array}{ccccccc}
 & & & & H^2(W, L^*) & & \\
 & & & & \uparrow{\scriptstyle s} & & \\
0 & \longrightarrow & \mathrm{Ext}(I[W/H], L^*) & \longrightarrow & H^2(W, L^*) & \stackrel{r}{\longrightarrow} & H^2(H, L^*) \\
 & & \uparrow & & \uparrow & & \uparrow{\scriptstyle \mathrm{Id}} \\
0 & \longrightarrow & \mathrm{Ext}(I_n[W/H, L^*) & \longrightarrow & \mathrm{Ext}^2(Z/nZ, L^*) & \longrightarrow & H^2(H, L^*) \\
 & & & & \uparrow & & \\
 & & & & \mathrm{Ext}(Z, L^*) = (0) & &
\end{array}
$$

where all rows are exact, and the middle vertical exact sequence arises from the exact sequence:

$$0 \longrightarrow Z \stackrel{n}{\longrightarrow} Z \longrightarrow Z/nZ \longrightarrow 0$$

and the identification of $\mathrm{Ext}^i(Z, L^*)$ with $H^i(W, L^*)$. A standard verification shows that $r$ is the restriction map and $s$ is the multiplication by $n$ map. Parts 1) and 3) follow immediately. As for 2), this is the description of $K(A)$ in [**Ro**]. Finally, to consider 4), let $\alpha \in \mathrm{Ext}(I_n[W/H], L^*)$ and let $\beta$ be its image in $\mathrm{Ext}(I[W/H], L^*)$. We have:

$$\begin{array}{ccccccccc}
0 & \longrightarrow & L^* & \longrightarrow & I[W/H]_\beta & \longrightarrow & I[W/H] & \longrightarrow & 0 \\
 & & \downarrow & & \downarrow & & \downarrow & & \\
0 & \longrightarrow & L^* & \longrightarrow & I_n[W/H]_\alpha & \longrightarrow & I_n[W/H] & \longrightarrow & 0 \\
 & & & & \downarrow & & \downarrow & & \\
 & & & & Z & & Z & & \\
 & & & & \downarrow & & \downarrow & & \\
 & & & & 0 & & 0 & &
\end{array}$$

where the top and bottom long rows represent $\beta$ and $\alpha$ respectively. By 1.1, $L_\alpha(I_n[W/H])^W$ is rational over $L_\beta(I[W/H])^W$ and 4) follows. Q.E.D.

## Section Two: Invariants of reductive groups

Let $G$ be a linear reductive connected algebraic group and $V$ a representation of $G$ which is finite dimensional over the ground field $F$. Assume that $V$ has a point with trivial $G$ stabilizer. Form the field $F(V) = q(F(V))$ and consider the invariant field of $G$ on $F(V)$ which we will denote by $F(V)^G$. The goal of this section is some general results about these invariant fields and, specifically, to show how division algebras and Brauer Severi varieties arise in the study of these invariant fields. We begin with a increasingly well known result which states that up to stable isomorphism, $F(V)^G$ is independent of the choice of $V$.

**Theorem 2.1 (Bogomolov [Bo]).** *Let $V$, $V'$ be two representations of $G$ with elements of trivial stabilizer. Then $F(V)^G$ and $F(V')^G$ are stably isomorphic.*

*Proof.* There is actually more than one proof of this result now in print ([**Sh**] as well as [**Bo**]). We give the following proof only because it is somewhat different from the others and so may give a different sort of insight into this very useful result.

It suffices to show that $F(V \oplus V')^G$ is rational over $F(V')^G$. Write $L = F(V')$, and let $W = LV \subseteq F(V \oplus V')$. It is easy to see that $W \cong L \otimes_F V$ and we use the isomorphism to identify $W$ with $L \otimes_F V$. In particular, $W$ is an $L$ vector space with dimension equal to the $F$ dimension, $n$, of $V$. If $\{y_1, \cdots, y_n\}$ is a $G$ invariant basis for $W$, then it is easy to see that the $y$'s form a transcendence basis for $F(V \oplus V')$ over $F(V')$ and so for $F(V \oplus V')^G$ over $F(V')^G$. Thus it suffices to construct this basis $\{y_i\}$. In other words, it suffices to construct $G$ invariant linearly independent $y_1, \cdots, y_n$.

There is a general method for constructing $G$ invariant elements we will now develop. Let $S$ be an $F$ algebra with a $G$ action. Assume that as a $G$ module,

$S \supseteq V_1$ where $V_1$ is isomorphic as a $G$ representation to $V^*$. In $S \otimes_F V$ there is a submodule $V_1 \otimes_F V$ isomorphic to $V^* \otimes_F V$ and the latter is isomorphic to End($V$). In End($V$) there is a $G$ invariant element Id which is the identity endomorphism. Let $y_1 \in S \otimes_F V$ be the corresponding element of $W$. We say $y_1$ is the $G$ invariant element associated with $S$ and $V_1$. We must find $n$ submodules $V_1, \cdots, V_n \subseteq L$, all isomorphic to $V^*$, so that the associated $y_1, \cdots, y_n \in L \otimes_F V' = W$ are linearly independent.

As $V$ has a point with trivial stabilizer, there is a $G$ invariant embedding $G \to V$. Thus there is a $G$ invariant surjection $T \to K$ where $T$ is a $G$ invariant localization of $F[V]$ and $K$ is the field of fractions of the affine ring of $G$. Assume we have constructed $V_1', \cdots, V_n' \subseteq K$, all isomorphic $V^*$, such that the associated $y_1', \cdots, y_n' \in K \otimes_F V'$ are linearly independent. Since $G$ is reductive, there are submodules $V_i \subseteq T$ such that the map $T \to K$ induces an isomorphism $V_i \cong V_i'$. Let $y_i \in T \otimes_F V$ be the elements associated with the $V_i$. It follows that the $y_i$ are linearly independent in $F(v) \otimes_F V$. It remains to construct the $V_i \subseteq K$.

As $V$ is a $G$ representation there is a $G$ invariant map $G \to GL(V)$. It follows that there is a $G$ invariant $R \to K$ where $R$ is the affine ring of $GL(V)$. $R$ has the form $F[x_{11}, \cdots, x_{nn}](1/f)$ where $f$ is the determinant of the $x$'s and the $x_{ij}$ correspond to a basis $v_1, \cdots, v_n$ of $V$. Set $V_1'' \subseteq R$ to be the $F$ span of $x_{i1}, \cdots, x_{in}$. Checking definitions shows that $V_i'' \cong V^*$, and the associated $y_i'' \in R \otimes_F V$ are just $y_i'' = (x_{i1} \otimes v_1) + \cdots + (x_{in} \otimes v_n)$. The $y_i''$ are linearly independent because $f$ is their determinant with respect to the basis $\{v_j\}$. If $V_i'$ is the image of $V_i''$ in $K$, and $y_i'$ are the associated elements in $K \otimes_F V$, then the $y_i'$ are the image of the $y_i''$. Hence the $y_i'$ are linearly independent and the result is proved. Q.E.D.

In this paper we are interested in studying the invariant fields $F(V)^G$ up to stable rationality. It will therefore make sense to choose for each group $G$ a $G$ representation $V$ with an element of trivial stabilizer and define $F(G)$ to be $F(V)^G$. Our results will state that $F(G)$ is stably isomorphic to some other field and thus will apply to $F(V)^G$ for any such $V$.

Let $V$ and $G$ be as above. Let $g$ be the Lie algebra of $G$, or in other language, the adjoint representation of $G$. Of course, $V \oplus g$, viewed as a $G$ representation, has an element with trivial stabilizer and it suffices to consider $F(V \oplus g)^G$. There is a standard trick for reducing $G$ invariants to invariants of subgroups which we will apply here. Let $T \subseteq G$ be a maximal torus, $N = N_G(T) \supseteq T$ the normalizer of $T$ in $G$, and $W = N/T$ the Weyl group, a finite group. This first lemma gives some well known but not easy to quote facts about $g$.

**Lemma 2.2.**

a) *There is a Zariski open $U \subseteq g$ such that for all $v \in U$, the centralizer $c_g(v)$ of $v$ in $g$ is a maximal toral subalgebra.*

b) *There is an open $U' \subseteq U \subseteq g$ such that for all $v \in U'$, the centralizer $C_G(v)$ of $v$ in $G$ (or equivalently the stabilizer $G_v$) is a maximal torus in $G$.*

*Proof.*

a) Let $z \subseteq g$ be the center of $g$. As $g$ is reductive, $z$ and $g/z$ are semisimple. Thus we may assume $g$ is semisimple. Choose $U$ such that for all $v \in U$, $c_g(v)$ has minimal dimension. Such a $U$ can be defined by the nonvanishing of coefficients of the characteristic polynomial of the adjoint operator adj($v$). By **[Bu]** p.23, $c_g(v)$ is a maximal toral subalgebra.

b) By **[H1]** p.89, the Lie algebra of $C_G(v)$ is $c_g(v)$. Thus the connected component $C_G(v)^o$ is abelian and semisimple if $v \in U$. In other language,$C_G(V)^o$ is a torus. Since $c_g(v)$ is maximal, $T = C_G(v)^o$ is a maximal torus. It remains to find an open $U' \subseteq U$ such that if $v \in U'$, $T = C_G(v)$.

For $v \in U$, the characteristic polynomial of adj($v$) has the form $t^m f(t)$ for $m = \dim(c_g(v))$. Define the open subset $U' \subseteq U$ by the condition that $f(t)$ has distinct roots. Let $v \in U$ and set $t = c_g(v)$. Recall that $g$ decomposes as $t \oplus (\oplus_\alpha g_\alpha)$ where the $\alpha$'s are the nontrivial roots of $t$ in $g$, and $g_\alpha = \{w \in g \mid \text{for all } v \in t, \operatorname{adj}(v)(w) = \alpha(v)w\}$ Also recall that $g_\alpha$ has dimension 1 for every nontrivial $\alpha$ (*e.g.* **[H1]** p.160). Since the $\alpha(v)$'s are the roots of $f(t)$, $U'$ is nonempty.

Now suppose $v \in U'$. If $\sigma \in C_G(v)$, then $\sigma$ normalizes $T = C_G(v)^o$ and so $\sigma \in N_G(T)$. If $\sigma \notin T$, $\sigma$ induces a nontrivial element of $W$. Now $W$ acts faithfully on the root system, so there is some root $\alpha$ such that $\sigma(\alpha) \neq \alpha$. But this is a contradiction since we have that $\alpha(v) \neq \sigma(\alpha)(v)$ and $\sigma(v) = v$. Thus $C_G(v) = T$. Q.E.D.

By a standard trick (compare **[LR]** p.489), the next result will reduce the invariant field of $G$ on $V \oplus g$ to an invariant field of $N = N_G(T)$. Let $T \subseteq G$ be a fixed maximal torus and $t \subseteq g$ the corresponding maximal toral subalgebra.

**Proposition 2.3.** $F(V \oplus g)^G \cong F(V \oplus t)^N$.

*Proof.* There is a natural map $\phi : F[V \oplus g]^G \to F[V \oplus t]^N$ induced by the restriction of a polynomial on $V \oplus g$ to $V \oplus t$. If $f \in F[V \oplus g]^G$ is in the kernel of $\phi$, then $f$ is $G$ invariant and $f(V \oplus t) = 0$. All maximal toral subalgebras are $G$ conjugate. Thus for a generic $v \in g$, there is a $\sigma \in G$ with $\sigma(v) \in t$. Hence for $v \in g$ generic, $f(w, v) = f(\sigma(w), \sigma(v)) = 0$ and so $f = 0$.

Thus there is an embedding $F(V \oplus g)^G \subseteq F(V \oplus t)^N$. We want this embedding to be an isomorphism. That is, we want the map of quotient spaces $(V \oplus t)/N \to (V \oplus g)/G$ to be generically finite of degree 1. Generically, closed points of both quotient spaces correspond to orbits (**[R]**). Consider an orbit $G(w, v)$ of $G$ on $V \oplus g$. For generic $(w, v)$ we can assume $v$ is in a maximal toral subalgebra. As these are all conjugate, we can assume that $v \in t$. The preimage of this orbit consists of all orbits $N(w', v') \subseteq G(w, v)$ with $v' \in t$. If $v$ is in the $U'$ defined in 2.2, and $\sigma \in G$ is such that $v' = \sigma(v) \in t$, then $C_G(\sigma(v)) = T$, $\sigma T \sigma^{-1} = T$ and so $\sigma \in N$. Thus $N(w', v')$ has the form $N(w'', v)$ and $w'' = \sigma(w)$ for $\sigma \in C_G(v) = T$. We conclude that $N(w'', v) = N(w, v)$ is the unique preimage and 2.3 is proved. Q.E.D.

Of course, $V \oplus t$ is an $N$ representation with an element of trivial $N$ stabilizer. By 2.1, $F(V \oplus t)^N$ is stably isomorphic to $F(V')^N$ for any finite dimensional $N$ representation with an element of trivial stabilizer. By choosing $V'$ wisely we can describe $F(V \oplus t)^N$ up to stable isomorphism.

Suppose $V'$ is an arbitrary $N$ representation, not necessarily having an element with trivial stabilizer. As a $T$ representation, $V'$ is a direct sum of irreducibles and so $V' \cong \oplus F v_i$ where for each $v_i$ there is a $\phi_i \in X(T) =$

$\mathrm{Hom}_{\mathrm{alg}}(T, F^*)$ with $\tau(v_i) = \phi_i(\tau)v_i$ for all $\tau \in T$. We call $V'$ *good* if we can choose the $v_i$ such that for all $\sigma \in N$, $\sigma(Fv_i) = Fv_j$ for some $j$. Such a set of $v_i$ we will call a *good basis* for $V'$. The $v_i$ will be a good basis if, for example, the $\phi_i$ are all distinct.

Assume now that $V'$ is good, with good basis $\{v_i\}$. We delay until later the (quite trivial) construction of good representations. Form the function field $F(V')$. In $F(V')^*$, let $P'$ be the multiplicative subgroup generated by $F^*$ and the $v_i$'s. By assumption, $P'$ is closed under the action of $N$. If $P = P'/F^*$, $P$ has a trivial $T$ action and so is a $W$ lattice. In fact, $P$ is a permutation $W$ lattice with basis the image of the $v_i$'s. Define $f' : P' \to X(T)$ by setting $f(v_i) = \phi_i$ and $f'(F^*) = 0$. In general, $f'(\alpha) = \phi$ if $\tau(\alpha) = \phi(\tau)\alpha$ for all $\tau \in T$ and so $f'$ is an $N$ morphism. $f'$ is surjective if and only if $V'$ has an element with trivial stabilizer. $f'$ induces a $W$ morphism $f : P \to X(T)$. Let $M \subseteq P$ be the kernel of $f$ and $M' \subseteq P'$ the kernel of $f'$. By definition, $T$ acts trivially on $M'$ and so there is an induced $W$ action on $M'$. Consider the exact sequence:

$$(1) \qquad 0 \longrightarrow F^* \longrightarrow M' \longrightarrow M \longrightarrow 0$$

of $W$ modules. Denote the extension (1) by $\alpha$. We call $M$ and $\alpha$ the $W$ lattice and extension induced by $V'$. Earlier, we defined an invariant field $F_\alpha(M)^W$ associated to such a lattice and extension, and we are about to observe that $F(V')^N$ is precisely $F_\alpha(M)^W$. However, for future use, we must state the following result in a bit more generality. Let $L$ be a field on which $W$ acts and define an action of $N$ on $L$ by letting $T$ act trivially. Then there is an induced action of $N$ on $L(V')$.

**Proposition 2.4.**

a) $L(V')^N \cong L_\alpha(M)^W$.

b) *There is an exact sequence of $W$ lattices:*

$$(2) \qquad 0 \longrightarrow M \longrightarrow P \longrightarrow X(T) \longrightarrow 0$$

*where $P$ is a permutation $W$ lattice.*

*Proof.* Part b) follows from the definition of $M$. As for a), let $L' \subseteq L(V')$ be the subfield generated by all monomials in the $v_i$ occurring in $M'$. It is easy to see that $L'$ is isomorphic to $L(M) = q(L[M])$ (as defined in section one) and that the induced action of $W$ on $L(M)$ is just the $\alpha$ twisted action. Since $L(V')^M = L((V')^T)^W$, it suffices to show that $L(V')^T = L'$.

That $L' \subseteq L(V')^T$ is clear. Assume $f = k/h \in L(V')^T$. Note that $L[V']$ is a polynomial ring and hence a UFD. We can therefore write $f = k/h$ where $k$, $h$ have no common factors. If $\tau \in T$, $\tau(f) = f$ so $\tau(h)k = \tau(k)h$. Hence $k$ divides $\tau(k)$ and $h$ divides $\tau(h)$. But each of the pairs have equal degrees so there is a $\psi \in X(T)$ with $\tau(h) = \psi(\tau)h$ and $\tau(k) = \psi(\tau)k$. It follows that $k = \Sigma k_i$ and $h = \Sigma h_i$ where the $k_i$'s and $h_i$'s are monomials in $P'$ and $\tau(k_i) = \psi(\tau)k_i$ and $\tau(h_i) = \psi(\tau)h_i$ for all $i$. It suffices to show $k_i/h \in L'$. But $1/(k_i/h) = \Sigma h_j/k_i \in L'$ and we are done. Q.E.D.

We must, of course, construct good $N$ representations. We will, in fact, show that for any permutation lattice $P$ and any $W$ map $f : P \to X(T)$, $f$ is induced by a good $N$ representation. We begin by observing that though

$$1 \longrightarrow T \longrightarrow N \longrightarrow W \longrightarrow 1 \tag{3}$$

does not split, there is a more limited splitting result that does hold. Let $\phi \in X(T)$ be a character and let $W_\phi \subseteq W$ be its stabilizer under the $W$ action. Then (3) restricts to an extension

$$1 \longrightarrow T \longrightarrow N_\phi \longrightarrow W_\phi \longrightarrow 1$$

and this induces an extension:

$$1 \longrightarrow T/\ker(\phi) \longrightarrow N_\phi/\ker(\phi) \longrightarrow W_\phi \longrightarrow 1 \tag{4}$$

**Lemma 2.5.** *The exact sequence (4) splits as a sequence of $W$ modules.*

*Proof.* First consider the case $G$ is semisimple. Since $\phi$ has a conjugate which is a dominant weight, there is a $G$ representation $U$ such that $U_\phi$ has dimension one ([**H1**] p.192). If we write $U_\phi = Fv$, then for each $\sigma \in N_\phi$, $\sigma(v) = fv$ for some $f \in F^*$. But $\phi$ is nontrivial so there is a $\tau \in T$ with $\phi(\tau) = 1/f$. In other words, for each $\sigma' T \in W_\phi$, there is a representative $\sigma \in N$ such that $\sigma(v) = v$. Furthermore, $\sigma$ is unique module $\ker\phi$. It is easy to check that the map $\sigma' T \to \sigma(\ker\phi)$ is the required splitting.

If $G$ is only reductive, then $G$ is an image of $G' \times T'$ where $G'$ is semisimple and $T'$ is a torus. Given the result for $G'$, it follows for $G' \times T'$ and then for $G$. Q.E.D.

The next result is our promised construction of good $N$ representations.

**Proposition 2.6.** *Suppose $P$, $Q$ are permutation $W$ lattices.*

a) *If $f : P \to X(T)$ is a $W$ morphism, there is a good representation $V_f$ inducing $f$.*

b) *Suppose there is a commutative diagram:*

$$\begin{array}{ccc} P & \xrightarrow{f} & X(T) \\ {\scriptstyle j}\downarrow & & \downarrow{\scriptstyle \mathrm{Id}} \\ P \oplus Q & \xrightarrow{g} & X(T) \end{array} \tag{5}$$

*where Id is the identity and $j$ is the natural inclusion. Then there are good $N$ representations $V_f \subseteq V_g$ and good bases $\{v_i\} \subseteq V_f$, $\{w_j\} \subseteq V_g$ such that the following holds. First, $\{v_i\} \subseteq \{w_j\}$, and furthermore, $V_f$, $V_g$ and these bases induce a diagram:*

$$\begin{array}{ccccccc} 0 & \longrightarrow & M' & \longrightarrow & P' & \xrightarrow{f'} & X(T) \\ & & {\scriptstyle h}\downarrow & & {\scriptstyle j'}\downarrow & & \downarrow{\scriptstyle \mathrm{Id}} \\ 0 & \longrightarrow & N' & \longrightarrow & (P \oplus Q)' & \xrightarrow{g'} & X(T) \end{array} \tag{6}$$

*where $j'$ is induced by the inclusion of $\{v_i\}$ in $\{w_j\}$; $P'$ and $(P \oplus Q)'$ are extensions of $P$ and $P \oplus Q$ by $F^*$; $h$ is the identity on $F^*$; and the right half of (6), modulo $F^*$, is just (5).*

*Proof.*

a) Let $\alpha_1, \cdots, \alpha_r$ be elements of $P$ such that the union of the $\alpha$'s and all their $W$ images form a basis of $P$. For each $i$, let $W_i'$ be the stabilizer of $\alpha_i$ in $W$ and let $W_i \supseteq W_i'$ be the stabilizer of $\phi_i = f(\alpha_i)$. Denote by $N_i$, $N_i'$ respectively the preimages of $W_i$ and $W_i'$ in $N$. Define an $N_i$ representation $V_i'$ by setting $V_i' = F\{W_i/W_i'\}$ as a $W_i$ representation and letting $T$ act on $V_i$ via $\phi_i$. This is well defined by 2.5. Finally, set $V_i$ to be the $N$-representation induced from $V_i'$. If $V_f$ is the direct sum of the $V_i$, then it is trivial to check that $V_f$ behaves as required.

b) Let $k$ be the map $g$ restricted to $Q$, so $g = f \oplus k$. Use the construction above to define $V_f$ and $V_k$, and then set $V_g = V_f \oplus V_k$. Choose good bases for $V_f$ and $V_k$ which induce $f$ and $k$, and take a good basis, $\{w_j\}$, for $V_g$ which is the union of these good bases. The multiplicative $N$ submodule generated by $F^*$ and $\{w_j\}$ is an extension of $P \oplus Q$ by $F^*$, so it makes sense to denote this submodule by $(P \oplus Q)'$. Of course, $P'$ denotes the multiplicative submodule generated by $F^*$ and the good basis of $V_f$. The inclusion of bases induces the map $j' : P' \to (P \oplus Q)'$, and it is easy to check that the rest of diagram (6) follows. Q.E.D.

Recall that $F(G)$ is an invariant field $F(V)^G$ where $V$ is a $G$ representation with trivial $G$ stabilizer. Combining 2.4 and 2.6 we have:

**Corollary 2.7.** *Let $G$ be a reductive connected linear algebraic group. Let $T \subseteq G$ be a maximal torus, $W = N_G(T)/T$ the Weyl group, and $X(T) = Hom(T, F^*)$, the character group viewed as a $W$ lattice. Choose any surjective $W$ map $f : P \to X(T)$ with $P$ a permutation lattice. Let $M$ be the kernel of $f$. Then $F(G)$ is stably isomorphic to $F_\alpha(M)^W$ where $\alpha$ is some extension of $M$ by $F^*$.*

To apply 2.7 it is most convenient to have $\alpha$ a split extension. Up to stable isomorphism, this can often be arranged as follows. Let $U$ be a faithful $W$ representation and let $L = F(U)$ have the obvious $W$ action. Suppose $M \subseteq P$ to be faithful $W$ lattices and assume $\alpha \in \mathrm{Ext}(M, F^*)$ is in the image of $\mathrm{Ext}(P, F^*)$.

**Lemma 2.8.**

*a) $L_\alpha(M)^W/F_\alpha(M)^W$ and $L(M)^W/F(M)^W$ are both rational extensions.*

*b) $L_\alpha(M)^W \cong L(M)^W$.*

*c) $F_\alpha(M)^W$ and $F(M)^W$ are stably isomorphic.*

*Proof.*

a) $L(M) = F(M)(U)$ and $F(M)/F(M)^W$ is $W$ Galois, so $L(M)^W/F(M)^W$ is rational by 1.2. The same argument applies to $L_\alpha(M)^W/F_\alpha(M)^W$.

To prove b), consider the following commutative diagram:

$$\begin{array}{ccc} \mathrm{Ext}(P, F^*) & \longrightarrow & \mathrm{Ext}(P, L^*) \\ \big\downarrow & & \big\downarrow \\ \mathrm{Ext}(M, F^*) & \longrightarrow & \mathrm{Ext}(M, L^*) \end{array}$$

Since $\mathrm{Ext}(P, L^*)$ is isomorphic by **[B]** p.61 to a direct sum of cohomology groups $H^1(H, L^*)$ for $H \subseteq W$, $\mathrm{Ext}(P, L^*) = (0)$. Thus $\alpha$ has zero image in $\mathrm{Ext}(M, L^*)$. It follows that the $\alpha$-twisted and usual $W$ action on $L(M)$ are isomorphic and so $L_\alpha(M)^W \cong L(M)^W$. Part c) follows from a) and b). Q.E.D.

Having shown that the fields $F(G)$ are stably isomorphic to fields of the form $F_\alpha(M)^W$, we will use 1.3 to show how the $F(G)$ can be described using Brauer Severi varieties. Our setup will be the following. Suppose $\phi : G' \to G$ is a surjective homomorphism of reductive connected algebraic groups with central kernel. Then there is an induced surjection $T' \to T$ of maximal tori in $G'$ and $G$ respectively, and a surjection $N' \to N$ of the normalizers of $T'$ and $T$. The induced map $N'/T \to N/T$ is an isomorphism which we use to identify $W$ with the Weyl group of $G'$. The map $T' \to T$ induces a $W$ lattice embedding $X(T) \to X(T')$. Let $P, Q$ be permutation $W$ lattices and $f$, $g$ $W$ module surjections such that the following diagram commutes:

$$\text{(8)} \qquad \begin{array}{ccccc} P & \xrightarrow{f} & X(T) & \longrightarrow & 0 \\ {\scriptstyle j}\big\downarrow & & \big\downarrow & & \\ P \oplus Q & \xrightarrow{g} & X(T') & \longrightarrow & 0 \end{array}$$

where $j$ is the natural inclusion. By 2.6 we can associate to (8) the diagram:

$$\text{(9)} \qquad \begin{array}{ccccccccc} 0 & \longrightarrow & M_\alpha & \longrightarrow & P' & \longrightarrow & X(T) & \longrightarrow & 0 \\ & & \big\downarrow & & \big\downarrow & & \big\downarrow & & \\ 0 & \longrightarrow & R_\beta & \longrightarrow & (P \oplus Q)' & \longrightarrow & X(T') & \longrightarrow & 0 \end{array}$$

where $F_\alpha(M)^W$ is stably isomorphic to $F(G)$ and $F_\beta(R, W)$ is stably isomorphic to $F(G')$. Using this setup we make two observations, the first one concerning the fact that we can sometimes ignore the "$\alpha$" twisting.

**Lemma 2.9.** *Suppose $R$ is a permutation lattice. Then $F(M)^W$ is stably isomorphic to $F_\alpha(M)^W$. More precisely, let $U$ and $L = F(U)$ be as in 2.8. Then $L(M)^W \cong L_\alpha(M)^W$, and both $L(M)^W/F(M)^W$ and $L_\alpha(M)^W/F_\alpha(M)^W$ are rational.*

*Proof.* This lemma is an immediate consequence of (9) and 2.8.

As an immediate consequence we have:

**Corollary 2.10.** *Assume the notation of 2.8. If $M$ is a faithful permutation lattice over $W$, then $F(G)$ is stably rational.*

*Proof.* $F(G)$ is stably isomorphic to $F_\alpha(M)^W$ which by 2.8 is stably isomorphic to $F(M)^W$. If $M$ is permutation, then $F(M)^W$ is stably isomorphic to $F(V)^W$ for any faithful $W$ representation $V$. As $W$ is a Weyl group, there is a representation $V$ such that $W$ is generated by reflections. By the Chevalley-Shepard-Todd theorem (*e.g.* **[Sp]** p.76), $F(V)^W$ is rational. Q.E.D.

More essentially, the diagram (9) and 1.3 yield the main result of this paper.

**Theorem 2.11.** *Suppose $\phi : G' \to G$ is a surjective homomorphism of linear connected reductive algebraic groups with central kernel cyclic of order $n$. Then $F(G')$ is stably isomorphic to $F(G)(A)$ for some central simple algebra $A/F(G)$ of exponent $n$.*

*Proof.* Of course if $M_\alpha$ and $R_\beta$ are as in (9), it suffices by 2.7 to show that $F_\alpha(M)^W(A)$ is stably isomorphic to $F_\beta(R)^W$ for some $A$. By enlarging $P$ if necessary, we can assume $M$ is $W$ faithful. Our conditions on $\phi$ imply that $X(T')/X(T)$ is cyclic of order $n$. Let $x \in X(T')$ be a preimage of a generator of $X(T')/X(T)$, and let $H \subseteq W$ be the stabilizer of $x$. In (8), we may take $Q = Z[W/H]$. Diagram chasing shows that $R_\beta/M_\alpha \cong I_n[W/H]$. Set $L' = F(M)$ with the $\alpha$ twisted action of $W$. The exact sequence $0 \to M_\alpha \to R_\beta \to I_n[W/H] \to 0$ induces an extension we call $\gamma$ of $I_n[W/H]$ by $L'^*$. By our definitions $F_\beta(R, W)$ is just $L'_\gamma(I_n[W/H])^W$ and so this theorem follows from 1.3. Q.E.D.

It would obviously be of interest to describe the Brauer group element $[A]$ that arises in 2.11. It is equivalent, in the notation of the proof of 1.3, to describe the cohomology class $\delta \in H^2(W, M_\alpha)$ that corresponds to the extension $0 \to M_\alpha \to R_\beta \to I_n[W/H] \to 0$. This is difficult because the structure of $M_\alpha$ is hard to get one's hands on. However, we will give some partial results, and then give a full description of $A$ for special $G$.

Continuing with the notation of diagrams (8) and (9) and the proof of 2.11, let $\mu$ be the canonical generator of $Q = Z[W/H]$ which we can assume satisfies $g(\mu) = x$. Choose $\mu' \in (P \oplus Q)'$ a preimage of $\mu$. Let $\{u_\sigma \mid \sigma \in W\}$ be a set of representatives of $W$ in $N'$. For each $\sigma \in W$ let $c(\sigma) \in P'$ be a preimage of $\sigma(x) - x$, with $c(1) = 0$. Set $d(\sigma) = u_\sigma(\mu') - \mu' - c(\sigma) \in (P \oplus Q)'$. Then $d(\sigma) \in R_\beta$ and $\delta(\sigma, \tau) = \sigma(d(\tau)) + d(\sigma) - d(\sigma\tau) \in M_\alpha$ is a cocycle representing $\delta$. This construction makes it clear that the image, $\delta'$, of $\delta$ in $H^2(W, M)$ can be described as follows.

**Lemma 2.12.** *Let $e(\sigma) = \sigma(x) - x \in X(T)$. Then $e(\sigma)$ defines an $e \in H^1(W, X(T))$ that generates the kernel of $H^1(W, X(T)) \to H^1(W, X(T'))$ and $\delta'$ is the coboundary of $e$ with respect to the exact sequence $0 \to M \to P \to X(T) \to 0$.*

To add anything further to our description of $\delta$ we must specialize our setup. Suppose $U$ is a faithful $W$ representation over $F$ and $L = F(U)$ is as in 2.8. Assume $M_\alpha$ and $R_\beta$ split in $\mathrm{Ext}_W(L^*, M)$ and $\mathrm{Ext}_W(L^*, R)$ respectively. As $L(V)^N$ is rational over $F(V)^N$ it suffices to describe the image of $\delta$ in $H^2(W, L^* \oplus M)$.

**Lemma 2.13.** *If $M_\alpha$ and $R_\beta$ are as above, then the image of $\delta$ in $H^2(W, L^* \oplus M)$ is equal to the image of $\delta' \in H^2(W, M)$. In other words, $\delta$ has no contribution from $H^2(W, L^*)$.*

*Proof.* By assumption there are inclusions $M_\alpha \to L^* \oplus M$ and $R_\beta \to L^* \oplus R$ and a commutative diagram:

$$\begin{array}{ccccccccc} 0 & \longrightarrow & M_\alpha & \longrightarrow & R_\beta & \longrightarrow & I_n[W/H] & \longrightarrow & 0 \\ & & \downarrow & & \downarrow & & \downarrow & & \\ 0 & \longrightarrow & L^* \oplus M & \longrightarrow & L^* \oplus R & \longrightarrow & I_n[W/H] & \longrightarrow & 0 \end{array}$$

where the bottom row is defined by the image of $\delta$. But the bottom row maps are the identity on $L^*$. Hence $\delta$ is trivial when restricted to $L^*$ and the lemma is proved. Q.E.D.

In essence, 2.13 says that under the stable isomorphism of $F_\alpha(M)^W$ and $F(M)^W$, $\delta'$ maps to $\delta$ and so the description of $\delta$ can be used to find the "A" in 2.11. We will use this approach in the case $G' = SL_n$.

In a different direction, assume the sequence $1 \to T \to N \to W \to 1$ splits, and fix an embedding $W \to N$. Using this embedding we can regard $P'$ as a $W$ module. In fact, the construction of $P'$ shows that $P' \cong F^* \oplus P$ as $W$ modules and so $M_\alpha$ is the split extension of $M$ by $F^*$. In the computation above we may choose $c(\sigma) \in P$. Then the image of $u_\sigma(u_\tau\mu' - \mu' - c_\tau) - u_\sigma\mu' - \mu' - c_\sigma - (u_{\sigma\tau}\mu' - \mu' - c_{\sigma\tau})$ in $F^*$ is just $u_\sigma(u_\tau\mu') - u_{\sigma\tau}\mu'$. To compute this expression let $C$ be the kernel of $N' \to N$ and $W' \subseteq N'$ the preimage of $W$. We can assume that the $u_\sigma$ are in $W'$. If $c(\sigma, \tau) \in C$ is defined by $u_\sigma u_\tau = c(\sigma, \tau)u_{\sigma\tau}$, then $c$ represents the cohomology class $c \in H^2(W, C)$ corresponding to $W'$. Now $x \in X(T')$ is a homomorphism $T' \to F^*$ which can be seen to induce an injection on $C$, and $u_\sigma(u_\tau\mu') - u_{\sigma\tau}\mu' = x(c(\sigma, \tau))$. We have computed that:

**Lemma 2.14.** *Under the above assumptions, $x^*(c)$ is the image of $\delta$ in $H^2(W, F^*)$.*

## Section Three: Some specific groups

### Case I: $SL_n$ and its images

In this section we will apply the results of section 2 to some specific groups. To begin, consider $SL_n(F)$ and note that the center of $SL_n$ is cyclic of order $n$. Fix a divisor $r$ of $n$ and set $C_r \subseteq SL_n$ to be the unique central subgroup which is cyclic of order $r$. Define $G_r = SL_n/C_r$. Of course, $G_n = PGL_n$ and we have a pair of surjections $SL_n \to G_r \to PGL_n$. Let $T'' \subseteq SL_n$ be a maximal torus, $T' \subseteq G_r$ and $T \subseteq PGL_n$ the corresponding maximal tori, $W$ the Weyl group of all the groups, and $X(T) \subseteq X(T') \subseteq X(T'')$ the character lattices.

The structure of $W$, $X(T)$, and $X(T'')$ can be read off from the root diagram, $A_{n+1}$, for $SL_n$ (*e.g.* **[Bu]** p.185). In fact, $W$ is $S_n$, the symmetric group on $n$ letters. Let $U$ be the real vector space with basis $e_1, \cdots, e_n$ and let $S_n$ act on the $e$'s and hence on $U$ in the obvious way. Let $E$ be the $S_n$ invariant hyperplane defined by $\Sigma x_i = 0$ where the $x_i$'s are the coordinates with respect to the $e_i$'s. Set $Q$ to be $E \cap (\Sigma Z e_i)$ which is an $S_n$ lattice. Denote by $\pi \in Q$ the element

$(e_2 - e_1) + \cdots + (e_n - e_1)$, and set $Q_r$ to be $Q + Z(r/n)\pi$. Note that $Q_r$ is again a $W$ lattice, and $X(T) \subseteq X(T') \subseteq X(T'')$ is isomorphic to $Q \subseteq Q_r \subseteq Q_n$.

Let $H \subseteq S_n$ be the stabilizer of $e_1$. $H$ is also the stabilizer of $\pi$. It is easy to compute that $Q \cong I[S_n/H]$. Furthermore, $Q_n$ is generated as an $S_n$ lattice by $(1/n)\pi$. One can then compute that $Q_n \cong Z[S_n/H]/(Z\xi)$ where $\xi$ is the sum of the cosets of $H$ under the usual identification.

We next will define permutation lattices and surjections onto the $Q$'s. Let $P$ be the permutation lattice with $Z$ basis the set $\{y_{ij} \mid 1 \leq i,j \leq n; i \neq j\}$ where $S_n$ acts by acting in the obvious way on the indices. Further, let both $P_1$ and $P_2$ be the isomorphic to $Z[S_n/H]$ where $P_i$ has canonical generator $\alpha_i$. We form the diagram:

$$\begin{array}{ccccccccc} 0 & \longrightarrow & M & \longrightarrow & P & \xrightarrow{\phi} & Q & \longrightarrow & 0 \\ & & \downarrow & & \downarrow & & \downarrow & & \\ 0 & \longrightarrow & M_r & \longrightarrow & P \oplus P_1 & \xrightarrow{\chi} & Q_r & \longrightarrow & 0 \\ & & \downarrow & & \downarrow & & \downarrow & & \\ 0 & \longrightarrow & M_n & \longrightarrow & P \oplus P_1 \oplus P_2 & \xrightarrow{\psi} & Q_n & \longrightarrow & 0 \end{array}$$

where $\chi$ extends $\phi$ and $\psi$ extends $\chi$; $\phi(y_{ij}) = x_i - x_j$; $\chi(\alpha_1) = (1/r)\pi$, and $\psi(\alpha_2) = (1/n)\pi$. Now $\psi$ restricted to $P_2$ is already a surjection with kernel isomorphic to $Z$. It follows that $M_n \cong Z \oplus P \oplus P_1$ and so $M_n$ is a permutation lattice.

Let $U$ be the faithful $S_n$ representation $F[S_n/H]$ and $L = F(U)$ as in 2.8. Using 2.9 and 2.11 we have:

**Proposition 3.1.** *Let $W = S_n$. $F(G_r)$ is stably isomorphic to $L(M_r)^W$, $F(SL_n)$ is stably isomorphic to $L(M_n)^W$, and $F(PGL_n)$ is stably isomorphic to $L(M)^W$. Furthermore, $F(G_r)$ is stably isomorphic to $F(PGL_n)(A)$ for some $A/F(PGL_n)$ central simple and of exponent $n/r$.*

We will finish our discussion of $G_r$ by determining the algebra $A$. Note that in $PGL_n$ the extension $1 \to T \to N \to S_n \to 1$ splits and so we can assume $W \subseteq N$. By 2.12 and 2.13, $A$ corresponds to the cohomology class $\delta \in H^2(S_n, M)$ which is the coboundary of $d(\sigma) = \sigma((r/n)\pi) - (r/n)\pi$. Let $\zeta \in H^2(W, M)$ be the coboundary of $e(\sigma) = \sigma((1/n)\pi) - (1/n)\pi$, so $\delta = r\zeta$. We form the crossed product $D' = \Delta(L(M)/L(M, S_n), S_n, \zeta)$. As $\zeta$ is trivial when restricted to $H$, $D'$ is split by $L(M)^H$. Thus $D'$ is Brauer equivalent to an algebra $D$ of degree $n$ with $L(M)^H$ as a maximal subfield. The description of Formanek and Procesi ([**P1**] and [**F**]) shows that $L(M)^W$ is isomorphic to the center of the generic division algebra $UD(F, n, 2)$ and that $D$ corresponds to $UD(F, n, 2)$. We have proved:

**Theorem 3.2.** *Let $Z$ be the center of the generic division algebra $UD(F, n, 2) = D$ and $A$ the division algebra in the class $r[D]$. Set $G_r$ to be $SL_n(F)/C_r$ where $C_r$ is the central cyclic subgroup of order $r$, let $V$ be a $G_r$ representation over*

*$F$ with an element with trivial stabilizer and $F(G_r)$ the invariant field of $G_r$ on $F(V)$. Then $F(G_r)$ is stably isomorphic to $Z(A)$.*

$D_r = D \otimes_Z Z(A)$ has exponent $r$ in the Brauer group. There is ([**S2**]) a "generic division algebra of degree $n$ and exponent $r$" which by [**S3**] is stably isomorphic to $Z(A)$. Thus Theorem 3.2 can be rephrased as saying:

**Corollary 3.3.** *$F(G_r)$ is stably isomorphic to the center of the generic division algebra over $F$ of degree $n$ and exponent $r$.*

Let $PO_n$ be the projective orthogonal group and $PSP_n$ the projective symplectic group. Assume $n$ is even. In [**P1**], the centers of "general division algebras with involution" were shown to be of the form $F(V)^G$ where $G = PO_n$ or $G = PSp_n$ and $V$ is the direct sum of at least two copies of $M_n(F)$. In [**BS**], these centers were shown to be of the form $Z(A)$ where $A$ was an algebra in the Brauer class of $2[D]$. Direct observation shows that $V$ has elements with trivial stabilizer. Thus we have:

**Corollary 3.4.** *Let $n$ be even. Then $F(G_2)$, $F(PO_n)$ and $F(PSp_n)$ are all stably isomorphic.*

## Case II: Spin groups of odd degree.

There are two connected semisimple groups with diagram $B_n$, namely, the spin group $\mathrm{Spin}_{2n+1}$ and the special orthogonal group $SO_{2n+1}$ (*e.g.* [**Bu**] p.191). They fit into a diagram $\mathrm{Spin}_{2n+1} \to SO_{2n+1}$ where the map (call it $\phi$) is a surjection. Further, $\phi$ has kernel $C$ central of order 2. We have need to recall the definition of Spin and $\phi$ later on. Let $T' \to T$ and $N' \to N$ be the corresponding maps of maximal tori and their normalizers. $W = N/T = N'/T'$ and $X(T) \subseteq X(T')$ can be described as follows ([**H**] p.64). Let $Q$ be an $S_n$ lattice with $Z$ basis $e_1, \cdots, e_n$ and the obvious $S_n$ action. Define $\tau_i \in \mathrm{Aut}_Z(Q)$ by setting $\tau_i(e_j) = e_j$ if $j \neq i$ and $\tau_i(e_i) = -e_i$. Let $H$ be the subgroup of $\mathrm{Aut}_Z(Q)$ generated by the $\tau$'s and $W$ the subgroup generated by $H$ and $S_n$. Then $H$ is abelian, of exponent 2, with basis the $\tau$'s and $H$ is normal in $W$. There is a split exact sequence $1 \to H \to W \to S_n \to 1$ and the conjugation action of $S_n$ on $\{\tau_1, \cdots, \tau_n\}$ is the usual one. Set $\pi = (1/2)(e_1 + \cdots + e_n)$ and $Q' = Q + Z\pi$. Then $Q$, $Q'$ are $W$ lattices and $X(T) \subseteq X(T')$ is isomorphic to $Q \subseteq Q'$.

Let $P$ be the following permutation $W$ lattice. $P$ has $Z$ basis $f_1, \cdots, f_n$, $h_1, \cdots, h_n$ such that $S_n$ permutes the $f$'s and $h$'s as usual. In addition, $\tau_i(f_j) = f_j$ and $\tau_i(h_j) = h_j$ for $j \neq i$; $\tau_i(f_i) = h_i$ and $\tau_i(h_i) = f_i$. Let $P_1 = Z[W/S_n]$ have canonical generator $\beta$. Form the diagram:

$$\begin{array}{ccccccccc} 0 & \longrightarrow & M & \longrightarrow & P & \xrightarrow{\phi} & Q & \longrightarrow & 0 \\ & & \downarrow & & \downarrow & & \downarrow & & \\ 0 & \longrightarrow & R & \longrightarrow & P \oplus P_1 & \xrightarrow{\chi} & Q' & \longrightarrow & 0 \end{array}$$

where $\phi(f_i) = e_i$, $\phi(h_i) = -e_i$, and $\chi(\beta) = \pi$. $M$ has $Z$ basis $(f_1 + h_1), \cdots, (f_n + h_n)$ and so is a permutation $W$ lattice. Furthermore, direct observation shows that $1 \to T \to N \to W \to 1$ splits.

Let $U = Fx_1 + \cdots + Fx_n$ be the faithful $W$ representation defined by the relations $\tau_i(x_j) = x_j$ if $j \neq i$, $\tau_i(x_i) = -x_i$, and $S_n$ acts on the $x$'s in the obvious way. Form $L = F(U)$. By 2.11 and 2.9, $F(SO(2n+1))$ is stably isomorphic to $L(M)^W$ and $F(\mathrm{Spin}(2n+1))$ is stably isomorphic to $L(M)^W(A)$ where $A$ has exponent dividing 2. In addition, $A$ corresponds to a cohomology class $\delta \in H^2(W, L^* \oplus M)$ and we can write $\delta = \delta_1 + \delta_2$ where $\delta_1 \in H^2(W, L^*)$ is described by 2.14 and $\delta_2 \in H^2(W, M)$ is described by 2.12.

Let us compute the $\delta_i$ explicitly. By 2.12 and 2.13, $\delta_2$ is the coboundary of the 1 cocycle $d(\sigma) = \sigma(\pi) - \pi \in Q$. Because of this, it is easy to compute that $\delta_2$ contains a 2 cocycle $\delta_2(\sigma, \sigma')$ which is zero on $S_n$ and satisfies $\delta_2(\tau_i, \tau_j) = 0$ if $i \neq j$ and $\delta_2(\tau_i, \tau_i) = f_i + h_i$.

To describe $\delta_1$, we begin by outlining a description of the map $\mathrm{Spin}_{2n+1} \to SO_{2n+1}$. Let $V''$ be the $F$ vector space with basis $a_1, \cdots, a_n, b_1, \cdots, b_n, c$. Define a quadratic form, $q$, on $V'''$ by specifying that $q(c,c) = -1$, $q(a_i, b_i) = 1$ and $q(x,y) = 0$ for any other choice of $x$ and $y$ from the basis. We can identify $SO_{2n+1}$ with the special orthogonal group on $V''$ with respect to this form. $T \subseteq SO_{2n+1}$ can be chosen to be the group of diagonal matrices with respect to this basis, and $W \subseteq SO_{2n+1}$ can be chosen so that $W$ fixes $c$, $S_n$ permutes the $a$'s and $b$'s as usual, $\tau_i$ is the identity on all $a_j$, $b_j$ for $j \neq i$, and $\tau_i$ transposes $a_i$ and $b_i$.

Let $C(q)$ be the Clifford algebra of $(V'', q)$, and identify $V''$ with a subspace of $C(q)$ in the canonical way. One defines an involution, $\alpha$, on $C(q)$ by setting $\alpha(v) = -v$ for all $v \in V''$. Let $\Gamma \subseteq C(q)^*$ be the multiplicative subgroup defined by: $x \in \Gamma$ if and only if $\alpha(x)Vx^{-1} = V$. For $x \in \Gamma$, let $\phi_x$ be the endomorphism of $V''$ defined by $\phi_x(v) = \alpha(x)vx^{-1}$. $\mathrm{Spin}_{2n+1}$ can be identified with $\{x \in \Gamma \mid \alpha(x)x = 1 \text{ and } \det(\phi_x) = 1\}$. The map $\mathrm{Spin}_{2n+1} \to SO_{2n+1}$ is the map $x \to \phi_x$. The kernel, $C$, of this map is generated by $-1 \in C(q)$.

We let $W' \subseteq \mathrm{Spin}_{2n+1}$ be the inverse image of $W$. According to 2.14, we must investigate the extension $1 \to C \to W' \to W \to 1$, use this extension to define a cocycle $\eta \in H^2(W, C)$, and then find the image of $\eta$ in $H^2(W, L^*)$ with respect to the map $x : T \to F^* \subseteq L^*$. Now $x$ must map $C$ isomorphically onto $\{1, -1\} \subseteq L^*$, and in $L^*$, $-1$ has a square root $\beta$. If we set $C' \subseteq L^*$ to be the subgroup generated by $\beta$, it suffices to consider the induced extension $1 \to C' \to W'' \to W \to 1$. Let $\tau_{ij} \in S_n \subseteq W$ be the $i,j$ transposition and $\tau_i \in H \subseteq W$ as usual. If $\tau'_{ij} = 1/2(a_i - a_j)(b_i - b_j) \in C(q)$, and $\tau'_i = (1/\sqrt{2})(a_i - b_i)c \in C(q)$, then $\tau'_{ij}$ and $\tau'_i$ are in $\mathrm{Spin}_{2n+1}$ and preimages of $\tau_{ij}$ and $\tau_i$ respectively. Hence $\tau'_{ij}$ and $\tau'_i$ are in $W'$. Set $\tau''_{ij} = \tau'_{ij}\beta \in W''$ and $\tau''_i = \tau'_i\beta \in W''$. A straightforward exercise shows that the $\tau''_{ij}$ generate a subgroup of $W''$ isomorphic to $S_n$, that $\tau_i''^2 = 1$, but that the commutator of $\tau''_i$ and $\tau''_j$, for $i \neq j$, is nontrivial. In other words, if $\eta'$ is the image of $\eta$ in $H^2(W, C')$, then $\eta'$ contains a 2 cocycle $\eta'(\sigma, \sigma')$ which is the identity of $S_n$ and satisfies $\eta'(\tau_i, \tau_i) = 1$ and $\eta'(\tau_i, \tau_j) = -1$ for $i \neq j$. Replacing $\eta'$ by $\delta_1$ in this last sentence yields a description of $\delta_1$.

Given all this information, $F(\mathrm{Spin}_{2n+1})$ can be described as follows. To begin with, note that $L(M)$ can be identified with the field $K'' =$ $F(x_1, \cdots, x_n, y_1, \cdots, y_n)$ where the $x$'s and $y$'s are a transcendence base and $W$ acts on $K''$ as follows. $S_n$ acts on the $x$'s and $y$'s as usual, $H$ is the identity on the $y$'s, but $\tau_i(x_i) = -x_i$ and $\tau_i(x_j) = x_j$ for $i \neq j$. Let $z_i = x_i^2$ and note that $K' = F(z_1, \cdots, z_n, y_1, \cdots, y_n)$ is the invariant field of $H$ on $K''$. Form the central simple algebra

$$A' = \Big(\bigotimes_{i<j}(z_i, z_j)\Big) \otimes \Big(\bigotimes_{1=1}^{n}(z_i, y_i)\Big)$$

where $A'$ has center $K'$ and $(x, y)$ is the generalized quaternion algebra (see introduction). The action of $S_n$ on $K'$ extends in an obvious way to $A'$. We set $A$ to be the invariant division algebra and note that the center of $A$ is $K$, the invariant field of $S_n$ on $K'$.

**Theorem 3.5.** *$F(SO_{2n+1})$ is stably isomorphic to $K$ and $F(\mathrm{Spin}_{2n+1})$ is stably isomorphic to $K(A)$.*

*Remark:* A result similar to, but not identical to, 3.5 was suggested to the author by I. Dolgachev and attributed by him to Bogomolov. The proof, as briefly outlined, was seemingly entirely different.

*Proof.* It is clear that $K \cong L(M)^W$. Thus it suffices to show that $A$, when identified with an element of $H^2(W, K''^*)$, corresponds to $\delta = \delta_1 + \delta_2$. This is tedious but routine, so we leave it to the reader. Q.E.D.

We will close this section with a few remarks on some other groups $G$. To begin with, if $G$ has diagram $C_n$, then $G$ is either the symplectic group $Sp_n$ or the projective symplectic group $PSp_n$. But $F(Sp_n)$ is known to be stably rational, and the reader can show this by applying 2.10. By **[BS]**, $F(PSp_n)$ is stably isomorphic to the "center of the generic division algebra with involution", and we cannot here improve on this description.

Finally, let us outline what the techniques of this paper say about groups with diagram $D_n$. The simply connected group with this diagram is the spin group $\mathrm{Spin}_{2n}$. This group has as an image the special orthogonal group $SO_{2n}$, and the kernel of this map is central of order 2. By 2.10 and 2.11, $F(SO_{2n}) = \hat{K}$ is stably rational and $F(\mathrm{Spin}_{2n})$ is stably isomorphic to $\hat{K}(\hat{A})$ for $\hat{A}/\hat{K}$ central simple of exponent 2.

Just as in case II, one can describe $\hat{A}$ precisely. The argument is very close to that of case II, so we will confine ourselves to highlighting the differences. Recall the notation of case II. If $\tau_i \in W$ and $S_n \subseteq W$ are as in case II, let $W' \subseteq W$ be the subgroup of $W$ generated by $S_n$ and all elements of the form $\tau_i\tau_j$. Then $W'$ is the Weyl group of $\mathrm{Spin}_{2n}$ and $SO_{2n}$. The surjection $\mathrm{Spin}_{2n} \to SO_{2n}$ induces character groups $X(T) \subseteq X(T')$ which are isomorphic to the lattices $Q \subseteq Q'$ considered as $W'$ lattices. Once again, the cohomology class $\delta \in H^2(W, L^*)$ can be written $\delta_1 + \delta_2$ where $\delta_1$ is described by 2.14 and $\delta_2$ by 2.12. It is easy to see that $\delta_2$ is the restriction to $W'$ of the corresponding cohomology class from case II. As for $\delta_2$, it once again is simpler to describe the image of $\delta_1$ in $H^2(W, C')$ where $C' \subseteq L^*$ has order 4. Direct computation shows that $\delta_1$ is also the restriction to $W'$ of the corresponding cohomology class from case II.

Hence $F(\mathrm{Spin}_{2n})$ can be described as follows. Let $K' =$ $F(z_1, \cdots, z_n, y_1, \cdots, y_n)$ be the rational extension of $F$ of transcendence degree $2n$ and let $S_n$ act on the $z$'s and $y$'s in the usual way. Form the central simple algebra:

$$A' = \left(\bigotimes_{i<j}(z_i, z_j)\right) \otimes \left(\bigotimes_{1=1}^{n}(z_i, y_i)\right)$$

and let the $S_n$ action extend to $A'$ in the obvious way. Set $K$ to be the invariant field of $S_n$ on $K'$ and $A/K$ the invariant division algebra. Note that $z = z_1 z_2 \cdots z_n \in K$ and set $\hat{K} = K(z^{1/2})$, $\hat{A} = A \otimes_K \hat{K}$. We have:

**Theorem 3.6.** *$F(SO_{2n})$ is stably isomorphic to $\hat{K}$ and $F(\mathrm{Spin}_{2n})$ is stably isomorphic to $\hat{K}(\hat{A})$.*

Among groups with diagram $D_n$ there are others we have not dealt with. In particular, $SO_{2n}$ has an image $PSO_{2n}$ which is the projective special orthogonal group. This is a subgroup of index 2 in the projective orthogonal group $PO_n$. We have already mentioned that by **[BS]** $F(PO_n)$ is stably isomorphic to the center of the generic division algebra with involution. The relationship of $F(PSO_{2n})$ to division algebras is a topic requiring further study.

## References


[A] Amitsur, S. A., *Generic splitting fields of central simple algebras*, Ann. Math. **62**, 8–43.

[Ar] Artin, *Brauer-Severi varieties*, appears in "Brauer groups in Ring Theory and Algebraic Geometry," Springer-Verlag, Berlin/Heidelberg/ New York, 1982 (LNM #917).

[BS] Berele, A. and Saltman, D. J., *The centers of generic division algebras with involution*, preprint.

[Bo] Bogomolov, F. A., *Rationality of some quotient varieties*, Math. USSR Sbornik **54**(2), 1986.

[Bu] Bourbaki, N. "Groupes et algebres de Lie," chpt. VII, Hermann, 1975.

[BD] Brocker, Theodor and Dieck, Tammo tom, "Representations of Compact Linear Groups," Springer-Verlag, New York/Berlin/Heidelberg/Tokyo, 1985.

[B] Brown, Kenneth S., "Cohomology of Groups," Springer-Verlag, New York /Heidelberg/Berlin, 1982.

[EM] Endo, S. and Miyata, T., *Invariants of finite abelian groups*, J. Math. Soc. Japan, **25**, 1973, 7–26.

[F] Formanek, E., *The center of* $3 \times 3$ *generic matrices*, Lin and Multlin. Alg. **7**, 1979, 203–212.

[H1] Humphreys, James E., "Linear Algebraic Groups," Springer-Verlag, New York/
Heidelberg/Berlin, 1975.

[H2] Humphreys, James E., "Introduction to Lie algebras and their Representations," Springer-Verlag, New York/Heielberg/Berlin, 1972.

[J] Jacobson, Nathan, "P. I. Algebras," Springer-Verlag, Berlin/ Heidelberg/New York, 1975 (LNM #441).

[LR] Luna, D. and Richardson, R. W., *A generalization of the Chevalley restriction theorem*, Duke Math. J., **46**(3), 1979, 487–496.

[M] Milnor, John, "Introduction to Algebraic $K$ Theory", Princeton University Press (Annals of Mathematics Studies), 1971.

[P1] Procesi, C., *Noncommutative affine rings*, Atti Acc. Naz. Lincei, s.VIII v.VIII f.g (1967), 239–255.

[P2] Procesi, C., *The invariant theory of* $n \times n$ *matrices*, Adv. in Math. **19**, 306–381.

[Ro] Roquette, P., *On the Galois cohomology of the projective linear group*$\cdots$, Math. Ann. **150**, 411–439.

[R] Rosenlicht, M., *A remark on quotient spaces*, An. Acad. Brasil, **35**, 1963.

[S1] Saltman, D. J., *Multiplicative field invariants*, J. of Alg. **106**, 1987, 221–238.

[S2] Saltman, D. J., *Indecomposable division algebras*, Comm. in Alg. **7**(8), 1979, 791–817.

[S3] Saltman, D. J., *Norm polynomials and algebras*, J. of Alg. **62**(2), 1980, 333–345.

[Sh] Shepherd-Barron, N. I., *The rationality of certain orbit spaces*, preprint.

[Sp] Springer, T. A., "Invariant Theory," Springer-Verlag, Berlin/Heidelberg/New York, 1977.

# Group Graded Rings, Smash Products and Additive Categories.

Liu Shaoxue
Beijing Normal University
Beijing, P.R. China

Fred Van Oystaeyen
University of Antwerp, UIA
Antwerp, Belgium

## Introduction.

For rings graded by a finite group the smash products of the rings with the grading groups play an important part in the duality theory that allows to relate properties of graded nature to ungraded properties. In this paper we propose to study similar topics but for rings graded by arbitrary groups and one of the new ingredients we introduce is the use of small additive categories associated to the smash products. First we derive the duality theorem for actions and coactions in Section 2, then we turn to study some properties of the Jacobson and the Baer radical in Section 3 and we provide some applications in Section 4. These applications center around the primitivity or simplicity of the smash product and a reinterpretation of the Wedderburn theorem for the small additive category associated to it. We have chosen a presentation that makes the material adaptable to the case of gradations by $G$-sets in the sense of [4], because we aim to come back to this case in forthcoming work.
The use of smash products in graded ring - or module theory, originating from a few observations of G. Bergman, is very wide-spread at this moment, we hope to have avoided essential overlaps with the recent work of a.o. M. Cohen, S. Montgomery [1], C. Năstăsescu and Rodino [2], D. Quin [5], ..., that inspired us for this paper. We use [3] as a basic reference on graded rings.



*F. van Oystaeyen and L. Le Bruyn (eds.), Perspectives in Ring Theory, 299–310.*

## 1. Preliminaries.

Let $G$ be any group and $A$ a $G$-graded ring, $A = \oplus_{g \in G} A_g$. The category of $G$-graded left $A$-modules will be denoted by $A$-gr; notation and terminology stems from [3].

Consider $M$ and $N$ in $A$-gr; we say that $N$ **weakly divides** $M$ if it is isomorphic to a graded direct summand of the direct sum $M^{(t)}$ of a finite number of copies of $M$. We say that $M$ and $N$ are **weakly isomorphic** in $R$-gr if each one weakly divides the other. A graded $A$-module $M$ is said to be **weakly $G$-invariant** if $M$ is weakly isomorphic to $M(\sigma)$ for all $\sigma \in G$, where $M(\sigma)$ is the shifted module obtained by taking the ungraded $A$-module $M$ with gradation defined by $M(\sigma)_\tau = M_{\tau\sigma}$ for all $\tau \in G$. For $M$ and $N$ in $A$-gr we let $\mathrm{HOM}_A(M,N)_\rho$ be the additive group of graded homomorphisms of degree $\rho \in G$ and $\mathrm{HOM}_A(M,N) = \oplus_{\rho \in G} \mathrm{HOM}_A(M,N)_\rho$. The morphisms in $A$-gr are then given by $\mathrm{Hom}_{A-\mathrm{gr}}(M,N) = \mathrm{HOM}_A(M,N)_e$, where $e$ is the neutral element of $G$.

Recall Theorem I.5.1. of [ 3 ], p. 43 : for a graded $A$-module $M$, $\mathrm{END}_A(M)$ is strongly graded by $G$ (i.e. a ring $R$ graded by $G$ is said to be strongly graded by $G$ if $R_\sigma R_\tau = R_{\sigma\tau}$ for all $\sigma, \tau \in G$) if and only if $M$ is weakly $G$-invariant. We say that $M \in A$-gr is **$G$-invariant** if for all $\sigma \in G$, $M \cong M(\sigma)$ in $A$-gr.

Consider the matrix ring $M_n(A)$ over the $G$-graded ring $A$. This matrix ring may be viewed as a graded endomorphism ring of a gr-free $A$-module if we equip it with a $G$-gradation depending on a set $\overline{\sigma} = \{\sigma_1, \ldots, \sigma_n\} \subset G$ as follows : $(M_n(A)(\overline{\sigma}))_\lambda = (A_{\sigma_i \lambda \sigma_j^{-1}})_{ij}$. If $A$ is strongly graded by $G$ then for any choice of $\overline{\sigma}$ the ring $M_n(A)(\overline{\sigma})$ is also strongly graded by $G$ and if $A$ is a crossed product $A_e * G$ then for any $\overline{\sigma}$, $M_n(A)(\overline{\sigma})$ is a crossed product over $M_n(A_e)$ too.

A graded ring $A$ is said to be **gr-semisimple** if $A$-gr a semisimple category i.e. if any $M \in A$-gr is a direct sum of gr-simple objects (being just simple objects in $A$-gr). Thus $A$ is gr-semisimple if and only if $A = L_1 \oplus \ldots \oplus L_m$ for some minimal graded left ideals of $A$; $A$ is **gr-simple** if it has such a decomposition with $\mathrm{HOM}_A(L_i, L_j) \neq 0$ for every $i,j = 1,\ldots,m$, or equivalently if $L_j \cong L_i(\sigma_{ij})$ for some $\sigma_{ij} \in G$. A gr-simple ring $A$ is **gr-uniformly simple** if it has a decomposition as above with $L_i \cong L_j$ in $R$-gr for $i,j = 1,\ldots,m$. If $A$ is gr-simple then $A_e$ is only semisimple in general but if $A$ is gr-uniformly simple then $A_e$ is a simple ring. $A$ gr-simple ring $A$ having a $G$-invariant graded simple object is necessarily gr-uniformly simple. Generally a $G$-graded ring $A$ is said to be **gr-uniformly primitive** if there exists a faithful $G$-invariant gr-simple module. In particular, if $A$ is gr-uniformly primitive and gr-simple then it is also gr-uniformly simple. Let us recall the graded version of Wedderburn's theorem as it is given in [3] p. 47. The $G$-graded ring $A$ is gr-simple, resp. gr-uniformly simple, if and only if $A \cong M_n(\Delta)(\overline{\sigma})$ for some graded-division ring $\Delta$ and a $\overline{\sigma} \in G^n$, resp. $A \cong M_n(\Delta)(\overline{\sigma}) = M_n(\Delta)$ where $\overline{\sigma} = \{e, \ldots, e\} \in G^n$.

That the gr-uniformly simple rings are **exactly** the gr-simple rings having a $G$-invariant gr-simple module follows after Theorem in Section 4 (this also makes the terminology in defining gr-uniformly primitive rings consistent and adequate).

We now recall the notion of smash products. Let $A$ be a $G$-graded ring with unit (as always unless otherwise mentioned). We define the smash product $A\#G$ as the free module $\oplus_{g\in G} Ap_g$ with multiplication given by : $(ap_h)(bp_g) = ab_{hg^{-1}}p_g$, for $g, h \in G$ and $a, b \in A$. It is clear that $A\#G$ is a ring but without unit when $G$ is infinite. We may rephrase the multiplication rule in $A\#G$ as follows : for $a_x \in A_x, b_y \in A_y$, $(a_xp_h)(b_yp_g) = a_xb_yp_g$ if $h = yg$ and $(a_xp_h)(b_yp_g) = 0$ if $h \neq yg$. Hence in any case $(A_xp_h)(A_yp_g) \subset A_{xy}p_g$.

The smash product $A\#G$ determines a small additive category $C_{\#}$ by taking the elements of $G$ for the objects of $C_{\#}$ and putting $\mathrm{Hom}(g, h) = A_{gh^{-1}}p_h$. Furthermore we may provide a matrix form representation for $A\#G$ by considering (possibly infinite) matrices with rows and columns indexed by $G$ with $\mathrm{Hom}(h, g)$ in the $(h, g)$-position, i.e. $A\#G \cong (\mathrm{Hom}(h, g))_{(h,g)} \cong (A_{hg^{-1}}p_g)_{(h,g)} \cong (A_{hg^{-1}})_{(h,g)}$, with matrix addition and multiplication, and where every matrix is assumed to have only a finite number of nonzero entries.

We write $A\#G$-mod for the category of all $A\#G$-modules $M$ with the property $(A\#G)M = M$ (we consider left modules unless otherwise stated), so an $M \in A\#G$-mod may be decomposed as $M = \oplus_{g\in G} e_{g,g}M$, where $e_{g,g} = 1_e p_g$ in the matrix presentation of $A\#G$.

**1.1. Lemma.** The set $\{e_{g,g} = 1_e p_g, g \in G\}$ is a system of orthogonal idempotent in $A\#G$ and $A\#G = \sum_{h,g\in G} e_{h,h}(A\#G)e_{g,g}$ (this says that $A\#G$ has local identities).

**Proof.** Straightforward, using the multiplication rules :

$$e_{h,h}(a_xp_g) = (1_ep_h)(a_xp_g) = a_xp_g \text{if } h = xg \text{ or } 0 \text{ otherwise}$$

$$(a_xp_g)e_{h,h} = (a_xp_g)(1_ep_h) = a_xp_h \text{if } g = h, \text{ or } 0 \text{ otherwise} \qquad \square$$

**1.2. Corollary.** If $(A\#G)M^* = M^*$ then $M^* = \oplus_{g\in G} e_{g,g}M^*$.

To a graded $A$-module $M = \oplus_{g \in G} M_g$ we associate the $A\#G$-module $M^* = (M_e, M_h, \ldots, M_g \ldots)^t$, using the matrix representation of $A\#G$ and matrix multiplication (on the left) on $M^*$. Conversely, every $A\#G$-module $N$ may be represented by $(e_{e,e}N, e_{h,h}N, \ldots, e_{g,g}N, \ldots)^t$ and we may associate to it the $G$-graded $A$-module $N_*$ given by $N_* = \oplus_{g \in G} e_{g,g}N$ and scalar multiplication by (homogeneous) elements of $A$ is the one induced in the obvious way by the scalar matrix multiplication on $N$. A graded morphism $\alpha : M \to M'$ in $A$-gr (hence of degree zero) takes $M_g$ to $M'_g$ for every $g \in G$ and thus we may view $\alpha$ as a morphism $M^* \to (M')^*$ ($A\#G$-linearity is easily checked). Conversely an $A\#G$-linear $\beta : N \to N'$ defines a graded morphism $\beta : N_* \to N'_*$. So we obtain :

**1.3. Proposition.** The correspondences $M \mapsto M^*, N \to N_*$, are inverse to each other and they define an isomorphism between $A$-gr and $A\#G$-mod. As a consequence, the representation type of the $G$-graded ring $A$ is the same as the representation type of $A\#G$.

## 2. Dualities.

The following lemma, pointed out to us by Luo Yun Lun, may exist in some generality in the literature, since we have no reference handy we include it here.

**2.1. Lemma.** Let $R$ be a ring, in general without 1, with a system of matrix units $\{e_{i,j}; i, j \in \mathcal{I}\}$, in general with $\mathcal{I}$ being infinite, such that the following condition holds : $R = \sum_{i,j \in \mathcal{I}} e_{i,i} R e_{j,j}$, then $R \cong M_{|\mathcal{J}|}(S)$ where $S = e_{1,1} R e_{1,1}$.

**Proof.** We define a map $\alpha : M_{|\mathcal{I}|}(S) \to R, \Sigma s_{i,j} \bar{e}_{i,j} \mapsto \Sigma e_{i,1} s_{i,j} e_{1,j}$ and the reader may check all claims. □

**2.2. Theorem (Duality for actions).** Let $S$ be a ring with unit and $G$ a group acting on $S$ by automorphisms, let the action be given by the group morphism $\varphi : G \to \mathrm{Aut} S$. The skew group ring $S *_\varphi G$ is the free $S$-module $S\{\sigma, \sigma \in G\}$ with multiplication : $(s_1\sigma_1)(s_2\sigma_2) = s_1 s_2^{\sigma_1^{-1}} \sigma_1\sigma_2$ where $s_1, s_2 \in S, \sigma_1, \sigma_2 \in G$. Then $(S *_\varphi G)\#G \cong M_{|G|}(S)$.

**Proof.** We aim to provide a complete system of matrix units i.e. a system of matrix units satisfying the condition of the lemma, say $\{e_{g,h}; g, h \in G\}$ such that $S = e_{e,e}((S *_\varphi G)\#G)e_{e,e} \cong S$.

Put $e_{g,k} = (1 * gh^{-1})p_h \in (S *_\varphi G)\#G$.

First it is easy to calculate :

$$e_{g,h}e_{x,y} = (1 * gh^{-1}p_h).(1 * xy^{-1})p_y = (1 * gh^{-1})(1 * xy^{-1})_{hy^{-1}}p_y$$

so if $h = x$ then $e_{g,h}e_{x,y} = (1 * gy^{-1})p_y = e_{g,y}$, and if $h \neq x$ then $e_{g,h}e_{x,y} = 0$.
Furthermore for $s \in S$; $g, h, k \in G$ we also calculate :
$e_{g,g}(s * h)p_k = (1 * e)p_g(s * h)p_k = (s * h)p_k$ if $g = hk$, and zero otherwise.
$(s * h)p_k e_{g,g} = (s * h)p_k(1 * e)p_g = (s * h)p_k$ if $k = g$, and zero otherwise.
Then $(S *_\varphi G)\#G = \sum_{g,h\in G} e_{g,g}((S *_\varphi G)\#G)e_{h,h}$ is easily verified and $e_{e,e}((S *_\varphi G)\#G)e_{e,e} \cong S$ follows from : $e_{e,e}(s * h)e_{e,e} = s * e$ if $h = e$ or zero otherwise. □

**2.3. Theorem (Duality for coactions).** Let $A$ be a $G$-graded ring with unit. We let $G$ act on $A\#G$ by putting : $(\sum_{h,k\in G} a_h p_k)^g = \sum_{h,k\in G} a_h p_{kg}$ ( this is indeed an action by automorphisms) and then we form the skew ring $(A\#G) * G$. We obtain : $(A\#G) * G \simeq M_{|G|}(A)$.

**Proof.** Multiplication in $(A\#G) * G$ is given by the rule :

$$((ap_g) * h)((bp_x) * y) = (ap_g)(bp_x)^{h^{-1}} * hy = (ap_g)(bp_{xh^{-1}}) * hy$$
$$= (ab_{g(xh^{-1})^{-1}})p_{xh^{-1}} * hy = (ab_{ghx^{-1}})p_{xh^{-1}} * hy$$

Put $e_{g,gh} = (1.p_g) * h$.
Now $e_{g,gh}e_{x,xy} = ((1p_g) * h)((1p_x) * y) = (1p_{xh^{-1}}) * hy = e_{xh^{-1},xy}$ if $gh = x$ (then it equals $e_{g,xy}$ too) but it is zero otherwise, hence $\{e_{g,gh}; g, h \in G\}$ is a system of matrix units.
Next one calculates in a straightforward way :

$$e_{x,x}((a_k p_h) * g) = ((1.p_x) * e)((a_k p_h) * g) = (a_k p_h) * g$$

if $x = kh$ and zero if $x \neq kh$;
$((a_k p_h) * g)e_{x,x} = ((a_k p_h) * g)((1p_{hg}) * e) = (a_k p_h) * g$ if $x = hg$ and zero if $x \neq hg$.
From these equalities one easily obtains the fact that $\{e_{g,gh}; g, h \in G\}$ is a complete system. Furthermore we may calculate $e_{e,e}((A\#G) * G)e_{e,e} = \sum_{g\in G}(A_g p_{g^{-1}}) * g = B$. If we define $\alpha : B \to A$, $(r_g p_{g^{-1}}) * g \mapsto r_g$ then $\alpha$ defines an isomorphism and then the claim has been established. Let us just check that $\alpha$ is a ring morphism :

$$\alpha[((r_g p_{g^{-1}}) * g)((r_h p_{h^{-1}}) * h)] = \alpha[(r_g r_h p_{h^{-1}g^{-1}}) * gh]$$
$$= r_g r_h = \alpha((r_g p_{g^{-1}}) * g)\alpha((r_h p_{h^{-1}}) * h)$$

□

## 3. Radicals of Smash Products.

Let $A$ be a $G$-graded ring with unit. If $I = I_e \oplus I_h \oplus \ldots \oplus I_g \oplus \ldots$ is a graded subring of $A$ then $I\#G$ is a subring of $A\#G$; if $I$ is a graded ideal then $I\#G$ is an ideal of $A\#G$.

**3.1. Lemma.** Let $M$ be a graded $A$-module, then we have :

$$\begin{aligned}(\mathrm{Ann}_A(M))\#G &= \mathrm{Ann}_{A\#G}(M^* \oplus M(h)^* \oplus \ldots \oplus M(g)^* \oplus \ldots) \\ &= \bigcap_{x\in G} \mathrm{Ann}_{A\#G}(M(x)^*)\end{aligned}$$

**Proof.** In matrix presentation the annihilation condition is given as

$$\begin{pmatrix} a_e & a_{h^{-1}} & \ldots & a_{g^{-1}} & \ldots \\ a_h & a_e & \ldots & a_{hg^{-1}} & \ldots \\ \vdots & & & & \\ a_g & a_{gh^{-1}} & \ldots & a_e & \ldots \end{pmatrix} \begin{pmatrix} M_x \\ M_{hx} \\ \vdots \\ M_{gx} \end{pmatrix} = 0$$

Let $\mathrm{Ann}_{A\#G}(M(x)) = (I^{(h)}_{hg^{-1}}(x))_{(h,g)}$, then $I^{(e)}_e(x) = \mathrm{Ann}_{A_e}(M_x)$, $I^{(e)}_{h^{-1}}(x) = \mathrm{Ann}_{A_{h^{-1}}}(M_{hx}), \ldots, I^{(h)}_e(x) = \mathrm{Ann}_{A_e} M_{hx}$.
Thus $\cap_{x\in G}\mathrm{Ann}_{A\#G}M(x) = (I^h_{gh^{-1}})_{(h,g)}$, where $:I^{(e)}_e = \cap_{x\in G}\mathrm{Ann}_{A_e}(M_x) = \mathrm{Ann}_{A_e}(M), I^{(h)}_h = \cap_{x\in G}\mathrm{Ann}_{A_h}(M_x) = \mathrm{Ann}_{A_h}(M), I^{(h)}_e = \cap_{x\in G}\mathrm{Ann}_{A_e}(M_{hx}) = \cap_{y\in G}\mathrm{Ann}_{A_e}(My) = \mathrm{Ann}_{A_e}(M)$. This learns that

$$\cap_{x\in G}\mathrm{Ann}_{A\#G}(M(x)) = (\mathrm{Ann}_{A_{hg^{-1}}}(M))_{h,g} = \mathrm{Ann}_A(M)\#G \qquad \square$$

**3.2. Corollary.** For $N \in A\#G$-mod we have :

$$\cap_{x\in G}\mathrm{Ann}_{A\#G}((N_*(x))^*) = \mathrm{Ann}_A(N_*)\#G$$

**3.3. Lemma.** For $M \in A$-gr we have the following equivalences :

**1.** $M$ is gr-simple if and only if $M(x)$ is gr-simple, for all $x \in G$.

**2.** $M$ is gr-simple if and only if $M^*$ is a simple $A\#G$-module.

**Proof. 1.** is easy **2.** follows from the isomorphism given in Proposition 1.3. $\square$

As a straightforward consequence of the foregoing corollary and lemma we obtain :

**3.4. Theorem.** $J(A\#G) = J^g(A)\#G$, where $J(-)$ denotes the Jacobson radical and $J^g(-)$ the graded Jacobson radical. We devote the sequel of this section to proving the equivalent of Theorem 3.4. for the Baer-radical of modules.

An $M \in A$-gr is called a **gr-prime** $A$-module if for every nonzero graded submodule $N$ of $M$ and every graded ideal $I$ of $A$, $IN = 0$ implies $I \subset \mathrm{Ann}_A(M)$. Note that $\mathrm{Ann}_A(M)$ is graded. For a ring $B$, possibly without unit, an $N \in B$-mod is **prime** if for every $B$-submodule $P \neq 0$ and ideal $J$ of $B$ such that $JP = 0$, $J \subset \mathrm{Ann}_B(P)$ holds. Note that in case $B$ has no unit, modules $X$ are always assumed to satisfy $BX = X$. The Baer radical of $B$, denoted by $I\!B(B)$, is the intersection of all $\mathrm{Ann}_B(N)$ for prime $B$-modules $N$. The graded Baer radical of a $G$-graded ring $A$ is the intersection of all annihilators (and these are graded) of gr-prime $A$-modules; we denote this radical by $I\!B^g(A)$. Clearly, if $M \in A$-gr is prime then each $M(x)$ is also gr-prime. We first proceed to prove that $M^* \oplus M(h)^* \oplus \ldots \oplus M(g)^* \oplus \ldots$ is then a prime $A\#G$-module. Observation : a graded submodule $N$ of a gr-prime $M$ is also gr-prime.

**3.5. Lemma.** Let $\{M_i, i \in \mathcal{I}\}$ be a family of gr-prime $A$-modules having the same annihilator $I$ in $A$, then $\oplus_{i\in\mathcal{I}} M_i$ is also gr-prime with annihilator $I$.

**Proof.** Put $P = \oplus_{i\in\mathcal{I}} M_i$, then $P_g = \oplus_{i\in\mathcal{I}}(M_i)_g$ for all $g \in G$. Let $J$ be a graded ideal of $A$ and $N$ a nonzero graded $A$-submodule of $P$ such that $JN = 0$. We aim to prove $J \subset \mathrm{Ann}_A P = \cap_{i\in\mathcal{J}} \mathrm{Ann}_A M_i = I$. The projection $P \to M_i$ is denoted by $\pi_i$; put $N_i = \pi_i N$, then $0 \neq N \subset \oplus_{i\in\mathcal{J}} N_i$. Obviously, $JN = 0$ implies $JN_i = 0$ for all $i \in \mathcal{I}$. Since at least one $N_i$ is nonzero and $M_i$ is a gr-prime module we do obtain $J \subset \mathrm{Ann}_A(M_i) = \mathrm{Ann}_A(P) = I$.

**3.6. Proposition.** If $M$ is a gr-prime module then $(\oplus_{x\in G} M(x))^* = P^*$ is a prime $A\#G$-module.

**Proof.** Put $\mathrm{Ann}_A(M) = \oplus_{h\in G} I_h$. By the above lemma and remarks :

$$\mathrm{Ann}_{A\#G}(P^*) = \mathrm{Ann}_{A\#G}\left(\begin{pmatrix} M_e \\ M_h \\ \vdots \\ M_g \\ \vdots \end{pmatrix} \oplus \begin{pmatrix} M_h \\ \\ \vdots \\ M_{g^h} \\ \vdots \end{pmatrix} \oplus \ldots \oplus \begin{pmatrix} M_g \\ M_{hg} \\ \vdots \\ M_{g^2} \\ \vdots \end{pmatrix} \oplus \ldots\right) =$$

$$= \mathrm{Ann}_{A\#G}(M, \ldots, M)^t = \mathrm{Ann}_A(M)\#G$$

Consider a nonzero $A\#G$-submodule $N = (N(e), N(h), \ldots, N(g), \ldots)^t$ of $P^*$. Put

$\mathrm{Ann}_{A\#G}(N) = J = (J^{(h)}_{hg^{-1}})_{(h,g)}$; we aim to establish $J \subset I\#G$.

We say that an $A\#G$-submodule $T$ of $P^*$ is a homogeneous submodule if $T$ may be written as $(T^{(e)}, T^{(h)}, \ldots, T^{(g)}, \ldots)^t$ where each $T^{(g)}$ is a graded $A$-submodule of $M$, for $g \in G$. Put $(T^{(g)})_h = T^{(g)}_{gh}$. Let $\overline{N}$ be the unique minimal homogeneous $A\#G$-submodule containing $N$. One easily checks $\mathrm{Ann}_{A\#G}(\overline{N}) = \mathrm{Ann}_{A\#G}(N)$ and therefore $J.\overline{N} = 0$. So we obtain :

$$0 = J.\overline{N} =$$

$$= \begin{pmatrix} J^{(e)}_e & J^{(e)}_{h^{-1}} & \cdots \\ J^{(h)}_h & J^{(h)}_e & \cdots \\ \vdots & \vdots & \end{pmatrix} \begin{pmatrix} N^{(e)}_e & + & N^{(e)}_h & + & \cdots & + & N^{(e)}_g & + & \cdots \\ N^{(h)}_h & + & N^{(h)}_{hh} & + & \cdots & + & N^{(h)}_{h_g} & + & \cdots \\ \cdots & \cdots & \cdots & \cdots & \cdots & \cdots & \cdots & \cdots & \cdots \\ 0 & + & 0 & + & \cdots & + & 0 & & \\ 0 & + & 0 & + & \cdots & + & 0 & & \end{pmatrix}$$

Some rows may be zero, but we can find a nonzero column, say the first one. Since $N^{(e)}_e \oplus N^{(h)}_h \oplus \ldots \oplus N^{(g)}_g \oplus \ldots$ is a nonzero graded $A$-module and $\sum_{h\in G} J^{(e)}_{h^{-1}}$ is a right ideal of $A$, the annihilation condition leads to $(\sum_{h\in G} J^{(e)}_{h^{-1}})(\sum_{h\in G} N^{(h)}_h) = 0$, hence $\sum_{h\in G} J^{(e)}_{h^{-1}} \subset \mathrm{Ann}_A M$ by the gr-prime condition on $M$. Applying a similar argument to the row $(J^{(h)}_h, J^{(h)}_e, \ldots, J^{(h)}_{hg^{-1}}, \ldots)$ we obtain that $\sum_{h\in G} J^{(h)}_{hg^{-1}} \subset \mathrm{Ann}_A M$ and therefore we arrive at $J \subset \mathrm{Ann}_A(M)\#G = \mathrm{Ann}_{A\#G}(P^*)$. □

**3.7. Theorem.** $\mathbb{B}(A\#G) = \mathbb{B}^g(A)\#G$.

**Proof.** It will suffice to establish that a prime $A\#G$-module $N$ corresponds to a gr-prime $A$-module $N_*$. Therefore consider a nonzero graded submodule $P$ of $N_*$ and let $IP = 0$ for some graded ideal $I$ of $A, I = I_e + I_h + \ldots$. Clearly we have that $P^*$ is an $A\#G$-submodule of $N$ and $I\#G$ is an ideal of $A\#G$ such that $(I\#G)P^* = 0$. Since $N$ is prime we obtain $(I\#G)N = 0$ hence $IN_* = 0$ or $I \subset \mathrm{Ann}_A(N_*)$. □

## 4. Some Applications.

We first use the connection between the graded ring and the small additive category associated to the smash product in order to derive some results of graded nature from the structure theorems for additive categories appearing in [6]. We recall :

**4.1. Theorem.** Let $C$ be a small additive category with object set $\Sigma$ and write ${}_\alpha C_\beta$ for $\mathrm{Hom}_C(\alpha, \beta)$. We may form the generalized matrix ring $({}_\beta C_\gamma)_{(\beta,\gamma)}$. If $H$ is any finite (consequently, any subset will do) subset of $\Sigma$ then we have : $\{({}_\alpha C_\beta); \alpha, \beta \in H\} \cap J\{({}_\alpha C_\beta); \alpha, \beta \in \Sigma\} = J\{({}_\alpha C_\beta); \alpha, \beta \in H\}$, i.e. if

$J\{({}_\alpha C_\beta);\alpha,\beta\in\Sigma\}=\{({}_\alpha J_\beta);\alpha,\beta\in\Sigma\}$ then $J\{({}_\alpha C_\beta);\alpha,\beta\in H\}=\{{}_\alpha J_\beta);\alpha,\beta\in H\}$.

**4.2. Theorem.** If $A$ is a $G$-graded ring (with unit) and $H$ is a subgroup of $G$ then $J^g(A^H)=J^g(A)\cap A^H=(J^g(A))^H$, where $A^H=\oplus_{h\in H}A_h$.

**Proof.** By the duality theorem we have $J^g(A^H)\#H = J(A^H\#H)$. By the above theorem : $J(A^H\#H)=J(A\#G)\cap(A^H\#H)=(J^g(A)\#G)\cap(A^H\#H)=(J^g(A))^H\#H$ and form $J^g(A^H)\#H = (J^g(A))^H\#H$ the equality $(J^g(A))^H = J^g(A^H)$ follows. □

Recall from [6] again :

**4.3. Theorem.** Consider the generalized matrix ring $({}_\alpha C_\beta)_{(\alpha,\beta)}$, for the small additive category $C$. The category $C$ is primitive if and only if $C$ is a dense subcategory of VD, the category of $D$-vector spaces over a division ring $D$ ("dense" means that ${}_\alpha C_\beta$ is a dense subset of $\mathrm{Hom}_D(\alpha,\beta)$, where $\alpha,\beta$ are $D$-vectorspaces up to identification of $C$ in $VD$, i.e. for any $n\in I\!N$ and $D$-independent elements $a_1,\ldots,a_n$ of $\alpha$, any elements $b_1,\ldots,b_n$ of $\beta$, there exists $\theta\in{}_\alpha C_\beta$ such that $a_i\theta=b_i, c=1,\ldots,n$.

From [6] we retain the following.

**4.4. Theorem.** Let $C$ be a simple additive category and for all $\alpha\in\Sigma, {}_\alpha C_\alpha$ is Artinian then $C$ is isomorphic to a full subcategory of the category FVD of all finite dimensional $D$-vector spaces over the division ring $D$.

**Remark.** A $C$-module $M$ may be written as the column $(M_\alpha,\alpha\in\Sigma)^t$ with matrix action of $({}_\alpha C_\beta)_{(\alpha,\beta)}$ defined on it. The functor $C\to VD$ takes $\alpha$ to $(M_\alpha)_D$ where $D=\mathrm{End}_{{}_\alpha C_\alpha}(M_\alpha)$ and ${}_\alpha a_\beta\in{}_\alpha C_\beta$ to $t_{{}_\alpha a_\beta}:M_\alpha\to M_\beta, m\mapsto_\alpha a_\beta.m$.

Before deriving the corresponding results on graded rings we need the following easy lemma.

**4.5. Lemma.** If the $G$-graded ring $A$ is a direct sum of minimal graded left ideals then $A\#G$ is a direct sum of minimal left ideals.

**Proof.** Put $A=L_1\oplus\ldots\oplus L_j\oplus\ldots, L_i=(L_i)_e\oplus(L_i)_h\oplus\ldots\oplus(L_i)_g\oplus\ldots$. Then $A\#G\cong(A_{hg^{-1}})_{(h,g)}=(\sum_i(L_i)_{hg^{-1}})_{(h,g)}$, and the latter clearly decomposes as a sum of minimal left ideals of $A\#G$, each one of the form

$$\begin{pmatrix} 0 & \dots & 0 & \dots & (L_i)_{g^{-1}} & 0 & \dots & 0 \\ \vdots & & \vdots & & \vdots & \vdots & & \vdots \\ \vdots & & \vdots & & (L_i)_{hg^{-1}} & \vdots & & \vdots \\ \vdots & & \vdots & & \vdots & \vdots & & \vdots \\ 0 & \dots & 0 & \dots & \dots & 0 & \dots & 0 \end{pmatrix}$$

□

**4.6. Theorem.** For a $G$-graded ring $A$ with unit the following assertions are equivalent :

**1.** $A$ is gr-uniformly primitive with all $A_h \neq 0$

**2.** $A\#G$ is a primitive ring

**3.** There exists a division ring $D$ and a $D$-vectorspace $M$ such that for all $h \in G$, $A_h$ is a dense subring of $\mathrm{End}_D(M)$.

**Proof. 1. ⇒ 2.** Let $M = M_e + M_h + \dots$ be a faithful $G$-invariant graded $A$-module amd let $\alpha_g : M \to M(h)$ be fixed isomorphisms for $h \in G$. Put $N = \{(m_e^{(e)} + \alpha_h m_e^{(e)} + \dots + \alpha_g m_e^{(e)}, \dots, \alpha_h m_e^{(h)} + m_e^{(h)} + \dots + \alpha_{hg^{-1}} m_e^{(e)} + \dots, \dots)^t;$ $m_e^{(e)}, m_e^{(h)}, \dots \in M_e\}$. It is clear that $N$ is an $A\#G$-submodule of $M^* + M(h^{-1})^* + \dots + M(g^{-1})^* + \dots$. Now if $I_h \subset A_h$ is such that $I_h(m_e^{(g)} + \alpha_h m_e^{(g)} + \dots) = 0$ then $I_h m_e^{(g)} = I_h \alpha_h m_e^{(g)} = \dots = 0$, and from this we deduce that : $\mathrm{Ann}_{A\#G}(N) = (\mathrm{Ann}_A(M))\#G = 0$, because the first term equals $\cap_{h\in G} \mathrm{Ann}_{A\#G} M(h)^*$. Consequently $N$ is a faithful $A\#G$-module and since $N \cong M^*$ it is also a $A\#G$-simple module (apply lemma 3.3.(2)) and therefore $A\#G$ is a primitive ring.

**2. ⇒ 3.** The primitive additive category $(A^{(h)}_{hg^{-1}})_{(h,g)}$ with object set $G$ is isomorphic to a dense additive category, over some division ring $D$, in view of Theorem 4.3. Paying attention to, and using the notation of, the remark preceding Lemma 4.5. we see that $(M_e)_D \cong (M_h)_D \cong \dots \cong (M_g)_D \cong \dots$ and we also have embeddings :

$$A_{hg^{-1}} p_h = A^{(h)}_{hg^{-1}} \hookrightarrow \mathrm{Hom}_D(M_h, M_g) = \mathrm{Hom}_D(M_e, M_e)$$

$$A_{(hx)(gx)^{-1}} p_{hx} = A^{(hx)}_{hg^{-1}} \hookrightarrow \mathrm{Hom}_D(M_{hx}, M_{gx}) = \mathrm{Hom}_D(M_e, M_e)$$

and $A^{(h)}_{hg^{-1}} = A^{(hx)}_{hg^{-1}}$ as a subring of $\mathrm{Hom}_D(M_e, M_e)$. It follows that $A_e \cong A_e^{(e)} = A_e^{(h)} = \dots$ is a dense subring of $\mathrm{Hom}_D(M_e, M_e)$ and the same holds for $A_{hg^{-1}} \cong A^{(e)}_{hg^{-1}} = A^{(h)}_{hg^{-1}} = \dots$.

**3.** $\Rightarrow$ **1.** Suppose $A = \oplus_{h \in G} A_h$ and all $A_h$ being dense subrings of $\mathrm{End}_D(V)$. Now we define $V_g$ equal to $V$ as a right $D$-vector space but defining $a_h.n_g = a_h(n_g)$ where on the right we view $a_h$ as the linear transformation $V \to V$ but putting $a_h(n_g)$ in $V_{hg}$. Put $M = \oplus_{g \in G} V_g$. Clearly $M$ is a graded $A$-module and it is $G$ invariant as well as faithfull. □

**Remark.** In the embedding of $A^{(h)}_{hg^{-1}}$ in $\mathrm{Hom}_D(M_e, M_e)$ one has to be careful how to write the order of multiplication in $\mathrm{Hom}_D(M_e, M_e)$ (or else write the opposite ring).

Considering gr-simple modules we may also obtain the following extension of the graded version of the Wedderburn theorem as given in [ 3 ], in particular in the gr-uniformly simple case.

**4.7. Theorem.** Let $A$ be a $G$-graded ring such that $A_e$ is left Artinian, then the following statements are equivalent :

**1.** $A$ is gr-quasi-simple (no graded ideals that are nontrivial) and it has a $G$-invariant gr-simple module.

**2.** $A\#G$ is simple.

**3.** For all $h \in G, A_h = M_n(D)$ for some $n$ not depending on $h$ and some division ring $D$.

**4.** $A$ is gr-uniformly simple with all $A_h \neq 0$.

**Proof. 1.** $\Rightarrow$ **3.** The foregoing theorem implies that $A$ is gr-uniformly primitive and each $A_h$ is dense in $\mathrm{End}_D(V)$ for some $D$-vectorspace $V$. But $A_e$ is then a left Artinian dense subring of $\mathrm{End}_D(V)$ and therefore $V$ is finite dimensional over $D$; that all $A_h$ are now isomorphic to $M_n(D)$ is obvious.

**3.** $\Rightarrow$ **2.** Trivial

**2.** $\Rightarrow$ **3.** Similar argumentation as in foregoing theorem but using Theorem 4.4. instead of Theorem 4.3. Since **4.** $\Rightarrow$ **1.** is trivial, the proof is finished if we establish **3.** $\Rightarrow$ **4.**.

**3.** $\Rightarrow$ **4.**. Let $C_i$ be the $i$-th column of $M_n(D)$; put $C_{i,h}$ equal to this column viewed as a left module in $A_h \cong M_n(D)$ and define $L_i = \oplus_{h \in G} C_{i,h}$. It is clear that $L_i$ is a minimal graded left ideal of $A$ and that $A = L_1 \oplus \ldots \oplus L_n$. Moreover $L_i \cong L_j$ (for all $i, j$) as graded $A$-modules. That each $L_i$ is $G$-invariant is obvious from the definition, so the claim follows. □

**4.8. Corollary 1.** The conditions in the theorem are equivalent to the following condition.

**4'.** $A$ is gr-simple having a unique, up to isomorphism, gr-simple $A$-module (this is then $G$-invariant too).

**2.** The Wedderburn theorem on p. 47 of [ 3 ] proves that a gr-uniformly simple ring $A$ is of the form $M_n(\Delta)(\bar{e})$ i.e. $(M_n(\Delta))_\lambda \cong M_n(\Delta_\lambda) \cong M_n(\Delta_e) = M_n(D), D = \Delta_e$ a division ring. From the above it follows thus that a gr-uniformly simple ring $A$ with $A_h \neq 0$ for all $h \in G$, has a $G$-invariant (and up to isomorphism unique) gr-simple module; the converse was mentioned in [ 3 ] but this implication was missing there.

**4.9. Note.** If for every $h \in G$ such that $A_h \neq 0$ we have $\Delta_h \neq 0$ then also $A \cong M_n(\Delta)(\bar{e})$ and it is strongly graded by the subgroup sup($\Delta$) of $G$ (those $g \in G$ such that $\Delta g \neq 0$). However, if $G \neq \text{Sup}(\Delta)$ then $A$ is not gr-uniformly simple. As an example consider the $\mathbb{Z}$-graded ring $A = M_2(k[X, X^{-1}])(0,1)$ where deg$X = 2$. One easily verifies that this ring is even strongly graded by $\mathbb{Z}$ but its part of degree zero equals $k \oplus k$ and as this is not simple $A$ cannot be gr-uniformly simple !

**References.**


[1] M. Cohen, S. Montgomery, *Group Graded Rings, Smash Products and Group Actions*, Trans. A.M.S. 284, 1984, 237-258.

[2] C. Năstăsescu, N. Rodinó, *Group Graded Rings and Smash Products*, Rend. Sem. Mat. Univ. Padova, 74, 1985.

[3] C. Năstăsescu, F. Van Oystaeyen, *Graded Ring Theory*, Math. Library vol. 28, North Holland, 1982.

[4] C. Năstăsescu, S. Raianu, F. Van Oystaeyen, *Graded Modules over G-sets, Applications*. To appear.

[5] D. Quin, *Group Graded Rings and Duality*, Trans. AMS 292, 1985, 155-167.

[6] Liu Shaoxue, *Jacobson Structure Theorems for Additive Categories*, to appear.

[7] Liu Shaoxue, *A Wedderburn-Artin Theorem for Additive Categories*, to appear.

[8] M. Van den Bergh, *On a Theorem of Mongtomery and Cohen*, Proc. AMS

# Reflexive Auslander-Reiten Sequences

Hiroyuki Tachikawa
Institute of Mathematics, The University of Tsukuba
305 Ibaraki, Japan

Let A be a finite dimensional algebra over field K and X a finitely generated A-module. X is said to be reflexive (resp. torsionless) if the canonical map $\sigma_X : X \to X^{**}$ is an isomorphism (resp. an monomorphism), where $X^* = \mathrm{Hom}_A(X,A)$ and $\sigma_X$: $x \to (f \to f(x))$, $f \in X^*$. It is well known that A is selfinjective if and only if every A-module is reflexive. Cf.[5,7,8].

In this paper we shall say an Auslander-Reiten sequence is reflexive if all modules appeared in it are reflexive, and in order to generalize the above result, we study algebras for which all Auslander-Reiten sequences $0 \to X \to Y \to Z \to 0$ with projective A-modules X are reflexive. In §2 we give a characterization of such algebras . In fact, such algebras are maximal quotient QF-3 algebras satisfying a condition that the dominant dimensions of endomorphism rings of those minimal generator-cogenerator modules are at least four. Further we shall prove that no Auslander algebra has such reflexive Auslander-Reiten sequences.

Hereafter we shall abbreviate "Auslander-Reiten sequence"



*F. van Oystaeyen and L. Le Bruyn (eds.), Perspectives in Ring Theory, 311–320.*

to "AR-sequence". E(X) denotes the injective envelope of X ,and D, $\tau$ and $\tau^{-1}$ means $Hom_K(?,K)$, DTr and TrD respectively.

§ 1 Reflexive AR-sequences and QF-3 Algebras

Let $0 \to A \to Q_1 \to Q_2 \to Q_3 \to ***$ be a minimal injective resolution of a left module A . Following [10] dom dim ${}_AA \geqq n$ if $Q_i$ is projective for $i \leqq n$. It is well known that A is QF-3 (resp. a maximal quotient QF-3 algebra) if and only if dom dim ${}_AA \geqq 1$ (resp. dom dim ${}_AA \geqq 2$). Cf.[9,10,11,12].

Lemma 1.1. Let $0 \to e\,A \to Y \to Z \to 0$ be an AR-sequence with an indecomposable projective right A-module e A, where e denotes a primitive idempotent of A. Let $\rho$ : rad A e $\to$ A e be the inclusion map and $\iota : (e\,A^*)^* \to A\,e^*$ the canonical isomorphism. Then it holds that

a) $\operatorname{Ker} \rho^* \iota\, \sigma_{e\,A} \simeq (\bar{A}\,\bar{e})^*$ and $\operatorname{Cok} \rho^* \iota\, \sigma_{e\,A} \simeq \operatorname{Ext}^1_A(\bar{A}\,\bar{e}, A)$,

b) $0 \to \operatorname{Ker} \rho^* \iota\, \sigma_{e\,A} \to \operatorname{Ker} \sigma_Y \to \operatorname{Ker} \sigma_Z \to \operatorname{Cok} \rho^* \iota\, \sigma_{e\,A}$
$\to \operatorname{Cok} \sigma_Y \to \operatorname{Cok} \sigma_Z$ is exact.

Proof. a) By applying $Hom_A(?,A)$ to an exact sequence $0 \to \text{rad}\,A\,e \overset{\rho}{\to} A\,e \to \bar{A}\,\bar{e} \to 0$ we have the following exact sequence

$$0 \to (\bar{A}\,\bar{e})^* \to A\,e^* \overset{\rho^*}{\to} (\text{rad}\,A\,e)^* \to \operatorname{Ext}^1_A(\bar{A}\,\bar{e}, A) \to 0.$$

So the conclusion is immediate.

b) Since rad A e is a unique maximal submodule of A e, a right A-homomorphism $\phi : e\,A \to A$ is not splitable monomorphism if and only if $\phi(e) \in$ rad A e. Hence $0 \to Z^* \to Y^* \to \text{rad}\,A\,e \to 0$ is exact and we have the following commutative diagram:

$$\begin{array}{ccccccccc}
0 & \to & e\,A & \to & Y & \to & Z & \to & 0 \\
 & & \downarrow{\scriptstyle \sigma_{e\,A}} & & & & & & \\
 & & e\,A^{**} & & \Big\downarrow{\scriptstyle \sigma_Y} & & \Big\downarrow{\scriptstyle \sigma_Z} & & \\
 & & \downarrow{\scriptstyle \iota} & & & & & & \\
 & & A\,e^{*} & & & & & & \\
 & & \downarrow{\scriptstyle \rho^{*}} & & & & & & \\
0 & \to & (\mathrm{rad}\ A\,e)^{*} & \to & Y^{**} & \to & Z^{**} & & .
\end{array} \qquad (1)$$

By the Snake lemma the conclusion is evident.

Proposition 1.2. Assume that Y is reflexive and Z is torsionless for every AR-sequence $0 \to e\,A \to Y \to Z \to 0$ with an indecomposable projective right A-module e A. Then dom dim ${}_{A}A \geq 2$, i.e., A is a maximal left quotient QF-3 algebra.

Proof. For a given AR-sequence e A is not injective. Hence it follows from Lemma 1.1 that $(\bar{A}\,\bar{e})^{*} = 0$ and $\mathrm{Ext}^1_A(\bar{A}\,\bar{e},A) = 0$ for every non-injective, projective right A-module e A. Therefore if $(\bar{A}\,\bar{f})^{*} \neq 0$ for a primitive idempotent f of A, then f A is projective and injective. This implies that if $\bar{A}\,\bar{f}$ appears in the socle of ${}_{A}A$, then D(f A) is projective and injective left A-module. Therefore $E({}_{A}A)$ is projective, i.e., A is left QF-3.

Now assume that $\bar{A}\,\bar{e}$ appears in the socle of $E({}_{A}A)/A$.

Suppose that e A is not injective. Then from our assumption it follows that $\mathrm{Ext}^1_A(\bar{A}\,\bar{e},A) = 0$.

Hence in the following diagram

$$\begin{array}{ccccccccc}
 & & & & & & \bar{A}\,\bar{e} & & \\
 & & & & \swarrow{\scriptstyle \Theta} & & \downarrow{\scriptstyle \theta} & & \\
0 & \to & A & \to & E({}_{A}A) & \to & E({}_{A}A)/A & \to & 0
\end{array}$$

any non-zero homomorphism $\theta$ can be lifted to a homomorphism $\Theta : \bar{A}\,\bar{e} \to E({}_AA)$. But this contradicts to that $E({}_AA)$ is an essential extension of ${}_AA$. Therefore every e A is injective. Since a projective left A-module D(e A) appears as a direct summand of $E({}_AA)$, we have dom dim ${}_AA \geqq 2$.

From now on we shall say that A has reflexive AR-sequences if all AR-sequences $0 \to e\,A \to Y \to \tau^{-1} e\,A \to 0$ with projectives e A's are reflexive.

Then we have

Corollary 1.3. If A has reflexive AR-sequences, then dom dim ${}_AA \geqq 2$.

Remark: The arguments in the proof of Proposition 1.2 seem to be appeared essentially in [6,9,12]. But the context of Proposition 1.2 is slightly different from them. So we don't omit the proof.

Further we would like to remark that conversely if dom dim ${}_AA \geqq 2$ and a primitive right ideal e A is not injective, then $(\bar{A}\,\bar{e})^* = 0$ and $\mathrm{Ext}_A^1(\bar{A}\,\bar{e}, A) = 0$. For if $(\bar{A}\,\bar{e})^* \neq 0$, then in the following commutative diagram:

$$\begin{array}{ccccc} 0 & \to & \bar{A}\,\bar{e} & \to & D(e\,A) \\ & & \downarrow \psi & \swarrow \Psi & \\ & & E({}_AA) & & \end{array}$$

a non-zero monomorphism $\psi$ is extended to a monomorphism $\Psi$, and as $E({}_AA)$ is projective, D(e A) is projective. Hence e A is injective. But this contradicts to our assumption. Next assume that $(\bar{A}\,\bar{e})^* = 0$ and $\mathrm{Ext}_A^1(\bar{A}\,\bar{e}, A) \neq 0$. Then the last condition implies that $\bar{A}\,\bar{e}$ appears as a submodule of

$E({}_A A)/A$. But from dom dim ${}_A A \geq 2$ it follows that $E(E({}_A A)/A)$ is projective and we have again a contradiction.

There is an interesting class of QF-3 algebras which is called Auslander algebras. An algebra A is said to be an Auslander algebra if and only if dom dim ${}_A A \geq 2$ and gl. dim ${}_A A \leq 2$.

It arises naturally a question whether Auslander algebras have reflexive AR-sequences. Unfortunately the next Proposition provides the negative answer.

Proposition 1.4. No Auslander algebra has reflexive AR-sequences.

Proof. Let A be an Auslander algebra and e A a non-injective primitive ideal. Without losing generality we may assume that $\tau^{-1} e\,A$ is torsionless.

Let $0 \to e\,A \to Q_o \to Q_1 \to \cdots$ be the minimal injective resolution of e A. Then

$$\mathrm{Hom}_A(D(A),Q_o) \to \mathrm{Hom}_A(D(A),Q_1) \overset{\alpha}{\to} \tau^{-1} e\,A \to 0$$

is exact and $\mathrm{Hom}_A(D(A),Q_i)$, i = 1,2, are projective.

Hence $0 \to \tau^{-1} e\,A^* \overset{\alpha^*}{\to} \mathrm{Hom}_A(D(A),Q_1)^* \to \mathrm{Hom}_A(D(A),Q_o)^*$ is exact. But since gl. dim $A \leq 2$, $\tau^{-1} e\,A^*$ is projective but a torsionless module $\tau^{-1} e\,A$ is reflexive if and only if $\mathrm{Ext}^1_A(\mathrm{Cok}\ \alpha^*, A) = 0$. Hence $0 \to \tau^{-1} e\,A^* \to \mathrm{Hom}_A(D(A),Q_1)^* \to \mathrm{Cok}\ \alpha^* \to 0$ is split and consequently $\tau^{-1} e\,A$ is projective. This is a contradiction.

Corollary 1.5. An algebra A has not reflexive

AR-sequences if $\tau^{-1}e\ A^*$ is projective.

We can state, however, the following path algebras defined by quivers with relations have reflexive AR-sequences:

Example 1

$\cdot \xrightarrow{\alpha_1} \cdot \xrightarrow{\alpha_2} \cdot \rightarrow \cdots \rightarrow \cdot \xrightarrow{\alpha_n} \cdot$ for $n \geq 3$ and $\alpha_{i+1}\alpha_i = 0$

for $i = 1,2,\ldots,n-1$.

Example 2

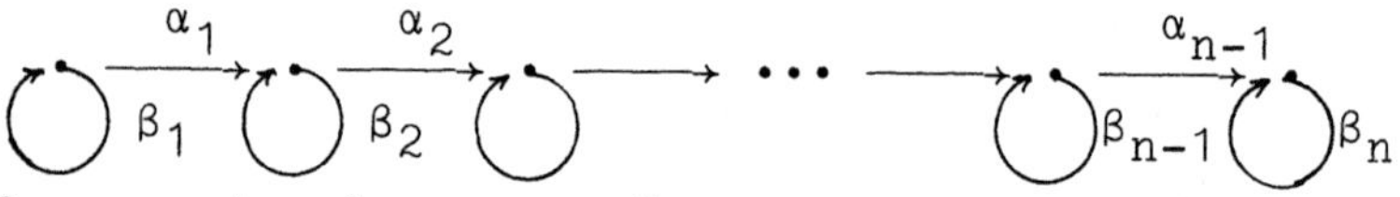


for $n \geq 4$ and $\alpha_{i+1}\alpha_i = 0$, $\beta_{i+1}\alpha_i = \alpha_i\beta_i$ for $i = 1,2,\ldots,n-1$ and $\beta_i^2 = 0$ for $i = 1,2,\ldots,n$.

## 2. Reflexivility of $\tau^{-1}e\ A$ and $\mathrm{End}_A(A \oplus DA)$

Lemma 2.1. Let M be a finitely generated, generator and cogenerator right A-module and R the endomorphism ring of M. Then the following conditions 1) and 2) are equivalent to each other.

1) $\mathrm{dom\ dim}\ R_R \geq n + 2$

2) $\mathrm{Ext}_A^i(M,M) = 0$ for $1 \leq i \leq n$.

For the proof we refer to (7.5) Theorem in [12].

Lemma 2.2. Let X and Y be finitely generated right A-modules. Let $\theta$ be a map : $Y \otimes X^* \rightarrow \mathrm{Hom}_A(X,Y)$ defined by $[\theta(y \otimes f)](x) = y\,f(x)$ for all $x \in X$. Then $\mathrm{Cok}\ \theta = \underline{\mathrm{Hom}}_A(X,Y) \simeq D\ \mathrm{Ext}_A^1(Y,\tau X)$ and $\mathrm{Ker}\ \theta \simeq D\ \mathrm{Ext}_A^2(Y,\tau X)$.

Corollary 2.3. Let X be a finitely generated right A-module. Then the following conditions 1) and 2) are equivalent to each other.

1) X is reflexive (resp. torsionless).

2) $\mathrm{Ext}_A^i(DA, \tau X) = 0$ for i = 1, 2 (resp. i = 1).

For the proofs of Lemma 2.2 and Corollary 2.3 we can refer to [2]. We would like to mention here that in order to obtain Corollary 2.3 how can we use Lemma 2.2.

Put Y = DA in Lemma 2.2. Then we have an exact sequence: $0 \to D\,\mathrm{Ext}_A^2(DA, \tau X) \to DA \otimes X^* \to \mathrm{Hom}_A(X, DA) \to D\,\mathrm{Ext}_A^1(DA, \tau X) \to 0$ .

On the other hand $D\,\mathrm{Hom}_A(X, DA) \simeq X$ and $D\,(DA \otimes X^*) \simeq X^{**}$, and $\theta$ corresponds to $\sigma_X$. Hence Corollary 2.2 is immediate.

We are now ready to prove

Theorem 2.4. An algebra A has reflexive AR-sequences if and only if the following two conditions 1) and 2) are satisfied:

1) $\mathrm{dom\,dim}\ {}_A A \geq 2$

2) $\mathrm{dom\,dim}\ R_R \geq 4$, where $R = \mathrm{End}_A(A_A \oplus DA_A)$.

Proof. "only if": 1) follows from Proposition 1.2. Since $\tau^{-1} e\,A$ is reflexive for a non-injective, projective right ideal e A, by Corollary 2.3 $\mathrm{Ext}_A^i(DA, \tau\,\tau^{-1} e\,A) = \mathrm{Ext}_A^i(DA, e\,A) = 0$ for i = 1,2. It follows that $\mathrm{Ext}_A^i(DA, A) = 0$ for i = 1,2 and consequently $\mathrm{Ext}_A^i(A_A \oplus DA_A, A_A \oplus DA_A) = 0$ for i = 1,2. Therefore by Lemmma 2.1 $\mathrm{dom\,dim}\ R_R \geq 4$.

"if": By Lemma 2.1 and Corollary 2.3 $\mathrm{Ker}\ \sigma_{\tau^{-1} e\,A} = 0$ and $\mathrm{Cok}\ \sigma_{\tau^{-1} e\,A} = 0$ for every non-injective primitive right ideal e A. Further by the remark after Corollary 1.3 $\mathrm{Ker}\ \rho^{*}\iota\,\sigma_{\tau^{-1}eA} = 0$ and $\mathrm{Cok}\ \rho^{*}\iota\,\sigma_{\tau^{-1}eA} = 0$ in the diagram (1) of the proof of Lemma 1.1. So the exact sequence in Lemma 1.2, a) induces that the middle term Y of AR-sequence $0 \to eA \to Y \to \tau^{-1}eA$ is reflexive.

Remark. As is shown by the following example, for an AR-sequence $0 \to eA \to Y \to \tau^{-1}eA \to 0$ with a primitive right ideal e A, Y is not necessarily reflexive even if $\tau^{-1}eA$ is reflexive.

Example 3. Let A be a path algebra defined by the following quiver with relations. Then Auslander-Reiten quiver is the diagram (2):

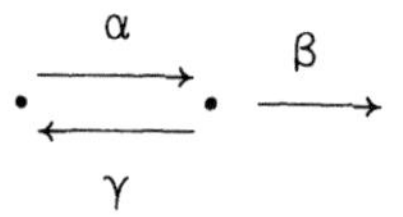


and $\beta\alpha = 0,\ \gamma\alpha = 0,\ \alpha\gamma = 0.$

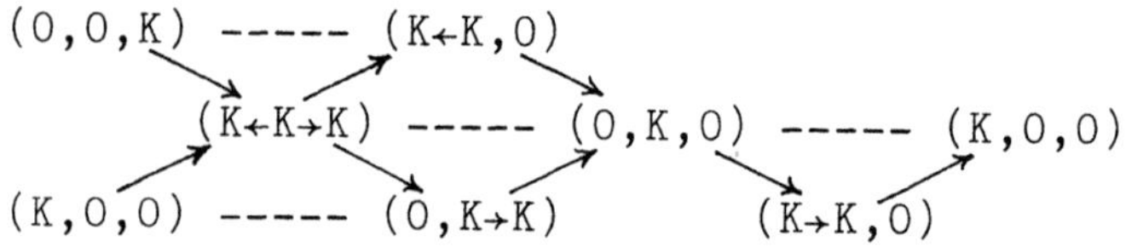


(2)

,where $0 \to (K\leftarrow K\to K) \to (K\leftarrow K,0)\oplus(0,K\to K) \to (0,K,0) \to 0$ is an AR-sequence with a primitive ideal $(K\leftarrow K\to K)$, and $(0,K,0)$ is reflexive, but $(K\leftarrow K,0)\oplus(0,K\to K)$ is not reflexive.

Though quoted already by H. Bass himself [4] this example shows that in order to hold Lemma 3.2 in [3] it needs some additional assumptions.

Finally the author wishes to express his thanks to

Drs. Wakamatsu and K. Yamagata for their valuable comments. Especially Example 3 is due to Yamagata.

References

[1] M. Auslander, Representation dimension of artin algebras, Queen Mary College Lecture Notes (1971).

[2] M. Auslander, Coherent functors, Proc. Conf. on Categorical Algebra, La Jolla(1966), Springer.

[3] H. Bass, Injective dimension in noetherian rings, Trans. Amer. Math. Soc. 102 (1962), 18-29.

[4] H. Bass, On the ubiquity of Gorenstein rings, Math. Zeit. 82 (1963), 8-28.

[5] J. Dieudonne, Remarks on quasi-Frobenius rings, Illinois J. Math. 12 (1958), 346-354.

[6] K. R. Fuller, Double centralizers of injectives and projectives over artinian rings, Illinois J. Math. 14 (1970), 658-664.

[7] J. Jans, Duality in rings, Proc. Amer. Math. Soc. 12 (1961), 829-835.

[8] K. Morita and H. Tachikawa, Character modules, submodules of a free module, and quasi-Frobenius rings, Math. Zeit. 65 (1956), 414-428.

[9] B. J. Mueller, The classification of algebras by dominant dimension, Canand. J. Math. 20 (1968), 394-409.

[10] H. Tachikawa, On dominant dimension of QF-3 algebras, Trans. Amer. Math. Soc. 112 (1964), 249-266.

[11] C. M. Ringel and H. Tachikawa, QF-3 rings, J. reine und angew. Math. 272 (1975), 49-72.

[12] H. Tachikawa, Quasi-Frobenius rings and Generalizations, QF-3 and QF-1 rings, Lecture Notes in Math. (Springer) 35 (1973)

[13] R. M. Thrall, Some generalizations of quasi-Frobenius algebras, Trans. Amer. Math. Soc. 64 (1948), 173-183.

# Noncommutative Invariant Theory

Yasuo Teranishi
Nagoya University
Nagoya, Japan

## 1. Introduction

Throughout this article $K$ will be a field of characteristic zero. Let $V$ be a finite dimensional vector space over $K$. Let

$$T(V) = K \oplus V \oplus V^{\otimes 2} \oplus \dots$$

denote the tensor algebra over $V$. The group of $K$-automorphisms $GL(V)$ can be identified with the group of homogeneous automorphisms of $T(V)$. If $G$ is a subgroup of $GL(V)$, there is an induced homogeneous action of $G$ on $T(V)$.

Let $I$ be a $GL(V)$-invariant two-sided ideal of $T(V)$. Then we get an induced homogeneous action of $G$ on the graded algebra $T(V)/I$. The fixed subalgebra of $T(V)/I$ under the action of $G$ will be denoted by $(T(V)/I)^G$. For example if $C$ is the commutator ideal of $T(V)$, $T(V)/C$ is the symmetric algebra on $V$ :

$$S(V) = K \oplus V \oplus S^2(V) \oplus \dots,$$

where $S^r(v)$ denotes the $r$-th symmetric tensor of $V$.

A classical theorem in invariant theory asserts that, if $G$ is linear reductive, $S(V)^G$ is finitely generated. A natural question arises about the finite generation of the graded algebra $(T(V)/I)^G$. In general the algebra $(T(V)/I)^G$ has not a finite number of generators. In his paper [5], Koryukin proved that if $G$ is linear reductive,



*F. van Oystaeyen and L. Le Bruyn (eds.), Perspectives in Ring Theory, 321–331.*

there is a finite subset of $T(V)^G$ which generates $T(V)^G$ using usual algebra operations and the action of the symmetric groups. In section 2, we prove a theorem (Theorem 2.1) by which we can reduce the problem to find $S$-generators of $T(V)^G$ to that of a commutative case. The Capelli identity plays an essential role.

Some works ([2], [4], [5]) of Dicks-Formanek, Kharchenko and Koryukin show that in almost all cases $T(V)^G$ is not finitely generated in the usual meaning. On the other hand Lane [6] and Kharchenko [3] proved that $T(V)^G$ is a free algebra.

The main problem in invariant theory is to establish so called the first and second fundamental theorems. The first fundamental theorem is an explicit list of invariants and the second fundamental theorem is a list of all the polynomial relations between the list of invariants given by the first fundamental theorem. In view of the Lane Kharchenko theorem, the first and second fundamental theorems for the algebra $T(V)^G$ is to find a free generating system of invariants. For the noncommutative ring of invariants of an $N$-ary form with noncommutative coefficients, one can explicitly obtain a free generating set of invariants (Teranishi [7]). In section 3, we give an algorithm for construction of many invariants when $V$ is an irreducible $SL(n, K)$-module by using the Young tableau method.

In section 4, we prove the dimension for the noncommutative invariants of binary ground form.

**Acknowledgement**

This paper was written while the author was visiting the University of Mannheim in West Germany. He likes to thank the Department of Mathematics and especially Prof. Dr. Popp for the hospitality.

## 2. $S$-generators for noncommutative invariants.

Let $V$ be a finite dimensional vector space over $K$ and let $\{e_1, \ldots, e_n\}$, $\{x_1, \ldots, x_n\}$ be a pair of dual basis for $V$ and $V^*$. Let

$$K < V >= K < x_1, \ldots, x_n >= T(X^*) \quad \text{and}$$
$$K[V] = K < x_1, \ldots, x_n >= S(V^*)$$

denote the free associative algebra of rank $n$ and the polynomial algebra repectively.

Let $f(x)$ be a polynomial in $n$ variables $x = (x_1, \ldots x_n)$ and let $y = (y_1, \ldots, y_n)$ be new variables. Consider the polarization

$$D_{yx}.f = \sum y_i \frac{\partial f}{\partial x_i}$$

The polar process is a derivation on $K[x_1, \ldots, x_n]$. For a polynomial $f(x)$ of degree $d$, let $x(1) = (x(1)_1, \ldots, x(1)_n), \ldots, x(d) = (x(d)_1, \ldots, x(d)_n)$ be new variables and consider the multilinear function

$$Pf(x(1), \ldots, x(d)) = D_{x(d)x} D_{x(d-1)x} \cdots D_{x(1)x} f(x)$$

$Pf$ is called the complete polarization of $f$.

For each $r \in N$, let $V(r)$ be $\oplus V$, the direct sum of $r$ copies of $V$, with the diagonal action of $GL(n, K)$. The polynomial ring on $V(r)$ is the polynomial ring in variables $X(j)_i, 1 \leq jr, 1 \leq i \leq n$.

$$K[V(d)] = K[x(1), \ldots, x(r)]$$

The ring $K[V(r)]$ is an $N^r$-graded algebra by giving the variables $x(j)_i$ degree $(0, \ldots, 0, |0, \ldots, 0)$ where the $j$-th coordinate is 1.

For each $\underline{d} = (d_1, \ldots, d_r) \in N^r$. let $K[V(r)]_{\underline{d}}$ denote the $K$-vector space of $K[V(r)]$ spanned by all homogeneous elements of multidegree $\underline{d}$. Consider the $GL(n, K)$-isomorphism

$$K[V(r)]_{\underline{d}} \simeq K[V]_{d_1} \otimes \ldots \otimes K[V]_{d_r}$$

In particular, if $d_1 = d_2 = \ldots = d_r = 1$, we have

$$K[V(r)]_{\underline{1}} \simeq K < V >_r, (\underline{1} = (1, \ldots, 1) \in N^r)$$

The process of the complete polarization gives an injective $GL(n,K)$-linear map

$$P : K[V]_d \to K[V(d)]_{\underline{1}} \quad (\underline{1} = (1,\ldots,1) \in N^d)$$

For each $\underline{d} \in N^r$,

$$\begin{aligned} K[V(d_1)]_{\underline{1}} \otimes \ldots \otimes K[V(d_r)]_{\underline{1}} &\simeq K < V >_{d_1} \otimes \ldots \otimes K < V >_{d_r} \\ &\simeq K < V >_{|\underline{d}|}, |\underline{d}| = d_1 + \ldots + d_r \end{aligned}$$

This together with the map

$$\otimes P : K[V]_{d_1} \otimes \ldots \otimes K[V]_{d^r} \to K[V(d_1)]_{\underline{1}} \otimes \ldots \otimes K[V(d_r)]_{\underline{1}}$$

gives rise to a $GL(n,K)$-linear map

$$\hat{P} : K[V(r)]_{\underline{d}} \to K < V >_{|\underline{d}|}$$

This map will be called the **complete polarization** in $K < V >$.

For each $d \in N$, consider the action of the symmetric group $S_d$ on $K < V >_d$ :

$$\sigma(x_1 \otimes \ldots \otimes x_d) = x_{\sigma(1)} \otimes \ldots \otimes x_{\sigma(d)}$$

Since this action of $S_d$ commutes with the action of $GL(n,K)$, the algebra $K < V >_d^G$ is closed under the action of $S_d$. Let $\{f_i\}_{i\in I}$ be a system of homogeneous invariants in $K < V >^G$ and suppose that $K < V >^G$ is generated by $\{f_i\}$ using the algebra operations together with the action of $S_d$ on $K < V >_d$ $(d = 1, 2, \ldots, )$. Then $\{f_i\}_{i\in I}$ is called a **homogeneous system of $S$-generators** of $K < V >^G$. If $K < V >^G$ has a homogeneous system of $S$-generators consisting of a finite number of invariants, we say that $K < V >^G$ is **$S$-finitely generated.** We now prove the following.

**Theorem 2.1.** Let $V$ be an $n$-dimensional vector space over $K$ and $G$ a subgroup of $GL(n,K)$. Let $\{f_i\}_{i\in I}$ be an $N^n$-homogeneous system of generators for the ring of invariants $K < V(n) >^G$. Then $\{\hat{P}f_i\}_{i\in I}$ is a homogeneous system of $S$-generators for $K < V >^G$, where $\hat{P}f_i$ is the complete polarization of $f_i$ in $K < V >^G$.

**Proof.** Let $f \in K < V >_d^G$ be an invariant of degree $d$ and consider the corresponding multilinear invariant $\widetilde{f}(x(1),\ldots,x(d))$. It follows from the Capelli identity (see Weyl [9]) that any invariant in the polynomial ring $K[x(1),\ldots,x(d)]$ can be obtained from $\{f_i\}_{i\in I}$ by means of succesive polarization and ring operations. Therefore $\tilde{f}$ is a linear combination of invariants having the following form

$$(*) \qquad D_1 f_{i_1} \dots D_m f_{i_m}$$

where $f_{i_1}, \dots, f_{i_m}$ is a subset of $\{f_i\}_{i\in I}$ and $D_1, \dots, D_m$ are successive polarizations. Since $\tilde{f}$ is multilinear with respect to vectors $x(1), \dots, x(d)$ and $f_i$ are $N^n$-homogeneous, we can assume that each term (*) is of multilinear form with respect to vectors $x(1), \dots, x(d)$. Then one sees that $D_1 f_{i_1}, D_2 f_{i_2}, \dots, D_m f_{i_m}$ are multilinear forms in some variables $x(j_1), \dots, x(j_\alpha)$, $x(k_1), \dots, x(k_\beta), \dots$, $x(l_1), \dots, x(l_\gamma)$, respectively and that

$$(x(j_1), \dots, x(j_\alpha), x(k_1), \dots, x(k_\beta, x(l_1), \dots, x(l_\gamma))$$

is a permutation of $(x(1), x(2), \dots, x(d))$. Therefore the element of $K < V >_d$ corresponding to $D_1 f_{i_1} D_2 f_{i_2} \dots D_m f_{i_m}$ can be obtained from $\hat{P} f_{i_1} \otimes \dots \otimes \hat{P} f_{i_m}$ by a place permutation. This completes the proof.

From this theorem, we obtain

**Koryukin's theorem.** If $G$ is linear reductive, $K < V >^G$ is $S$-finitely generated.

The following theorem follows from the special Capelli identity.

**Theorem 2.2.** Let $V$ be an $n$ dimensional vector space and $G$ a subgroup of the special linear group $SL(n, K)$. Let $\{f_i\}_{i\in I}$ be a homogeneous system of generators for the ring of invariants $K[V(n-1)]^G$. Let $\hat{P} f_i$ be the completely polarizations of $f_i$ in $K < V >$. Then $\{\hat{P} f_i\}_{i\in I}$ together with the standard polynomial of degree $n$

$$S_n(x_1, \dots, x_n) = \sum_\sigma sgn.\sigma x_{\sigma(1)} \otimes \dots \otimes x_{\sigma(n)}$$

constitute a homogeneous system of $S$-generators of $K < V >^G$.

**Example.** Let $G = \{1, \sigma\}$ be a cyclic group of order 2. Consider an action of $G : \sigma(x) = y, \sigma(y) = x$ and $\sigma(z) = -z$. Then $K[x, y, z]^G$ is generated by $x+y, xy, z^2$. On the other hand, consider the ring $K[x(1), y(2), z(1), x(2), y(2), z(2)]^G$. Then this ring is generated by invariants $x(i) + y(i), i = 1, 2, x(i)y(j) + y(i)x(j), z(i)z(j)$, $(x(i)y(j) - x(j)y(i))z(i), 1 \leq i, j \leq 2$. Since $S_3(x, y, z)$ is obtained from $(xy - yx)z$ by place permutations, it follows from Theorem 3.2. that $K < x, y, z >^G$ is $S$-generated by $x + y, xy + yx, (xy - yx)z$ and $z^2$. However $K < x, y, z >^G$ is not finitely generated, since $\sigma$ is not a scalar multiplication.

**3. A construction of (noncommutative) invariants.**

We recall that the irreducible $SL(n,K)$-modules are parametrized by Young diagrams with $< n$ rows. To be more precise, there is a one to one correspondence between irreducible $SL(n,K)$-modules and Young diagrams $\lambda = (\lambda_1,\ldots,\lambda_{n-1})$ with $< n$ rows. If $M_\lambda$ is the irreducible modul;e corresponding to $\lambda$, $M_\lambda$ is, as an $SL(n,K)$-module, isomorphic to $c_\lambda V^{\otimes|\lambda|}$ where $|\lambda| = \lambda_1 + \lambda + \lambda_{n+1}$ and $c_\lambda$ is the Young idempotent corresponding to $\lambda$. Hereafter in this section we identify $M_\lambda$ with $c_\lambda V^{\otimes|\lambda|}$. Then the map $f \to c_\lambda f$ defines a surjective $SL(n,K)$-homomorphism

$$p_r : V^{\otimes|\lambda|} = T_{(V)_{|\lambda|}} \longrightarrow M_\lambda$$

In this section, we construct $SL(n,K)$-invariants in the tensor algebra $T(M_\lambda)$. By an invariant of degree $d$ and weight $k$ we mean an element in $T(M_\lambda)_d$ which satisfies for any $g \in GL(n,K)$

$$g.f = (\det g)^k f$$

By a standard Young tableau of degree $d$ and weight $k$, we mean a Young tableau of the form

$$\begin{matrix} m_{11} & m_{12} & \ldots & m_{1k} \\ m_{21} & m_{22} & \ldots & m_{2k} \\ \vdots & \vdots & & \vdots \\ n_{n1} & m_{n2} & \ldots & m_{nk} \end{matrix}$$

with the following properties :

(1) The numbers are nondecreasing from left to right across along each row and from top to bottom down each column

(2) The numbers are contained in the set $\{1,2,\ldots,d\}$ and each number from $\{1,2,\ldots,d\}$ appears exactly $r$-times in the tableau.

Apparently, $nk = dr$. Suppose that in the $j$-th $(1 \leq j \leq k)$ column $(m_{1j}, m_{2j},\ldots,m_{nj})$ of a standard Young tableau of degree $d$ and weight $k$, the number 1 appears $\alpha(1,j)$ times, the number 2 appears $\alpha(2,j)$ times, etc. By the definition, we have

$$\alpha(1,j) + \ldots + \alpha(d.j) = n, \text{ for } j = 1,2,\ldots,k \text{ and}$$
$$\alpha(i,1) + \ldots + \alpha(i,k) = r, \text{ for } i = 1,2,\ldots,d$$

We set :

$$
\begin{aligned}
&< m_{1j}, m_{2j}, \ldots, m_{nj} > \\
&= \sum_{\sigma} sgn(\sigma)(e_{\sigma(1)} \otimes \ldots \otimes e_{\sigma(\alpha(1,j))}) \\
&\left.\begin{aligned}
&\otimes (e_{\sigma(\alpha(1,j)} + 1 \otimes \ldots \otimes e_{\sigma(\alpha(1,j)+\alpha(2,j)}) \\
&\vdots \\
&\otimes (e_{\sigma(\alpha(1,j)+\ldots+\alpha(d-1,j)+1}) \otimes \ldots \otimes e_{\sigma(n)})
\end{aligned}\right\} d
\end{aligned}
$$

where $\sigma$ is over all the permutations on $\{1,2,\ldots,n\}$ and $e_1,\ldots,e_n$ is a fixed basis of $V$.

Since

$$< m_{1j}, m_{2j}, \ldots, m_{nj} >= \sum_{\sigma} sgn(\sigma) e_{\sigma(1)} \cdots e_{\sigma(n)}$$

$< m_{1j}, m_{1j}, \ldots, m_{nj} >$ is invariant under the action of $SL(n,K)$.
Let $\otimes^d T(V)$ denote the $d$ times tensor algebra of the algebra $T(V)$ whose multiplication is given by

$$({}^d\otimes a_i).({}^d\otimes b_i) = {}^d\otimes(a_i \otimes b_i), a_i, b_i \in T(V)$$

Then $< m_{1j}, m_{2j}, \ldots, m_{nj} >$ is an element in the algebra ${}^d\otimes T(V)$. If we set

$$F_Y =< m_{11} m_{21}, \ldots, m_{n1} > . < m_{12} m_{22}, \ldots, m_{n2} > . < m_{1k} m_{2k}, \ldots, m_{nk} >$$

one finds that $F_y$ is an $SL(n,K)$-invariant in the space ${}^d\otimes T(V)_{|\lambda|}$. Let $f_y$ be the element of ${}^d\otimes T(M_\lambda)$ defined as

$$f_y = {}^d\otimes p_r.F_y, \text{ since } {}^d\otimes T(V)_{|\lambda|} = T(T(V)_{|\lambda|})_d$$

then the construction above shows that $f_y$ is an invariant of degree $d$ and weight $k$. Thus for a given standard Young tableau of degree $d$ and weight $k$, we can construct an invariant of degree $d$ and weight $k$.

**Theorem 3.1.** Let $M_\lambda$ be an irreducible $SL(n,K)$-module and let $Y$ be a standard Young tableau of degree $d$ and weight $k$. Then $f_y$ is an invariant of degree $d$ and weight $k$.

As an application, we consider the invariants of degree 2.

**Proposition 3.2.** Let $M_\lambda$ be the irreducible $SL(n,K)$-module corresponding to a Young diagram $\lambda$ with $< n$ rows. Then

(1) If $M_\lambda$ is not isomorphisc to its dual $M_\lambda^*$ as an $SL(n,K)$-module, then there is no invariant of degree 2.

(2) If $M_\lambda$ is isomorphic to its dual $M_\lambda^*$ as an $SL(n,K)$-module, then the space of invariants of degree 2 is one dimensional and any invariant of degree 2 is, up to constant, given by $f_y$ where

$y =$

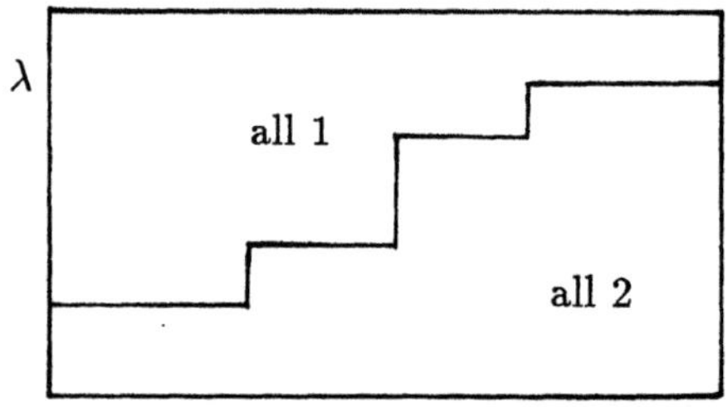


**Proof.**

$$\begin{aligned}\dim .T(M_\lambda)_2^{SL(n,K)} &= \dim .\mathrm{Hom}_{SL(n,K)}(M_\lambda \otimes M_\lambda, 1) \\ &\quad (1 = \text{the trivial rep. of } SL(n,K)) \\ &= \dim .\mathrm{Hom}_{SL(n,K)}(M_\lambda, M_\lambda^*)\end{aligned}$$

By Schur's lemma

$$\dim .T(M_\lambda)_2^{SL(n,K)} = \begin{cases} 1, & \text{if } M_\lambda = M_\lambda^* \\ 0, & \text{if } M_\lambda = M_\lambda^* \end{cases}$$

The rest of the proof follows from Theorem 3.1.

## 4. Noncommutative invariants of binary forms.

In this section, we consider noncommutative invariants of a binary form. Consider the generic binary form (ground form)

$$F = \sum_{i+j=r} \binom{r}{i} a_{ij} x^i y^j$$

where $\binom{r}{i}$ is the usual binary coefficient and the coefficients are indeterminates. If a linear transformation $x \to ax + by, y \to cx + dy$, transforms $f$ to $f' = \sum \binom{r}{i} a'_{ij} x^i y^j$, $a_{ij} \to a'_{ij}, 0 \le i,j \le n$, defines a linear representation of $GL(2,K)$. Let $V$ be the standard $SL(2,K)$-module with a basis $a_1, a_2$. The map $a_1^i a_2^j \to a_{ij}$ induces an $SL(2,K)$-isomorphism

$$0 : K < S^r(V) > \xrightarrow{\sim} K < a_{ij} >$$

Let

$$y = \begin{matrix} m_{11}\ m_{12}, \dots, m_{1k} \\ m_{21}\ m_{22}, \dots, m_{2k} \end{matrix}$$

be a standard Young tableau of degree $d$ and weight $k$. Let $f_y$ be the corresponding invariant in $K < S^r(V) >$ constructed in §3. Then the element defined by $f_y(a_{ij} = \varphi(f_y)$ is an invariant in $K < a_{ij} >$.

**Definition.** A standard tableau of degree $d$ and weight $k$ is called indecomposable if

**(1)** it is a column strict Young tableau

**(2)** there is no number $k'(1 \leq k' < k)$ such that the sub-tableau

$$\begin{matrix} m_{11}m_{12}, \dots, m_{1k}, \\ m_{21}m_{22}, \dots, m_{2k} \end{matrix}$$

is a stardard tableau with lower degree and lower weight.

The following theorem gives the first and second fundamental theorem for the fixed algebra $K < a_{ij} >^{SL(2,K)}$.

**Theorem 4.1.** Let $\Lambda$ be the set of indecomposable Young tableaus. Then the set of invariants $f_y(a_{ij}), y \in \Lambda$ is a free generating set of $SL(2,K)$-invariants of $K < a_{ij} >$.

For the proof, see Theorem 3.3 in [7].

Recall the Cayley-Sylvester formula (see [8]) :

$$\dim . K[a_{ij}]_d^{SL(2,K)} = (1-t)\frac{(1-t^{r+1})\dots(1-t^{r+d})}{(1-t)\dots(1-t^d)}\Bigg|_{\frac{dr}{2}}$$

In the noncommutative case, the dimension formula has the following form.

**Proposition 4.2.** (cf. 5.3. Proposition [1])

$$\dim . K < a_{ij} >_d^{SL(2,K)} = (1-t)\frac{(1-t^{r+1})^d}{(1-t)^d}\Bigg|_{\frac{dr}{2}}$$

**Proof.** The character of the group $\{\begin{pmatrix} \epsilon & 0 \\ 0 & \epsilon^{-1} \end{pmatrix}, |\epsilon| = 1\}$ on $S^r(V)$ is equal to

$$\begin{aligned} &\epsilon^r + \epsilon^{r-2} + \ldots + \epsilon^{2-r} + \epsilon^{-r} \\ &= \epsilon^{-r} \frac{1 - \epsilon^{2(r+1)}}{1 - \epsilon^{-2}} \end{aligned}$$

By Weyl's integration formula, we have

$$\begin{aligned} &\dim.K < a_{ij} >_2^{SL(2,K)} \\ &= \frac{1}{4\pi\sqrt{-1}} \int_{|\epsilon|=1} \epsilon^{-rd} \left( \frac{1 - \epsilon^{2(r+1)}}{1 - \epsilon^2} \right)^d \frac{(1 - \epsilon^2)(1 - \epsilon^{-2})}{\epsilon} d\epsilon \\ &= \frac{1}{2\pi\sqrt{-1}} \int_{|\epsilon|=1} \epsilon^{-rd} \left( \frac{1 - \epsilon^{2(r+1)}}{1 - \epsilon^2} \right)^d \frac{1 - \epsilon^2}{\epsilon} d\epsilon \end{aligned}$$

Since the inside function of the integral is holomorphic except $= 0$, the integral above is equal to

$$\begin{aligned} &\mathrm{Res}_{\epsilon=0} e^{-rd} \left( \frac{1 - \epsilon^{2(r+1)}}{1 - \epsilon^2} \right)^d \frac{1 - \epsilon^2}{\epsilon} \\ &= (1 - t) \left( \frac{1 - t^{r+1}}{1 - t} \right)^d \Bigg|_{\frac{dr}{2}} \end{aligned}$$

This completes the proof.

**References.**


[1] G. Almqvist, W. Dicks and E. Formanek, *Hilbert series of fixed free algebra and noncommutative classical invariant theory*, J. Algebra 93(1985).

[2] W. Dicks and E. Formanek, *Poincaré series and a Problem of S., Montgomery*, Linear and Multilinear Algebra 12 (1982), 21-30.

[3] V.K. Kharchenko, *Algebra of invariants of free algebras*, Algebra i Logica 17(1978), 478-487.

[4] V.K. Kharchenko, *Noncommutative invariants of finite groups and Noetherian varieties*, J. Pure Appl. Alg. 31 (1984), 83-90.

[5] A. N. Koryukin, *Noncommutative invariants of reductive groups*, Algebra i Logica 23(1984), 419-429.

[6] D.R. Lane, *Free algebras of rank two and their automorphisms*, Ph. D. Thesis, London, 1976.

[7] Y. Teranishi, *Noncommutative classical invariant theory*, preprint.

[8] T. A. Springer, *Invariant Theory*, Lecture Notes in mathematics No. 585, Springer, 1977.

[9] H. Weyl, *The Classical Groups*, Princeton Univ. Press, 1946.

# The Brauer Severi scheme of the trace ring of generic matrices

Michel Van den Bergh*
Department of Mathematics and Computer Science
University of Antwerp, UIA

## Introduction

If $D$ is a division algebra with index $n$ with center some field $K$ then the Brauer Severi scheme of $D$ is defined as the scheme representing the left ideals of rank $n$ of $D$. This definition was subsequently extended to Azumayaalgebras and even to some classes of orders. In the last case the Brauer Severi scheme was defined as a certain connected component of the scheme representing the functor of ideals of minimal rank that are locally split off. See e.g. [A1]. Our aim in this note is to give a general framework for the study of Brauer Severi schemes. We define $\mathrm{Bsev}_n(A, R)$ for an $R$-algebra $A$ as the scheme representing the coideals of rank $n$ of $A$. It is easy to see that the component containing the generic fiber of $\mathrm{Bsev}_n(A, R) \to \mathrm{Spec}R$ coincides with the Brauer Severi schemes described above. In the first part of this note we state some general functorial properties of the Brauer Severi scheme. Furthermore we analyze the Brauer Severi scheme of the free algebra over the ground field. This is a nice scheme with an open covering of affine spaces.
The second part of this note is devoted to the Brauer Severi scheme of the trace ring of the ring of generic matrices. It turns out that it is almost equal to the Brauer Severi Scheme of the free algebra. As a consequence we obtain a new proof of Saltman's result that the Brauer Severi scheme of the generic division algebra is rational.

* The author is supported by an NFWO grant



*F. van Oystaeyen and L. Le Bruyn (eds.), Perspectives in Ring Theory, 333–338.*

## Generalities

In the following $R$ will always be a commutative $k$-algebra where $k$ is some basering, A will be an $R$-algebra.

**1. Definition :** Denote by $\mathcal{B}sev_n(A,R)$ the set of isomorphism classes of surjective maps $A \to P$ of $A$-modules where $P$ is projective of rank $n$ as $R$-module.

It is clear that the functor $S \to \mathcal{B}sev_n(A \otimes_R S, S)$ defines a covariant functor from the category of commutative $R$-algebras to Sets. Moreover this functor can obviously be extended to a functor from Spec$R$-schemes to Sets. By a slight abuse of notation we will denote this functor also by $\mathcal{B}sev_n(A,R)$

**2. Proposition :** The functor $\mathcal{B}sev_n(A,R)$ is representable by a scheme $\mathrm{Bsev}_n(A,R)$ over Spec$R$.

**Proof :** Clearly $\mathcal{B}sev$ is a subfunctor of the grassmanian functor which is represented by a scheme over $R$. It suffices to show that $\mathcal{B}sev$ is a *closed* subfunctor. This is easy.

If $T$ is a $k$-algebra we want to give a characterization of the $T$ points of $\mathrm{Bsev}_n(A,R)$.

**3. Lemma :** The $T$-points of $\mathrm{Bsev}_n(A,R)$ are in one-one correspondence with triples $(\theta, x, P)$ where $P$ is a $T$-module locally free of rank $n$, $\theta$ is a ringmorphism $A \to End_T(P)$ such that $\theta(R) \subset T$ and $x \in P$ such that $\theta(A)Tx = P$.

**Proof :** Suppose that we are given a $T$-point Spec$T \to \mathrm{Bsev}_n(A,R)$. Composition with the canonical map $\mathrm{Bsev}_n(A,R) \to$ Spec$R$ gives a map Spec$T \to$ Spec$R$ and hence a ring map $p : R \to T$. Now the $T$-points giving rise to such a map are in one-one correspondence with the $T$-points of $\mathcal{B}sev(p^*A, T)$, hence with surjective $p^*A$-module maps $\psi : p^*A \to P$ where $P$ is a $p^*A$-module locally free of rank $n$ as $T$-module. The action of $p^*A$ on $P$ determines an action of $A$ on $P$ and hence a morphism $\theta : A \to End_T(P)$. We claim that the following diagram is commutative :

$$\begin{array}{ccc} A & \xrightarrow{\theta} & End_T(P) \\ \uparrow & & \uparrow \\ R & \xrightarrow{p} & T \end{array}$$

Let $r \in R$, $q \in P$. Then $\theta(r)(q) = p^*(r)q = p(r)q$.
Thus $\theta(R) \subset T$. Let $x = \psi(1)$. Then clearly $P = \psi(p^*A) = p^*Ax = ATx$ since $p^*A$ is generated by the images of $A$ and $T$. Because of the fact that the actions of $A$ and $End_T(P)$ are compatible we obtain that $\theta(A)Tx = P$.

By inverting the above construction we associate with a $(\theta, x, P)$ a $T$-point of $\mathrm{Bsev}_n(A, R)$.

**4. Remark :** Instead of considering triples $(\theta, x, P)$ we could also consider triples $(\theta, i, P)$ where $i$ is the map $T \to P$ determined by $x$. It is then easy to verify that $i$ must be split. I.e. $P$ always contains a 1-dimensional free direct summand.

**5. Proposition :** (functoriality)

(1) If $S$ is a commutative $R$-algebra then there is a natural isomorphism of Spec$S$-schemes between $\mathrm{Bsev}_n(A \otimes_R S, S)$ and $\mathrm{Bsev}_n(A, R) \times_{\mathrm{Spec} R} \mathrm{Spec} S$.

(2) If $B$ is a quotient of $A$ then there is a canonical closed embedding of $\mathrm{Bsev}_n(B, R)$ in $\mathrm{Bsev}_n(A, R)$.

(3) f $R$ is a $R'$ algebra then there is a canonical closed embedding of $\mathrm{Bsev}_n(A, R)$ in $\mathrm{Bsev}_n(A, R')$.

**Proof :**

(1) General nonsense.

(2) If $B \otimes_R S \to P$ represents an $S$-point of $\mathrm{Bsev}_n(B, R)$ then the composition $A \otimes_R S \to B \otimes_R S \to P$ represents a point of $\mathrm{Bsev}_n(A, R)$. Conversely an $S$-point of $\mathrm{Bsev}_n(A, R)$ is obtained from a (unique) point of $\mathrm{Bsev}_n(B, R)$ if the map $A \otimes_R S \to P$ factors through $B \otimes_R S$. This is a closed condition.

(3) We use the criterion of lemma 3. A triple $(\theta, x, P)$ that represents a $T$-point in $\mathrm{Bsev}_n(A, R)$ will represent a $T$-point of $\mathrm{Bsev}_n(A, R')$ if and only if $\theta(R') \subset T$. This is a closed condition.

Any $k$-algebra may be written as a $k$-quotient of $F_I$ where $F_I$ is the free $k$-algebra on the variables $(X_i)_{i \in I}$, $I$ being some index set. From proposition 4 it follows that $\mathrm{Bsev}_n(A, R)$ is a closed subscheme of $\mathrm{Bsev}_n(F_I, k)$. It is therefore interesting to study the latter scheme. For simplicity's sake we will assume that $k$ is an algebraically closed field.
Assume that $(\theta, x, V)$ is a $k$-point of $\mathrm{Bsev}_n(F_I, k)$ and let there be given a sequence $M$ of couples $(a_1, i_1), \ldots, (a_{n-1}, i_{n-1})$ where $1 \leq a_j \leq j$, $i_j \in I$, $(a_j, i_j) \neq (a_{j'}, i_{j'})$ if $j \neq j'$. Clearly if $I$ is finite then the number of such sequences is finite.
We will call a basis $(e_1, \ldots, e_n)$ of $V$, $M$-special if and only if :

$$\begin{aligned} e_1 &= x \\ e_{j+1} &= \theta(X_{i_j}) e_{a_j} \end{aligned}$$

From the fact that $\theta(F_I)x = V$ it is easily deduced that there is always some sequence $M$ for which $V$ has a $M$-special basis.
Now we claim (for a given sequence M) :

(1) The set of points in $\mathrm{Bsev}_n(F_I, k)$ that are $M$-special form an affine space. If $I$ is finite then the dimension of this space is $(|I|-1)n^2+n$.
(2) The condition of being $M$-special is open.

Proof of (1) : If $(\theta, x, V)$ is $M$-special then there is a canonical basis $(e_1, \ldots, e_n)$. Hence we can identify $V$ with $k^n$ via this basis. The map $\theta$ is completely determined by the matrices $\theta(X_i)$. Clearly the $a_j$'th column of $\theta(X_{i_j})$ must be equal to

$$\begin{pmatrix} 0 \\ \vdots \\ 1 \\ \vdots \\ 0 \end{pmatrix} \longleftarrow j+1'\text{th place}$$

but otherwise the coefficients are free. If $|I|$ is finite then the number of coefficients that are free is equal to $|I|n^2-(n-1).n=(|I|-1)n^2+n$.

Proof of (2) : For any $M$ there is a set of monomials $(Y_2, \ldots, Y_n)$ such that a $M$-special basis must be of the form $x, Y_2x, \ldots, Y_nx$. Hence $(\theta, x, V)$ is $M$-special if and only if $x, Y_2x, \ldots, Y_nx$ are linearly independent. This is an open condition. Hence we have proved :

**6. Theorem :** If $k$ is an algebraically closed field then $\mathrm{Bsev}_n(F_I, k)$ has an open covering with affine spaces. If $|I|$ is finite then this covering is finite and the dimension of these affine spaces is $(|I|-1)n^2+n$

**7. Corollary :** If $A$ is finitely generated over $k$ then $\mathrm{Bsev}_n(A, R)$ is of finite type.

Let us from now always assume that $k$ is algebraically closed. In that case the $k$-point of $\mathrm{Bsev}_n(A, R)$ are represented by pairs $(\theta, x)$ where $\theta$ is a $k$-algebra map $A \to M_n(k)$ and $x \in k^n$ such that $\theta(A)x = k^n$. Two such couples represent the same point if there is a $\mu \in GL_n(k)$ such that $\theta' = \mu\theta\mu^{-1}$ and $x' = \mu x$

**8. Proposition :** Let $(\theta, x)$ represent a $k$-point $p$ of $\mathrm{Bsev}_n(A, R)$. The $T_p$ fits in an exact sequence of the form

$$0 \to M_k(k) \to Der^\circ_\theta(A, M_n(k)) \times k^n \to T_p \to 0$$

where $Der^\circ_\theta(A, M_n(k))$ denotes the $\theta$-derivations mapping $R$ into $k$.
In particular $dim T_p$ is independent of $x$.

**Proof :** An element of $T_p$ is given by a point $\mathrm{Spec}k[\epsilon]/\epsilon^2 \to \mathrm{Bsev}_n(A, R)$ such that restriction to $\mathrm{Spec}k \to \mathrm{Bsev}_n(A, R)$ gives the point $p$. Now let $(\theta+\epsilon\pi, x+\epsilon y)$

be such a point. Then from the fact that $\theta + \epsilon\pi$ is a ring morphism we deduce that $\pi(ab) = \theta(a)\pi(b) + \pi(a)\theta(b)$. Hence $\pi$ is a $\theta$ derivation. The fact that $\theta + \epsilon\pi$ maps $R$ to $k[\epsilon]$ is expressed by the fact that $\pi(R) \subset k$.
$(\theta + \epsilon\pi, x + \epsilon y)$ and $(\theta + \epsilon\pi', x + \epsilon y')$ represent the same point if there is a $t \in M_n(k)$ such that

$$(1 + \epsilon t)(\theta + \epsilon\pi)(1 - \epsilon t) = \theta + \epsilon\pi'$$
$$(1 + \epsilon t)(x + \epsilon y) = x + \epsilon y'$$

Or

$$\pi' - \pi = \theta t - t\theta$$
$$y' - y = tx$$

Hence there is an exact sequence

$$M_n(k) \xrightarrow{i} Der^\circ_\theta(A, M_n(k))) \times k^n \xrightarrow{j} T_p \longrightarrow 0$$

where $i$ is given by $t \to (\theta t - t\theta, tx)$ and $j$ is given by $(\pi, y) \to (\theta + \epsilon\pi, x + \epsilon y)$. We claim that $i$ is injective. Suppose that $\theta t - t\theta = 0$ and $tx = 0$. Then $\forall a \in A$ : $\theta(a)tx = t\theta(a)x = 0$. Hence $tk^n = 0$ and therefore $t = 0$.

## Traces

If $A$ can be embedded in a matrix ring over a commutative ring such that $Tr(A) \subset A$ where $Tr$ denote the reduced trace map on this matrix ring then $A$ can be equipped with a trace map which is the restriction of $Tr$ to $A$. In that case it is natural to make the following definition :

**9. Definition :** Let $t : A \to R$ be a trace map. Then if $S$ is an $R$ algebra we define $\mathcal{B}sev_n(A \otimes_n S, S, t) = \{(\theta, x, P) \mid \theta$ commutes with traces where $End_S(P)$ is equipped with the reduced tracemap $\}$
Again the assignment $S \to \mathcal{B}sev_n(A \otimes_R S, S, t)$ defines a functor from Sch/$R$ to Sets which we denote by $\mathcal{B}sev_n(A, R, t)$.
The following is easy to verify

**10. Proposition :** $\mathcal{B}sev_n(A, R, t)$ is a closed subfunctor of $\mathcal{B}sev_n(A, R)$ and hence is representable by a closed subset $\mathrm{Bsev}_n(A, R, t)$ of $\mathrm{Bsev}_n(A, R)$.

**11. Proposition :** Properties (1),(2) and (3) of Proposition 5 hold also for $\mathrm{Bsev}_n(A, R, t)$.

In order to get a better understanding of the relation between $\mathrm{Bsev}_n(A, R)$ and $\mathrm{Bsev}_n(A, R, t)$ look at the following situation. $R$ is a domain with quotient field $K$, $D$ is a central simple algebra of index $n$ over its center $K$ and $A$ is an $R$-order in $D$. By this we mean that $A$ is finitely generated over $R$ and that $KR = D$.

We will further assume that $A$ is closed under the reduced trace map $t : D \to K$. Then $\mathrm{Bsev}_n(A, R, t) \times_{\mathrm{Spec}R} \mathrm{Spec}K \simeq \mathrm{Bsev}_n(A \otimes_R K, K, t) = \mathrm{Bsev}_n(D, K) \simeq \mathrm{Bsev}_n(A, R) \times_{\mathrm{Spec}R} \mathrm{Spec}K$ Hence $\mathrm{Bsev}_n(A, R, t)$ contains the irreducible component of $\mathrm{Bsev}_n(A, R)$ containing the generic fibre.

### Generic Matrices

Let $\mathbb{G}_{m,n}$ denote the ring of $m$ $n \times n$ generic matrices. Then the tracering $\mathbb{T}_{m,n}$ is obtained by adjoining traces to $\mathbb{G}_{m,n}$. Denote with $Z_{m,n}$ the center of $\mathbb{T}_{m,n}$ and let $t$ be the natural tracemap $\mathbb{T}_{m,n} \to Z_{m,n}$.

**12. Proposition :** $\mathrm{Bsev}_n(\mathbb{T}_{m,n}, Z_{m,n}, t) = \mathrm{Bsev}_n(F_m, k)$ ($F_m$ is the free algebra on $m$ indeterminates).

**Proof :** Let $(\theta, x, P)$ be a $T$-point of $\mathrm{Bsev}_n(F_m, k)$. Then $\theta$ factors through $\theta' : \mathbb{G}_{m,n} \to End_T(P)$. This map can be lifted to a map $\theta'' : \mathbb{T}_{m,n} \to End_T(P)$. Then the triple $(\theta'', x, P)$ determines a $T$-point of $\mathrm{Bsev}_n(\mathbb{T}_{m,n}, Z_{m,n}, t)$. Conversely suppose that $(\phi, x, P)$ represents a $T$-point of $\mathrm{Bsev}_n(\mathbb{T}_{m,n}, Z_{m,n}, t)$. If $\phi'$ is the restriction of $\phi$ to $\mathbb{G}_{m,n}$ then $(\phi', x, P)$ represents a $T$-point of $\mathrm{Bsev}_n(\mathbb{G}_{m,n}, k)$. To see this we have to show that $\phi'(\mathbb{G}_{m,n})Tx = P$. But $P = \phi(\mathbb{T}_{m,n})Tx = \phi'(\mathbb{G}_{m,n})\phi(Z_{m,n})Tx = \phi'(\mathbb{G}_{m,n})Tx$. Lifting $\phi'$ to $F_m$ gives a $T$-point of $\mathrm{Bsev}_n(F_m, k)$. It is easy to see that the two constructions are each others inverse.

**13. Corollary :** The Brauer Severi variety of the generic division algebra (the quotient field of $\mathbb{T}_{m,n}$) is rational over the ground field.

**Proof :** Follows from Theorem 6 and Proposition 12.

### References

[A1] M. Artin,Left ideals in max orders,LNM 917,pp 182-193,Springer Verlag (1981).

# Some Problems on Associative Rings

F. Van Oystaeyen
Dept. of Mathematics
University of Antwerp, UIA, Belgium.

## Introduction

In this compilation I have collected some problems, ring theoretical in nature, but having a distinct flavor of other fields within Algebra. Central topics touched upon include Brauer groups of rings, orders in central simple algebras, projective representations of finite groups, graded or filtered rings, generalized crossed products and certain groups associated to those ... This incomplete list may suggest that the material is rather inhomogeneous but I believe that the problems mentioned here are linked by more than just my accidental interest, admitting that there may be more than one connected component. Just to avoid complete chaos I organized the problems in a few separate sections even if the existing interrelations make such a partition look artificial at times, so it is best to view the partition in sections in your fuzziest topology.

## Brauer Groups.

[1] Consider a commutative ring $R$ graded by an abelian group $G$. The Brauer group of $G$-graded algebras is obtained by taking equivalence classes of $G$-graded Azumaya algebras central over $R$; this group is denoted $Br^g(R)$. For $\mathbb{Z}$-gradings this Brauer group is well-studied e.g. in [CVO] but for more general groups there remain many unsettled questions.

One of the applications obtained in [CVO] for $\mathbb{Z}$-graded rings comes down to a



*F. van Oystaeyen and L. Le Bruyn (eds.), Perspectives in Ring Theory, 339–357.*

description of the usual Brauer group of a graded ring (satisfying some regularity conditions) in terms of several graded Brauer groups $Br^g(R_{(n)})$ for certain $n \in \mathbb{N}$, where $R_{(n)}$, is the $\mathbb{Z}$-graded ring obtained by "blowing-up" in the gradation i.e. $(R_{(n)})_{mn} = R_m$ for all $m \in \mathbb{Z}$. The relation obtained is of the type (*): $Br(R) = \cup_{n \in \mathbb{N}} Br^g(R_{(n)}$ (and one may restrict to a finite union) which states that any Azumaya algebra over $R$ may be made into a $\mathbb{Z}$-graded Azumaya algebra up to blowing-up of the gradation on $R$. One of the easiest cases where (*) holds is obtained by considering graded fields (every homogeneous element is invertible) i.e. $k[x, x^{-1}]$ for $k$ a field, $x$ a variable.

For an arbitrary abelian group $G$ a $G$-graded field is always of the form $kG^c$, the twisted group ring with respect to a 2-cocycle $c$ representing $[c] \in H^2(G, k^*)$. Recall that $kG^c$ is the vectorspace $k\{u_\sigma, \sigma \in G\}$ with multiplication defined by $u_\sigma u_\tau = c(\sigma, \tau) u_{\sigma\tau}$for all $\sigma, \tau \in G$. The first case to consider is $c = 1$, i.e. $R = kG$. The natural equivalent of blowing-up of the gradation is obtained by viewing $RG$ as an $E$-graded ring where $E$ is an abelian extension of $G$ such that $[E : G] < \infty$. The problem is to describe $Br(kG^c)$, or starting with $Br(kG)$ as a first attempt, in terms of $E$-graded Azumaya algebras for a class of groups $E$ as above; in other words $Br(kG^c) = \cup_{E \in F} Br^{gE}(kG^c_{(E)})$, where $kG^c_{(E)}$ is $kG^c$ viewed as an $E$-graded ring and $Br^{gE}$ is the $E$-graded Brauer group. $F$ is a suitable class of abelian extensions of $G$ and one hopes $F$ may be taken to be finite if $G$ is finitely generated. In the latter case a possible approach seems to be following : fix a presentation $\mathbb{Z}^m \to G \to 0$ and try to relate the Azumaya algebras over $kG^c$ and $k\mathbb{Z}^m$ in such a way that the blowing-up procedure for $\mathbb{Z}^m$ (this is easily understood) yields a finite number of extensions $E$ of $G$ obtained from some diagrams like :

$$\begin{array}{ccccccccc} 0 & \longrightarrow & K & \longrightarrow & \mathbb{Z}^m & \longrightarrow & G & \longrightarrow & 0 \\ & & \downarrow & & \downarrow & & \downarrow & & \\ 0 & \longrightarrow & K' & \longrightarrow & \mathbb{Z}\frac{1}{n_1} \oplus \ldots \oplus \mathbb{Z}\frac{1}{n_m} & \longrightarrow & E & \longrightarrow & 0 \end{array}$$

with $K' \cap \mathbb{Z}^m = K$.

In the finitely generated case the problem seems to split naturally in two parts : the case of a finite group and the case of a torsionfree group. The latter is an easy case because it reduces to the case $G = \mathbb{Z}^m$, so all the problems are really hidden in the finite case.

Note that, if char $k = p \neq 0$ and $G$ is a $p$-group, then $Br(kG^c)$ equals $(Br(k))^m$ where $m$ is the number of maximal ideals in $kG^c$, hence the problem reduces to $p'$-groups in this case.

Finally let us point out that a calculation of $Br(kG^c)$ in terms of $Br(k)$and (projective) characters of $G$ would also solve our problem (and more !). The possibility of projective characters appearing provides a link to projective representation theory : the $E$-gradations on the Azumaya algebras which are minimal, in the sense that the considered Azumaya algebra cannot be considered as $E'$-graded Azumaya algebra for a smaller $E' \subset E$, provide a link with the theory of strongly graded (Azumaya) algebras.

Some references for results connected to this problem : [CVO], [NVO], [DI], [KO].

[2] Clifford-Schur Groups.

For a field $k$, the Schur subgroup $S(k)$ of $Br(k)$ is obtained by looking at the elements in $Br(k)$ represented by an epimorphic image of $kG$ for some finite group $G$ i.e. by a central simple algebra appearing as a simple component in some decomposition of a $k$-separable groupring of a finite group. Each such $k$-algebra is a Clifford system for $G$ over $k$. In general, the Schur group is not too well understood, the best results available hold over local or global fields, cf. [$Y$]. The link with representation theory of finite groups is an obvious one. In relation with some problems on projective representations of finite groups, the Clifford-Schur group is a natural extension of the foregoing concept. This subgroup $CS(k)$ of $Br(k)$ consists of those elements of $Br(k)$ represented by an epimorphic image of $kG^c$, for some finite group $G$ and $c \in H^2(G, k^*)$, as defined in Problem [1]. For any subgroup $H$ of $H^2(G, k^*)$ we define the $H$-Schur groups $CS^H(k)$ to be the subgroup of $Br(k)$ consisting of elements represented by epimorphic images of twisted group rings representing an element of $H$. On the other hand, to any class of groups $\phi$ which is closed under products we associate $CS_\phi(k)$ consisting of the elements of $CS(k)$ that can be obtained from a group in $\phi$. We write $CS_{nil}(k)$ for the subgroup determined by the nilpotent groups, $CS_{ab}(k)$ if abelian groups are being considered and $CS_p(k)$ when $p$-groups are being used.

The theory works well also in the case where $k$ is a connected commutative ring, cf. [CVO2]. For numberfields $CS(k) = Br(k)$ holds, for fields containing all roots of unity we have $Br(k) = CS(k) = CS_{ab}(k)$. When roots of unity are present $CS_{ab}(k)$ consists of the "symbols" i.e. tensor productsof cyclic crossed products. The equality $CS(k) = CS_{ab}(k)$ depends on the Mercurjev-Suslin theorem and the result $Br(k) = CS(k) = CS_{ab}(k)$ is equivalent to this theorem. It would therefore be very interesting to find another proof for $CS(k) = CS_{ab}(k)$ in case $k$ contains the roots of unity. It seems that such result will follow from a projective version of the Brauer-Witt theorem on representations of finite groups; here the main problem is to carry out the Brauer induction method for projective representations.

Moreover, so far we have no example where $Br(k) \neq CS(k)$. In fact the equality $Br(k) = CS(k)$ would be rather surprizing because it would provide a cohomological description (in an odd way) of $Br(k)$ in terms of $H^2(G, k^*)$ i.e. central cohomology.

A candidate for a counter example to $Br(k) = CS(k)$ is a generic crossed product for a non-abelian group over a purely transcendental extension of $Q$, but it seems to be difficult to check whether something is **not** in $CS(k)$ perhaps because we do not have yet enough information about the algebras representing elements of $CS(k)$ at hand. Some other problems : $CS_p(k) = CS(k)_p$ ? (the latter is the $p$-torsion part of $CS(k)$; it seems to be easy to derive $CS_{nil}(k)_p = CS(k)_p$).

It is possible to look at epimorphic images of $kG^c$ for profinite groups $G$ without assuming that $kG^c$ is a P.I. algebra (but of course considering only P.I.-quotients; are there new elements in this profinite Clifford-Schur group not in $CS(k)$ ? A lifting result for a simple projective representation cf. [Na.V.O.] yields that there is a finite dimensional separable extension $l$ of $k$ (free too, even if $k$ is only a connected commutative ring) such that $CS^{<c>}(k)$ maps to $S(l)$ under $- \otimes_k l$. If $H$ is finitely generated,is there an $l/k$ as above such that $- \otimes_k l$ maps $CS^H(k)$ to $S(l)$ ? What is $\text{Ker}(CS^H(k) \to S(l))$ in that case.

For $H = < c >$ can one then choose $l$ such that $\Psi_l : CS^H(k) \to S(l)$ is injective ? Do algebras from $CS(k)$ satisfy the property of the local invariants established in [B.S.] for algebras of the Schur group ? Our interest in the Clifford-Schur groups is a natural extension of the projective representations we wanted to study in [CVO3], [Na.V.O.], but when we started writing [C.V.O.2] we discovered the papers [O], [L.O.] which we had to credit for some of the fundamental results over fields; perhaps the fact that one of the papers is in German explains why it seems to have passed rather unnoticed whereas the connections with the Mercurjev-Suslin result makes this an interesting source of problems in my opinion.

[3] Crossed Products over Fields.

Let $k$ be a field, $G$ a group acting on $k$ as a group of automorphisms i.e. there is given an automorphism $\phi : G \to Autk$. To a 2-cocycle $c : G \times G \to k^*$ we now associate a crossed product algebra $(k, \phi, c) = kG$ by defining multiplication on the vectorspace $ku_\sigma$ as follows : $u_\sigma u_\tau = c(\sigma, \tau) u_{\sigma\tau}$ for all $\sigma, \tau \in G$ and $u_\sigma \lambda = \phi_\sigma(\lambda) u_\sigma$ for all $\sigma \in G$. Assume that $k *_c G$ is such that it is a P.I. algebra; for conditions on c or $G$ leading to this situation cf. [VO1]. First question : when is $[k : k^G] < \infty$; this is the case when $k *_c G$ is semiprime e.g. when char $k = 0$, cf. [VO.1] but it might follow in general from $k *_c G$ being a P.I. algebra. Second question : is $Im\phi$ a locally finite group; this happens e.g. if $c$ is $\phi$ invariant in the sense that $\phi_\gamma c(\sigma, \tau) = c(\sigma, \tau)$ for all $\sigma, \tau, \phi \in G$. For general $c$ it is not immediately possible to realize $Im\phi$ as a subgroups of units of a P.I. algebra in order to evoke the special case of the Burnside problem as dealt with by C. Procesi in [P]. Third problem : Is there always a central extension of $k *_c G$ which is a finite module over the centre; this is true again when $k *_c G$ is semiprime. The final problem is to derive properties about the exponent of $k *_c G$ in the case where it is central simple algebra. If we restrict attention to char $k = 0$ then $k *_c G$ will be an Azumaya algebra provided $[G : Z(G)] < \infty, | G' | < \infty$ where $G'$ is the commutator subgroup of $G$, cf. loc. cit. When one restricts the problem to twisted group rings $kG^c$ it does not become trivial, not even for abelian groups, to determine the exponent of $kG^c$ in the Brauer groups of its centre. Obviously the order of $[c]$ in $H^2(G, k^*)/H^2_{sym}(G, k^*)$ will appear, where $H^2_{sym}$ consists of the classes of symmetric cocycles e.g. $c(\sigma, \tau) = c(\tau, \sigma)$ for $\sigma, \tau \in G$ in the abelian case. For a finite group $G$ acting in a Galois-manner on a central simple algebra $A$ with centre $k$ the determination of the centre will present no problem as it reduces to $k$ again. Determination of the exponent of $A *_c G$ is then the problem I introduced in the problem section of the 1983 meeting at U.I.A.; using Morita equivalence of $A *_c G$ and $A^G$ this problem reduces to rather straightforward calculations concerning ramifications of maximal orders over the valuations of $k$ from $A^G$ to $A$ and in many specific cases one should obtain an answer to the problem. Of more interest to us is the case where $G$ is infinite but it acts finitely on $A$ i.e., $\phi : G \to Aut_k A$ has finite image. Now the centre of $A *_c G$ depends on the ray classes for $c$, cf. [VO2] and $A *_c G$ is an Azumaya algebra and not simple in general.

[4] Graded Division Rings that are Central Simple Algebras.

A $G$-graded ring $D$ is said to be a gr-division ring when all homogeneous elements of $D - 0$ are invertible in $D$. In general, for arbitrary $G$, the structure of gr-division rings is not well-known e.g. if $G$ is polycyclic-by- finite then $D$ is Noetherian but surely one hopes there would be more structure available on $D$. Of course $D_e$ is always a division algebra ($e \in G$ the neutral element) and $D$ is a generalized crossed product over $D_e, D = D_e *_c G$ for a certain action $\phi : G \to AutD_e$, $[c] \subset H^2(G, UZ(D_e))$. This allows to derive some properties in a fairly straightforward way e.g. : if $G$ is finite then $D$ is Artinian, if $G$ is poly-infinite-cyclic then $D$ is a Noethenian Ore domain cf. [NaVO], and if $D$ satisfies the identities of $n \times n$-matrices and $| G' |^{-1} \in D$ then a result of [VO1] implies that $D$ is an Azumaya algebra. Consequently, for abelian $G$, $D$ is an Azumaya algebra whenever it satisfies the identities of $n \times n$-matrices; this may be viewed as an indication of the extra structure hidden in the graded information. Now let us look at gr-division algebras that are central simple algebras.

For example, if $k$ contains $n^{th}$-roots of unity then we may view $M_n(k)$ as a twisted group ring with respect to $G = \mathbb{Z}/n\mathbb{Z} \times \mathbb{Z}/n\mathbb{Z}$ for a certain 2-cocycle $c \in H^2(G, k^*)$.

If the gr-division algebra $D$ is central simple then we have $D = M_r(\Delta)$ where $\Delta$ is a skewfield. The main problem is to determine $r$, or the Schur index of $D$, or $\sqrt{dim_k \Delta}$ where $k = Z(D)$, in terms of [c] and $Im\phi$ where [c] and $\phi$ determine the structure of $D = D_e *_c G$. I suppose one can try to use the representation theory of $D$, using some graded-techniques, to get at the simple modules of $D$ and the dimension of $\Delta$. Note : what are good necessary conditions on $c$ and $G$ (or $Im\phi$), for example sufficient and necessary conditions if one can find these, in order that $D$ is a central simple algebra (centre may be larger than $k$) where $D_e$ is a skewfield e.g. start from $D_e = k$. Once a good approach to this last problem exists the solution of the other problems concerning the Schur index should also follow. If one applies some results of [VO1] it follows that one may restrict attention to finitely generated groups $G$ and then also to polycyclic-by-finite groups $G$ again by results of loc.cit.

[5] Index of Twisted Group Rings.

This problem is in fact a special case of some foregoing questions but here we may benefit from a relation with the theory of projective representations. Troughhout [5] let $G$ be a finite group, $R$ a commutative connected ring such that $| G |^{-1} \in R$. Let $A$ be an Azumaya algebra over $R$, $[\alpha] \in H^2(G, U(R))$ and consider $B = AG^{\alpha}$, the twisted group ring. Now $B$ is an Azumaya algebra, see [4] and references given there. Let $i(A)$ be the period of $A$ in $Br(R)$, $i(B)$ the period of $B$ in $Br(Z(B))$. Is it true that $i(A)/i(B)$ and $i(B)/i(A)exp(G)$. If $A$ is a commutative field, these relations are easily checked ! A more exact question is to determine $i(B)$ in terms of $exp(G)$ and projective representation properties of $G$ for $\alpha$. Using projective characters of $G$ it is clearly seen that $i(B) = 1$ when $k$ contains the values of the characters afforded by the simple components of $kG^{\alpha}$, where $k = R$ is a field. This may be generalized for $R$ as introduced above after having generalized some results about projective characters and splitting, this will appear in a recent paper by E. Nauwelaerts and myself. In the split case the proposed relations may be verified. In the general case one may split $B$ first and then try to use the degrees $[k(\chi):k]$ for characters $\chi$ afforded by $B$ (first in the field case then for connected commutative rings) and relate these to $i(B)$. Since $A \otimes_R RG^\alpha \cong AG^\alpha$ the essential problem will be settled if we can deal with the case $A = R$.

[6] Clifford Groups.
Put $B_n = (\mathbb{Z}/2\mathbb{Z})^n$, let $\alpha$ represent a class $[\alpha] \in H^2(B_n, k^*)$ where $k$ is a field with char $k \neq 2$. A projective representation of $B_n$ corresponding to $[\alpha]$ is given unambiguously by a representation of $kB_n^\alpha$, the twisted group ring as defined before, see [VO2]. We say that $kB_n^\alpha$ determines a Clifford representation, cf. [LVO], in case the subgroup of $\alpha$-regular elements in $B_n$ is minimal i.e. it is trivial when $n$ is even and equal to $\mathbb{Z}/2\mathbb{Z}$ in case $n$ is odd, (recall that $x \in B_n$ is $\alpha$-regular when $\alpha(x,y) = \alpha(y,x)$ for all $y$. We say that $G$ is a Clifford group over $k$ if $kG$ splits into a direct sum of Clifford representations, $kG = \oplus_{i=1}^r k_i B_{n_i}^{\alpha_i}, k_i$ field extensions of $k$. It seems that the class of Clifford groups over a given field is not so easily determined. Is this class closed under products ? There is a sort of generic Clifford group, cf. [LVO], namely the groups $G_n$ generated by $a_1, ..., a_n, b$ with $a_i^2 = 1$, $[a_i, a_j] = b$ for all $i \neq j$. It follows that $b^2 = 1$ and that $b \in Z(G_n)$. If $k = \mathbb{R}$, examples of Clifford groups are obtained by taking groups such that the dimension of irreducible representations is at most 4 but with at most one block of dimension 2. Main problem : obtain structure results for Clifford groups. Note : the term "Clifford group" is ambiguous, it has another meaning in the literature connected to Clifford algebras and generalizations.

## Orders.

[7] Extensions of Tame Orders and Associated Groups.
Let $B$ be a tame order over a Noetherian integrally closed domain $R$ and let $A$ be a ring containing $B$. For a left $B$-module $M$ we let $M^* = Hom_R(M, B)$ be the right $B$-module obtained by dualizing. A $B$-bimodule $P$ contained in $A$ is said to be divisorial if $P = P^{**}$ and there exists a $B$-subbimodule $Q$ of $A$ such that $Q = Q^{**}$ and we have : $(QP)^{**} = (PQ)^{**} = B$. The divisorial $B$-bimodules in $A$ form a group $D_B(A)$ with respect to $P.\ Q = (PQ)^{**}$.
We let $X^1(R)$ be the set of height one prime ideals of $R$, to $X^1(R)$ we correspond

the kernel functor $\kappa_1$ on $B$-mod, $\kappa_1 = \wedge\{\kappa_P, p \in X^1(R)\}$ i.e. its Gabriel filter $\ell(\kappa_1)$ has a basis of ideals of $B$ containing an $r \in R$ not contained in any $P \in X^1(R)$. This localization-theoretic aspect deserves some interest here because on $R$-lattices $Q_{\kappa 1}(M) \simeq M^{**}$ holds (here we consider $B$-modules $M$ which are also $R$-lattices, this makes sense because $B$ itself is a divisorial $R$-lattice). The relative Picard groups at $\kappa_1$ have been introduced in [VVO] as a natural generalization of the class group; indeed : $Pic(R, \kappa_1) = Cl(R)$, if $B$ is an Azumaya algebra then $Pic(B, \kappa_1) = Cl(R)$ and this holds even for so-called reflexive Azumaya algebras (cf. [VVO], [CVO]). In the context of tame orders we put $\Pi ic(B) = Pic(B, \kappa_1)$. We obtain a group morphism $D_B(A) \to \Pi ic(B)$ by just taking isomorphism classes of $B$-bimodules. We say that $B \subset A$ is an arithmetical situation $(A.S.)$ if the following conditions hold :

1. $R$ is integral over $R \cap Z(A) = C$
2. $C \hookrightarrow Z(A)$ satisfies P.D.E. (no blowing up)
3. $A$ is a tame order over $Z(A)$ (in particular : $Z(A)$ is a Krull domain)

Our main examples, cf. [TVO], of the A.S.-situation are the following ones (these are of main interest to us) :

E.1. $A$ is a prime ring divisorially graded by a finite group $G$ such that $| G |^{-1} \in A$ over $B$ i.e. $A_e = B, (A_\sigma A_\tau)^{**} = A_{\sigma\tau}$ for all $\sigma, \tau \in G$.

E.2. $B$ is a maximal order, $A$ is divisorially graded by a torsion free abelian group $G$ such that $B = A_e$. In fact this only garantuees the "weak" A.S. where 1. is replaced by the condition : $C \hookrightarrow R$ satisfies P.D.E. (this usually is good enough).

E.3. Both $A$ and $B$ are tame orders over Noetherian integrally closed domains $Z(A), Z(B)$ resp. and $A$ is divisorially graded by any group $G, A_e = B$.

For orders over Krull domains the class groups (reflexive Picard groups) are more natural invariants than the Picard groups, so divisorially graded rings arise in a more natural way than strongly graded rings. Note that the Noetherian assumption makes the $A_\sigma$ into $R$-lattices so the interplay between $Q_{\kappa 1}$ and $(-)^{**}$ may be used effectively.

In [Mi], Y. Miyashita associated several interesting groups to a strongly graded extension $B \hookrightarrow A$. Because we wanted to study these groups and the effect of their vanishing for orders we had to provide first a reflexive counter-part of Miyashita's results. This is the content of [TVO] and some extra results appear in [LVVO].

Let $R_B(A)$ be the set of isomorphism classes $[\phi]$ of maps $\phi : P \to Q$ where $P$ is a $B$-bimodule, $Q$ is an $A$-bimodule, such that $(A \otimes_B P)^{**} \to Q, a \otimes p \longmapsto a\phi(p)$, is an isomorphism (using bimodule isomorphisms fitting into obvious commutative diagrams in order to define the class $[\phi]$ of $\phi$). If we put : $[\phi].[\phi'] = [(\phi \otimes \phi')^{**}]$ then $R_B(A)$ is a group with the canonical $B \hookrightarrow A$ representing the identity element. The group $R_B(A)$ may be related to $Aut_B(A), \Pi ic_C(A)$, and also to $D_B(A), \Pi ic_C(B)$ by suitable long exact sequences. The grading group $G$ acts on $\Pi ic_C(B)$ and we can consider the fix-group for this action or the "fix-group up to simularity" cf. [TVO]. Furthermore, let $\Gamma_B(A)$ be the set of graded isomorphism classes of divisorially graded rings of type $G, \otimes_{\sigma \in G} H_\sigma$ say, where $H_e = B$ and the ring structure is defined by a factor set $\{h_{\sigma,\tau} : (H_\sigma \otimes H_\tau)^{**} \to H_{\sigma\tau}; \sigma, \tau \in G\}$, consisting of $B$-bimodule isomorphisms. The product of such a class $[H, h]$ and another one $[H', h']$ is given by : $H''_\sigma = (H_\sigma \otimes_B A_{\tau-1} \otimes_B H'_\sigma)^{**}$, $h''_{\sigma,\tau} = (h_{\sigma,\tau} \otimes j_{\sigma^{-1},\tau^{-1}} \otimes h'_{\sigma,\tau})^{**}$, where $j_{\sigma,\tau}$ stands for the "factor set" determining $A$. It follows that $\Gamma_B(A)$ is an abelian group. Finally let $\beta_B(A)$ be defined by the exactness of the sequence : $\Pi ic_c(B)^{(G)} \to \Gamma_B(A) \to \beta_B(A)$. This group generalizes the part of the "Brauer group of $B^G$" split by $B$, i.e. if $B$ is commutative and $A = B * G$, $G$ a Galois group of automorphisms for $B/B^G$ then $\beta_B(A)$ would indeed reduce to the part of the Brauer group of $B^G$ split by $B$. All the newly introduced groups fit into a long exact sequence generalizing the Case-Harrison-Rosenberg sequence (cf. [CHR]) and also te reflexive version of this sequence obtained for the reflexive Brauer group of a Krull domain cf. [VVO]; cf. [Mi] for the strongly graded case, and also [Ka].

Let us mention the particular case where $Z(B) \subset Z(A)$, because then this long exact sequence takes the very nice form :

$0 \to Aut_B(A)^G \to R_B(A)^G \to CCl(B) \to \Gamma_B^G(A) \to \beta_B(A) \to H^1(G, CCl(B)) \to H^3(G, \mathrm{U}(R))$,

where $CCl(B)$ is the central class group of the order $B$, cf. [L], [VO3]. The calculation of these groups and criteria for their vanishing provide a wealth of interesting problems; where $Pic$ and $Cl$ are invariants of the $H^1$-type, the groups $\Gamma_B(A)$ and $\beta_B(A)$ are of the $H^2$-type. Some problems :

1. Calculate the groups introduced above in case $B$ is a reflexive Azumaya algebra or in the slightly more general case where $CCl(B) = Cl(R)$, in particular when $Cl(R) = 1$.

2. Calculate the groups for the extension $B \hookrightarrow End_B(P)$ where $P$ is a faithful finitely generated projective $B$-bimodule (e.g. $B \hookrightarrow M_n(B)$ yields a not too difficult example). Conversely, is it possible to derive from certain vanishing properties of those groups, in paticular $\beta_B(A)$, some properties about the extension $B \hookrightarrow A$ ? In particular can we characterize the case where $A = End_B(P)$ ?

3. Specify problem 1. to the case where $B = R$ is a Krull domain; there is particular

interest in this when $Z(A)$ is Noetherian but $G$ is arbitrary as it is already clear from Example E.3. given above.

4. Let $B$ be a maximal order over a ring of integers $R$ in a number field and let $A$ be a tame order over $R[X_1, X_1^{-1}, ..., X_n, X_n^{-1}]$. This brings us to some very concrete calculations and we arrive at an interplay between orders over number rings and orders over rings of a more geometrical type, linked by $\mathbb{Z}^n$-gradation. The Picard- and class-groups of the rings considered play are important part here but it looks like one can use the construction of generalized Rees rings cf. [VO4], [L], [VVO] to control the class groups and reduce to the case $Pic(R) = 1$ up to some descent problems.

It is clear that one may try to obtain various structure results from conditions on the $H^2$-type invariants of orders; this theory is wide-open and it was the topic for future research after [TVO], so far not much happened, but I still hope to return to these problems eventually.

[8] Orders over subrings of integrally closed Noetherian domains.

8.a. Let $R$ be a subring of a ring of integers in a number field $K$. Since $R$ is a Noetherian domain all of its localizations are determined by certain prime ideals. If $c$ is the conductor ideal for $\overline{R}$ in $R$ then prime ideals not containing $c$ correspond to localizations yielding discrete valuation rings. So if $\kappa$ is a kernel functor on $R$-mod determined by prime ideals not containing $c$, i.e. $I \in \mathcal{L}(\kappa)$ (the Gabriel filter of $\kappa$) if $I \not\subset P$ for every one of the given prime ideals $P \in X(c) \subset SpecR$, then we calculate $Pic(R, \kappa) = Cl(Q_\kappa(\overline{R}))$. For prime ideals containing $c$, the problem reduces to the case of a Noetherian local domain $R$ dominated by a discrete valuation ring $R$. For a $\kappa$ determined by such prime ideals (containing $c$) we thus obtain for the relative group (as defined earlier, cf. [VVO]) that it will be a finite product of cyclic groups the orders of which relate to the "ramification" of $R_P$ in $\overline{R}_P$ (some power of the maximal ideal of $R_P$ is in $\overline{R}_P$ !). The general case is a mixture of the two cases mentioned and it is easily described (FVO, unpublished). Let us write $\kappa_c$ for $\wedge\{\kappa_p, P \supset c\}$, then $Pic(R, \kappa_c) = 1$ entails $R = \overline{R}$; moreover, if all relative Picard groups vanish then $R$ has to be a principal ideal domain (just an exercise). Now consider an $R$-order $\Lambda$; if $Pic(\Lambda, \kappa_c) = 1$ is then $\Lambda$ a maximal order over $R = \overline{R}$ ? In general, what properties of $\Lambda\overline{R}$ can one derive from knowledge of $Pic(\Lambda\kappa_c)$ ? The Brauer group of a ring like $R$ is equally interesting and there exist many problems concerning the crossed product structure or the structure of an Azumaya algebra in terms of smashed products with respect to Hopf algebras, even in case $R = \overline{R}$

(cf. [C] for the case of quaternion algebras).

8.b. If one considers Noetherian subrings of affine integrally closed domains over fields, the problems of 8.a. carry over to the geometrical situation. Then we are looking for relations between varieties and their normalizations expressed in terms of relative Picard groups (Brauer groups) of the coordinate ring and their vanishing properties. Of course, by passing to orders over the coordinate rings one obtains the problem worded in the terminology of noncommutative algebraic geometry, [VVO2]. Here one may obtain a good idea of what is going on by focussing on curves and their desingularizations. In this one dimensional case the use of the relative Picard group seems to globalize the information otherwise obtained by using Mayer-Vietoris sequences associated to the quotients of the conductor ideal.

[9] Orders as Galois objects.

A Galois object may be described roughly by saying that it is an algebra $A$ over $R$ such that there is a Hopf algebra $H(A,R)$ measuring $A$ over $R$ in the sense of M. Sweedler [S]. Various extensions and restrictions may be applied to this rough definition. The $R$-orders in a central simple algebra $\sum$ which are Galois objects must form a nice class; in particular if one imposes certain desirable conditions on $H(A,R)$. If $R$ is a Dedekind domain, some results obtained in [L2] give an indication of what kind of results one can obtain (results of [L2] do generalize to Krull domains !). Azumaya orders are Galois objects in a very strict sense but onc can go a few steps further and ask to determine the maximal orders that are dimodule-algebras i.e. having a compatible $G$-action and $G$-grading, or in other words, what are the Galois object $R$-orders for the Hopf algebra $RG\sharp RG^*$. In general one might hope to obtain information about smash products of orders $A$ and corresponding Hopf algebras $H(A,R)$ in order to obtain some results from a duality theory à la Cohen-Montgomery [CM]. Recently, L. Childs provided smash-product structure results for some Azumaya algebras over number rings which are not crossed products in the usual sense, [C]. It is a natural desire to try to obtain every Azumaya algebra as such a smashed product but the notion of "Galois object" and smashed products corresponding to it turn out to be too strong (a wise lesson S. Caenepeel and I learned in 1982 while optimistically proceeding in this direction). So as a final problem here we mention : is there a good definition of "weak" Galois object with respect to a (good) Hopf algebra such that every Azumaya algebra (e.g. over a number ring) is equivalent to a smashed product over a weak Galois object ? This would lead to a nice (weak-) smashed product theory for the Brauer group using some type of cohomology of Hopf algebras, of such general applicability that it would deserve to be viewed as the correct extension over a commutative ring (more restrictive conditions allowed) of the crossed product theory in the Brauer group of a field.

[10] Regularity of Orders, etc...

The notion of commutative regular ring (variety or scheme) may be generalized to certain regularity conditions for orders (non-commutative varieties or schemes) in different ways. One may use derivations, cf. [VVO3] (but the regularity condition obtained is not very useful), Brown-Hajarnavis regularity cf. [BH], or properties of Brauer-Severi schemes associated to the order, or Cohen-Macaulay type properties. Following Vasconcellos [V] we say that an $R$-order $\Lambda$ is a moderated Gorenstein order if its self-injective dimension equals the Krull dimension $n$ and $\Lambda - \text{Ext}\ \dim(\Lambda/Pi) = n$ for all prime ideals $P_1, ..., P_k$ lying over the maximal ideal of the local (by assumption !) ring $R$. If moreover, the global dimension of $\Lambda$ also equals $n$ then the order is said to be moderated regular. Vasconcellos showed that moderated Gorenstein rings over local rings are Cohen-Macaulay modules over the centre. It follows that moderated regular orders are tame orders and the converse holds if the order has global dimension equal to 1 or 2. From an algebraic point of view the moderated regular orders have several desirable good properties but the main problem is to derive the regularity properties fitting into a "geometric" theory of orders; for example : how can one define the notion of "tangent space" in the moderated regular case and can one relate it to well-behaving non-commutative differentials ? Evidently this problem is more a matter of providing a definition than anything else, but in order to select a suitable definition it is necessary to know where the theory of regular orders has to lead (from the geometrical point of view) and I am lost here.

A non-commutative curve may be viewed as the spectrum of a prime P.I. algebra $\Lambda$ of dimension one; it is birational to a so-called abstract (projective) non-commutative curve constructed by picking suitable maximal orders over discrete valuation rings in the central simple algebra $Q(\Lambda)$, such that the central valuations satisfy the approximation property. In terms of these maximal orders a Riemann-Roch theorem for non-commutative curves has first been derived in [VDVGVO]; however a Riemann-Roch result in terms of the arithmetical theory of central simple algebras had been known to E. Witt; cf. [W], and it reappeared in the litterature in several different forms since then. Extension of the Riemann-Roch result to higher dimensional varieties is possible but a concrete approach to non-commutative surfaces along the lines of O. Zariski's treatment of the commutative theory, cf. [Z], seems to run into difficulties because of a lack of a good intersection theory and a suitable non-commutative equivalent of Bertini's theorem. Can this be avoided ? Can such a theory be developed ? Perhaps it is still worthwhile to elaborate further the theory of orders over a surface by using primes (a generalization of the concept of a valuation to the non-commutative situation) associated to curves through a "non-singular" point on the surface. Some notions concerning primes and pseudo-places can be found in [VG], [VO5]; for some ideas on how a non-commutative algebraic geometry might look the reader may consult [VVO2] and several papers of M. Artin, C. Procesi, W. Schelter indicated in the references of loc.cit. For graded (regular) orders see also recent work of M. Artin [A].

## Graded Rings

[11] Normalizing Rees Rings.

Let $G$ be a finitely generated group, $R$ a strongly graded ring of type $G$, $R = \oplus_{\sigma \in G} R_\sigma$ and $R_\sigma R_\tau = R_{\sigma\tau}$ for all $\sigma, \tau \in G$. In this situation each $R_\sigma$ is finitely generated and projective as a left (and right) module. If $\sigma_1, ..., \sigma_r$ generate $G$ and $x^j_{\sigma_i}, j = 1, ..., n_i$ generate $R_{\sigma_i}, i = 1, ..., r$, then the $R_e$-ring generated by $x^j_{\sigma_i}, i = 1, ..., r, j = 1, ..., n_i$ is just $R$. If we can choose the $\sigma_1, ..., \sigma_r$ and the $x^j_{\sigma_i}$ such that for all $i, j$ we have that $rx^j_{\sigma_i} = x^j_{\sigma_i} r$ for all $r \in R_e$ then we say that $R$ is a **centralizing strongly graded** ring. If we can arrange the choice of the $x^j_{\sigma_i}$ such that we have $R_e x^j_{\sigma_i} = x^j_{\sigma_i} R_e$ (as a set) for all i,j, then we say that $R$ is a **normalizing** strongly graded ring. Recall that for a strongly graded ring $R$ each $R_e$-bimodule $R_\sigma, \sigma \in G$, is invertible and its inverse bimodule is $R_{\sigma^{-1}}$, the isomorphism class $[R_\sigma]$ of $R_\sigma$ is an element of the Picard group $Pic(R_e)$. The natural map $\varphi : Pic(R_e) \to Aut(Z(R_e)), [R_\sigma] \to \varphi_\sigma$ is defined as follows : for $[P] \in Pic(R_e), End_{R_e}({}_{R_e}P) \cong R_e^0, End_{R_e}(P_{R_e}) \cong R_e$ and $End_{R_e-R_e}(P) \simeq Z(R_e)$, i.e. a bimodule endomorphism of $P$ is given as the multiplication by an element of $Z(R_e)$, so if $c \in Z(R_e)$ then there is a unique $\varphi_\sigma(c)$ such that $R_\sigma c = \varphi_\sigma(c) R_\sigma$ and this defines $\varphi_\sigma : Z(R_e) \to Z(R_e), c \longmapsto \varphi_\sigma(c)$. Obviously in the centralizing case we have that $\varphi_\sigma = Id$ for each $\sigma \in G$. More generally, we say that $R$ is **quasi-inner graded** when $\varphi$ maps all $\sigma \in G$ to the identity under the map $G \to Pic(R_e) \to Aut(Z(R_e)), \sigma \longmapsto [R_\sigma] \to \varphi_\sigma$.

In the normalizing strongly graded case the gradation is not necessarily quasi-inner.

In both the centralizing and the normalizing case we may write $R = R_e[x^j_{\sigma_i}, i, j]$ in an unambiquous way because every element of $R$ may be expressed as a sum of monomials with coefficient from $R_e$ on the left.

**Proposition.**

Let $R_e$ be a prime P.I. algebra and $R$ a normalizing strongly graded ring of type $G$ such that the gradation is quasi-inner then $R$ is a centralizing strongly graded ring.

**Proof.**

First we evoke Proposition 1.4. and 1.6. of [NNVO] to conclude that $S^{-1}R = S_e^{-1}R = Q^g$ is a gr-simple gr-Artinian ring, where $S_e = \{s \in R_e, s \text{ regular in } R_e\}, S = \{s \in R, s \text{ regular in } R \text{ and homogeneous}\}$, and $(Q^g)_e = Q_e = Q_{cl}(R_e)$ is simple Artinian. From the cited results it also follows that $Q^g$ is a crossed product $Q^g = Q_e * G$, so that exist regular elements in $G$ of degree $\sigma$ for every $\sigma \in G$. Each one of the normalizing elements $x^j_{\sigma_i}$ is regular in $R$; indeed if $z = x^j_{\sigma_i}$ would be such that $zr = 0$ for some $r \in R$ then $zr_\tau = 0$ for all homogeneous components $r_\tau$ of $r$ but if $r_\tau \neq 0$ then $R_{\sigma_i^{-1}}zr_\tau = 0$ and $R_{\sigma_i^{-1}}zR_er_\tau = 0$ where $R_{\sigma_i^{-1}}zR_e$ is a nonzero ideal of $R_e$. Now from $R_{\sigma_i^{-1}}zR_er_\tau R_\tau^{-1} = 0$ it follows that $R_{\sigma_i^{-1}}z = 0$ or $r_\tau R_\tau^{-1} = 0$ but both are impossible unless $z = 0$ or $r_\tau = 0$, a contradiction. Let $\psi^j_{\sigma_i} = \psi^j_i$ be the automorphism of $R_e$ defined by $\psi^j_i(z) = x^j_{\sigma_i}z(x^j_{\sigma_i})^{-1}$ and the extension of $\psi^j_i$ to $Q_e$ is again denoted by $\psi^j_i$. Since $Z(Q_e) = Q(Z(R_e))$ by Posner's theorem, each $\psi^j_i$ fixes $Z(Q_e)$ hence these are inner automorphisms of $Q_e$. Let $y^j_i \in Q_e$ be such that $\psi^j_i(z) = y^j_i z(y^j_i)^{-1}$ and up to multiplying by some element of $Z(R_e)$ we may take $y^j_i \in R_e$. So for all $x \in R_e$ we have obtained : $\psi^j_i(z) = y^j_i x(y^j_i)^{-1} = x^j_{\sigma_i}x(x^j_{\sigma_i})^{-1}$. Consequently $(y^j_i)^{-1}x^j_{\sigma_i} \in Z_R(R_e)$ and therefore $x^j_{\sigma_i} \in R_eZ_R(R_e)$. Here, note also that $\psi^j_i$ and $\psi^k_i$ differ only by an inner automorphism of $R_e$ (look at any regular element of degree $\sigma_i^{-1}$, say $t_i$, and the automorphism induced by $x^j_{\sigma_i}t_ix^k_{\sigma_i}$).

We arrived at $R_{\sigma_i} \subset R_eZ_R(R_e)_{\sigma_i}$, and if we decompose the $R_e$-generators for $R_{\sigma_i}$ as $R_e$-combinations of elements in $Z_R(R_e)_{\sigma_i}$ then we see that we are in the centralizing situation and clearly $R$ is an extension of $R_e$ in the sense of C. Procesi, [P].

**Remark.**

If $R$ is a P.I. algebra too then $G$ is polycyclic-by-finite and $R$ is a Goldie ring, cf. [FVO1].

Let us now return to the general case and list some problems for normalizing strongly graded rings; some of these problems are based on facts known in the centralizing case, some derive from facts for finite normalizing extensions, cf. [BDR],

[HR].

Throughout $R$ is normalizing strongly graded over $R_e$.

a. Suppose that $R$ is a P.I. ring; if $R_e$ is a Jacobson ring, resp. a Hilbert algebra, and $G$ is finite, is $R$ a Jacobson ring, resp. a Hilbert algebra. Considering the Jacobson ring property in case $G$ is torsion free abelian may also turn out to be interesting.

b. If $G$ is finite such that $(G)^{-1} \in R_e$ and $R_e$ is an Azumaya algebra is $R$ also an Azumaya algebra. This is known for quasi-inner gradations, cf. [VO6].

c. Let $NCl(R)$ be the normalizing class groups of $R$. cf. [L], [VVO]. Is it true that $NCl(R) = NCl(R_e)/Im(\phi)$ where $\phi : G \to Pic(R_e)$ is the structure morphism of the gradation.

d. To $x^j_{\sigma_i}$ we may associate unambiguously $\psi^j_i mod Inn R_e = \overline{\psi}^j_i$ so we have a commutative diagram :

$$\begin{array}{ccccccc}
G & \longrightarrow & \mathrm{Aut}(R_e) & \hookrightarrow & \mathrm{Pic}(R_e) & \qquad \sigma_i \longrightarrow & \overline{\psi}^j_i \\
\phi \swarrow & & \downarrow \mathrm{res} & \swarrow & & & \\
\mathrm{Pic}(R_e) & \longrightarrow & \mathrm{Aut}(Z(R_e)) & & & \qquad [R_{\sigma_i}] \longrightarrow & \varphi_i
\end{array}$$

If $P$ is a prime ideal of $R$, are the minimal prime ideals of $R_e$ lying over $P \cap R_e$ conjugated i.e. conjugated by $R_\sigma(-)R_{\sigma^{-1}}$. If $P \subset Q \subset R$ and $P \cap R_e = Q \cap R_e$, does it follow that $P = R$.

It $p$ is a prime ideal of $R_e$ is there a prime ideal $P$ of $R$ such that $p$ is minimal over $P \cap R_e$. All these questions may be dealt with in the finite case, but we are interested in the situation where $G$ is infinite, e.g. polycyclic-by-finite.

e. For each $\sigma_i$ select some fixed $x^{j(i)}_{\sigma_i} = x_i$ and consider the graded ring $S_j = R_e[x_i, i = 1, ...r]$. Since for any $j, x^j_{\sigma_i} \in Q_e x_i$ as noted before, we obtain that $Q^g(R) = Q_e[x_i, i = 1, ...r] = Q^g(S_j)$.

Assume now that $Q_e$ is a central localization of $R_e$ e.g. if $R_e$ is a prime P.I. ring. In this case $R \subset Q(Z(R_e))S_j$ and $Z(S_j) \subset Z(R)$. If $S_j$ is an Azumaya algebra for some $j(i)$ then $R = S_j \otimes_{Z(S_j)} Z_R(S_j) = S_j \otimes_{Z(S_j)} Z(R)$ is an Azumaya algebra too.

If $S_j$ is an Azumaya algebra for a choice of $j (= \{j(i), i = 1, ..., r\})$ is each other $S_k$ also an Azumaya algebra ? Similar problems arise for maximal orders or tame orders and here one may use results of [NaVO2] or [NNVO] depending on the restrictive conditions put on the grading group. Note that the gradations on the rings $S_j$ satisfy (a strong form of) condition (E) (as studied in [NaVO2]). Both $Z(R)$ and $Z(S_j)$ contain the fixed ring $Z(R_e)^G$ and they have the same localization with respect to $(Z(R_e)^G)^*$, so the rings $R$ and $Z(R)S_j$ must be very close but it is not clear that equality should hold generally (it holds if $R$ or some $S_j$ is an Azumaya algebra).

[12] Relative Azumaya Algebras.

Let us consider prime algebras over a Noetherian centre $R$. The relative Azumaya we aim to consider are then defined by the property that for some subset $X(\kappa)$ in Spec(R) each localization $A_P, P \in X(\kappa)$, is an Azumaya algebra over $R_P$. The set $X(\kappa)$ is associated to a torsion theory or kernelfunctor $\kappa$ i.e. $X(\kappa)$ consists of the prime ideals not contained in the Gabriel filter $L(\kappa)$) of $\kappa$. In fact, geometrically speaking, the sets $X(\kappa)$ are just the generically closed sets in Spec($R$) (this is a generalization of the notions of Zariski-open set). A particular example, showing how $X(\kappa)$ may differ from an open set, is obtained by taking $X(\kappa_1) = \{P \in Spec(R), P$ has height one$\}$, the $\kappa_1$-Azumaya algebras are the reflexive Azumaya algebras studied in [VVO], [Yu]. For $\kappa \neq \kappa_1$ the $\kappa$-Azumaya algebras need not be maximal orders in central simple algebras, nevertheless they share many properties with these classical maximal orders. There remain many unsettled questions concerning the ring theoretical structure of relative Azumaya algebras. Let me here raise a purely torsion theoretic one : if $A$ is a prime P.I. ring such that each kernel functor $\sigma \geq \kappa$ on $A$-mod is a central kernel functor (i.e. $L(\sigma)$ has a basis of centrally generated ideals) is $A$ then a $\kappa$-Azumaya algebra ? We rediscover relative Azumaya algebras as rings of sections over generically closed sets of locally separable Algebras and in case projective varieties are being considered a very interesting class of such rings is provided by the $\kappa_1$-Azumaya algebras where $\kappa_+$ is the torsion theory associated to the filter generated by the powers of the ideal $R_+$, the positive part of a positively graded ring. Is a $\kappa_+$-Azumaya algebra a maximal order if $R$ is a positively graded Krull domain ? This question has a positive answer if $dimR \leq 3$.

[13] Let $A$ be a simple Artinian ring with an exhaustive separated filtration $FA$ such that $G(A)$ is a prime Noetherian ring. It is possible to prove that each homogeneous element of $G(A)$ which is regular is in fact a unit of $G(A)$ but in view of the failure of the graded version of Goldie's theorems (if one does not impose some extra condition) it does not follow that $G(A)$ is a direct sum of gr-simple (gr-Artinian)

algebras. If $G(A)$ is a P.I. ring then this is true; is it true in the situation we have here ? Is $F_0A$ a maximal order in $A$ (this holds too if $A$ or $G(A)$ is a P.I. ring or if one put a mild extra condition on the gradation of $G(A)$, or if $G(A)_0$ is prime instead of just semiprime; note that from $G(A)$ being prime it follows only that $G(A)_0$ is semiprime)?

If $G(A)$ is semiprime then one can use microlocalization at the multiplicative sets associated to minimal prime ideals of $G(A)$ and establish, cf. [VO7] for the P.I. case, that $F_0A$ is a finite intersection of maximal orders over discrete valuation rings. It seems one can extend, up to modification, this result to the case where $G(A)$ is a P.I. ring but $A$ neednot be. The general problem of relating $F_0A$ to $A$ in case $A$ is only simple Noetherian, allowing the restriction that it is also an Ore domain and $G(A)$ s prime (semiprime) Noetherian, is still uninvestigated.

## References


**[CVO]** S. Caenepeel, F. Van Oystaeyen; *Cohomology and Brauer Groups of Graded Rings*; Math. Monographs, M. Dekker, New York, 1988.

**[NVO]** C. Năstăsescu, F. Van Oystaeyen; *Graded Ring Theory*; Math. Library vol. 28, North Holland, Amsterdam, 1982.

**[DI]** F. De Meyer, E. Ingraham. *Separable Algebras over Commutative Rings*, Lect. Notes in Math. 181, Springer Verlag, Berlin 1971.

**[KO]** M.A. Knus, E. Ojanguren, *Théorie de la Descente et Algèbres d'Azumaya*, Lect. Notes in Math. 389, Springer Verlag, Berlin 1974.

**[Y]** S. Yamada. *The Schur Subgroup of the Brauer Group*, Lect. Notes in Math. 397, Springer Verlag, Berlin 1974.

**[CVO2]** S. Caenepeel, F. Van Oystaeyen. *Azumaya Algebras as Clifford Systems or the Clifford-Schur Subgroup of the Brauer Group.*

**[BS]** Benard, M. Schacher. *The Schur Subgroup II*, J. of Algebra 22, 1972, 378-385.

**[CVO3]** S. Caenepeel, F. Van Oystaeyen. *A Note on Generalized Clifford Algebras and Representations*, Comm. in Algebra, to appear in '88.

**[NaVO]** E. Nauwelaerts, F. Van Oystaeyen; *Projective Representations and the Brauer Splitting Theorem over Rings*, J. of Algebra.

**[O]** H. Opolka. *A Note on Projective Representations over Fields with Many Roots of Unity.*

[**LO**] F. Lorenz, H. Opolka. *Einfache Algebren und Projektive Darstellungen über Zahlkörpern*, Math. Z 162, 1978, 175-182.

[**VO1**] F. Van Oystaeyen, *Graded P.I. Rings and Generalized Crossed Product Azumaya Algebras.* Chinese Annals of Math. 8 B(1), 1987, 13-21.

[**VO2**] F. Van Oystaeyen, *Azumaya Strongly Graded Rings and Ray Classes*, J. of Algebra 103, 1986, 228-240.

[**P**] C. Procesi. *Rings with Polynomial Identities*, Monographs in Pure and Applied Math., Marcel Dekker, New York, 1973.

[**VO3**] F. Van Oystaeyen; *Note on the Central Class Group of Orders over Krull Domains*, in "Methods in Ring Theory", D. Reidel, Dordrecht, 1984.

[**LVO**] L. Le Bruyn, F. Van Oystaeyen, *A Note on Clifford Representation*, Bull. Soc. Math. Belg; to appear.

[**VVO**] F. Van Oystaeyen, A. Verschoren; *Relative Invariants of Rings. The Commutative Theory.*, Math. Monographs, M. Dekker, New York 1983.

[**TVO**] B. Torrecillas, F. Van Oystaeyen; *Divisorially Graded Rings Related Groups and Sequences*, J. of Algebra, 105, No. 2, 1987, 411-428.

[**Mi**] Y. Miyashita, *An Exact Sequence Associated with a Generalized Crossed Product*, Nagoya Math. J. 49, 1973, 21-51.

[**LVVO**] L. Le Bruyn, M. Van den Bergh, F. Van Oystaeyen; *Graded Orders* Birkhauser Verlag, to appear in 1988.

[**CHR**] S. Chase, H. Harrison, A. Rosenberg., *Galois Theory and Cohomology of Commutative Rings*, Mem. Am. Math. Soc. 52, 1965, 1-19.

[**Ka**] T. Kanzaki, *On Generalized Crossed Products and Brauer Groups*, Osaka J. Math. 5, 1968, 175-188.

[**L**] L. Le Bruyn, *Maximal Orders over Krull Domains, Ph. D. Thesis*, Antwerp 1983.

[**VO4**] F. Van Oystaeyen, *Arithmatically Graded Rings and Generalized Rees Rings* J. of Algebra, 82, 1, 1983, 185-193.

[**VVO2**] F. Van Oystaeyen, A. Verschoren; *Non-commutative Algebraic Geometry*, Lect. Notes in Algebra 887, Springer Verlag, Berlin, 1978.

[**S**] M.E. Sweedler, *Hopf Algebras*, Benjamin, New York, 1969.

[**L2**] L. Le Bruyn, *Universal Bialgebras Associated to Orders*, Comm in Algebra, 1981.

[CM] M. Cohen, S. Montgomery, *Group Graded Rings, Smash Products and Group Actions*, Trans. A. M. S. 282, No. 1, 1984, 237-257.

[VVO3] F. Van Oystaeyen, A. Verschoren, *Derivations of P.I. Algebras and Smoothness of Noncommutative Curves*, J. Pure Applied Algebra 29, 1983, 169-176.

BH] K. Brown, Majarnavis.

[V] Vasconcellos.

[W] E. Witt, *Riemann-Rochser Satz und z-Funktion in Hyper-komplexen*, Math. Ann. Bd. 110, 1934, 12-28.

'DVGVO] J.P. Van Deuren, J. Van Geel, F. Van Oystaeyen, *Genus and a Riemann-Roch Theorem for Noncommutative Curves*, Sem. Dubreille 1979, LNM 867, Springer Verlag, Berlin 1979, pp. 295-318.

[Z] O. Zariski, *An Introduction to the Theory of Algebraic Surfaces*, Lect. Notes in Math. 83, Springer Verlag, Berlin, 1969.

[VG] J. Van Geel, *Places and Valuations in Noncommutative Ring Theory*, Lect. Notes M. Dekker, 71, New York, 1981.

[VO5] F. Van Oystaeyen, *Prime Spectra in Noncommutative Algebra*, Lect. Notes in Math. 444, Springer Verlag, Berlin 1975.

[A] M. Artin. *Invent. Math.*

[BDR] J. Bit-David, J.C. Robson, *Normalizing Extensions II*, in Lect. Notes in Math. 825, Springer Verlag, Berlin 1981, 6-9.

[HR] A. Heimicke, J.C. Robson.

[VO6] F. Van Oystaeyen, *On Clifford Systems and Generalized Crossed Products*, J. of Algebra 87, 2, 1984, 396-415.

[NaVO2] E. Nauwelaerts, F. Van Oystaeyen, *Finite Generalized Crossed Products over Tame and Maximal Orders*, J. of Algebra 101(1), 1986, 61-68.

[MNVO] H. Marubayashi, E. Nauwelaerts, F. Van Oystaeyen, *Graded Rings over Arithmetical Orders*, Comm. in Algebra 12(6), 1984, 745-775.

[Yu] S. Yuan, *Reflexive Modules and Algebra Class Groups over Noetherian Integrally Closed Domains*, J. of Algebra 32, 1974, 405-417.

[VO7] F. Van Oystaeyen, *A Note on Filtrations on Simple Algebras*, to appear.

# DUBROVIN VALUATION RINGS

Adrian R. Wadsworth*
Department of Mathematics
University of California at San Diego
La Jolla, California 92093
U.S.A.

ABSTRACT. A few years ago Dubrovin defined a new notion of valuation ring for division rings and matrices over division rings. These valuation rings have better extension properties than the usual valuation rings on division rings. In addition, a Dubrovin valuation ring $B$ of a division ring $D$ is uniquely determined up to conjugacy by the valuation ring $B \cap Z(D)$ if $[D:Z(D)] < \infty$. In this paper some of the basic results on Dubrovin valuation rings will be reviewed, and a theorem will be described which uses Henselization to relate these rings back to the more traditional valuation rings in division rings.

## 1. VARIOUS KINDS OF VALUATION RINGS

Valuation theory has long been a basic tool in commutative algebra. Over the years there have been several attempts to extend the theory to a noncommutative setting, with varying degrees of success. We will examine here a notion of valuation ring on division rings or their matrix rings proposed by Dubrovin in [D1] and [D2], which seems particularly promising in view of the good extension properties these rings enjoy. No proofs will be given, but we will indicate where proofs can be found. The discussion and the references will reflect my bias toward division rings finite-dimensional over their centers. But we will not restrict exclusively to such division rings until §2. For comparison purposes, before considering Dubrovin's rings we will recall a bit about the two kinds of noncommutative valuation rings that have been studied most extensively.

* Supported in part by the N.S.F.



*F. van Oystaeyen and L. Le Bruyn (eds.), Perspectives in Ring Theory, 359–374.*

Let $R$ be a subring of a division ring $D$. We call $R$ an *invariant valuation ring* of $D$ if for every $d \in D^* = D-\{0\}$,

$$\text{(a)} \quad d \in R \quad \text{or} \quad d^{-1} \in R;$$

$$\text{(b)} \quad d R d^{-1} = R .$$

(The invariance in the name refers to property (b) -- invariance under inner automorphisms of $D$.) For such an $R$, the left ideals are the same as the right ideals, and are linearly ordered by inclusion. The unique maximal left (and right) ideal of $R$ is its Jacobson radical, $J(R)$, which is given by

$$J(R) = \{d \in R \mid d^{-1} \notin R \ \text{ or } \ d = 0\} . \tag{1}$$

The residue ring

$$\overline{R} := R/J(R) \tag{2}$$

is a division ring, and $R$ has a value group

$$\Gamma_R = D^*/R^* \qquad \text{(where } R^* = \text{units of } R\text{)}, \tag{3}$$

which is a totally ordered group ( $aR^* \geq bR^*$ iff $aR \subseteq bR$ ). The value group classifies the ideals of $R$, just as for a commutative valuation ring. The projection map $v\colon D^* \to \Gamma_R$ defines a valuation on $D$ in the sense of Schilling's book [S]. Conversely, every valuation on $D$ gives rise to an invariant valuation ring. Such valuation rings can be traced back to Hasse's work in [H] on orders in central simple algebras over local fields. Since then they have found notable applications in the construction of division algebras with nontrivial $SK_1$ (cf. [P], [$E_1$], [DK]), and in the construction of noncrossed products (cf. [$JW_1$], [Sa], [T], [K]). The Malcev-Neumann division algebras considered by Tignol and Amitsur in [TA] and [T] each have an invariant valuation ring. Recent work on such valuation rings in finite-dimensional division algebras can be found in [TW], [$JW_2$], and [PY, §3].

In the commutative theory valuations and places are two aspects of the same thing. But for division rings the two notions diverge. For places on division rings the corresponding type of valuation ring is a *total valuation ring*, i.e., a subring $R$ of a division ring $D$, such that

$$d \in R \ \text{ or } \ d^{-1} \in R, \text{ for each } d \in D^* .$$

In such a ring the Jacobson radical $J(R)$ is the unique maximal left (and also right) ideal, and is described by (1) above. The left ideals are linearly ordered by inclusion, but need not be the same as the right ideals. $R/J(R)$ is a division ring, and the associated place is the obvious map $D \to R/J(R) \cup \infty$. Not every total valuation ring is invariant, as shown by examples in [Ma, §4]. Even if $[D:Z(D)] < \infty$, where $Z(D)$ is the center of $D$, $D$ may have a noninvariant total valuation ring. This was shown first by Gräter in [$G_1$], correcting an error in [VG, Th. 1.8, p.32]. For a "value group" of a total valuation ring $R$, Mathiak has proposed

$$D^*/\bigcap_{d\in D^*} dR^*d^{-1}, \tag{4}$$

a group whose natural partial ordering is usually not a linear ordering. (This group is isomorphic to the group of order-isomorphisms of the set $\{dR \mid d \in D^*\}$ induced by left multiplications by elements of $D^*$.) See Mathiak's book [Ma] for a detailed account of total valuation rings, including applications to geometry and an extensive bibliography.

These two types of valuation rings on division rings are by no means the only ones that have been investigated. One can find, e.g., in Van Geel's book [VG] and the references given there, a number of other noncommutative valuation-like conditions.

Yet another definition of noncommutative valuation ring has been proposed by Dubrovin in [$D_1$] and [$D_2$]. It is the purpose of the rest of this paper to describe Dubrovin's rings, and to convince the reader that they form an interesting class of rings.

The rings considered by Dubrovin arise from a notion of place in the category of simple Artinian rings. Specifically, a subring $B$ of a simple Artinian ring $S$ is said to be a *Dubrovin valuation ring* of $S$ if $B$ has an ideal $I$ such that

(a) $B/I$ is a simple Artinian ring;

(b) for each $s \in S - B$ there exist $b_1, b_2 \in B$ such that $b_1 s, s b_2 \in B - I$.

It turns out that $I = J(B)$ and that $B$ is a left and right order in $S$ [$D_1$, §1, Th.4]. Hence, $B$ is a prime left and right Goldie ring. In addition, the center $Z(B)$ ($= B \cap Z(S)$) is a valuation ring of the field $Z(S)$. The following further properties noted by Dubrovin give some justification for calling such a ring $B$ a valuation ring:

(i) (Bezout) Every finitely generated left ideal of B is principal.

(ii) (semihereditary) Every finitely generated left ideal of B is a projective B-module.

(iii) B has the "k-chain property:" There is an integer $k$ such that for any $n > k$ and any $a_1, \ldots, a_n \in B$, the left ideal $Ba_1 + \ldots + Ba_n$ is generated by $k$ of the $a_i$.

(iv) The two-sided ideals of B are linearly ordered by inclusion, though the left ideals are in general not linearly ordered.

(v) (composition of places) If A is a subring of B with $J(B) \subseteq A$, then A is a Dubrovin valuation ring of S iff $A/J(B)$ is a Dubrovin valuation ring of $B/J(B)$.

(vi) (overrings and localization) For every ring T with $B \subseteq T \subseteq S$, T is a Dubrovin valuation ring of S, $J(T)$ is a prime ideal of B, T is the left (and right) localization of B with respect to the elements of B regular mod $J(T)$, and $B/J(T)$ is a Dubrovin valuation ring of $T/J(T)$. Furthermore, if $[S:Z(S)] < \infty$, then there is a one-to-one correspondence between prime ideals of B, prime ideals of $Z(B)$, and overings T of B.

(vii) (Morita invariance) If $S \cong M_n(D)$, i.e., $n \times n$ matrices over a division ring D, then $B \cong M_n(A)$, where A is a Dubrovin valuation ring of D. Furthermore, for any natural number k, $M_k(B)$ is a Dubrovin valuation ring of $M_k(S)$; if $e \in B$ is idempotent $(e \neq 0)$, then $eBe$ is a Dubrovin valuation ring of $eSe$.

See [$D_1$, §1,Th.4,Th.7; §2,Th.4] and [$D_2$, §1,Prop 2; §2,Th.1] for proofs of these properties. Of course, the right-hand versions of (i), (ii), and (iii) also hold.

Every Dubrovin valuation ring B has a residue ring $B = B/J(B)$, which is a simple Artinian ring. It is easy to check that B is a total valuation ring iff B is a division ring. One can also obtain a kind of value group for B as follows: First set

$$st(B) = \{s \in S^* \mid sBs^{-1} = B\} \tag{5}$$

(where S is the ring of quotients of B), the stablizer of B under the conjugation action of $S^*$. Then define the value group $\Gamma_B$ of B by

$$\Gamma_B = st(B)/B^*. \tag{6}$$

There is a natural bijection between $\Gamma_B$ and the set of fractional ideals $\{Bs \mid s \in S^* \text{ and } Bs = sB\}$. The linear ordering of this set by inclusion (cf. property (iv) above) induces on $\Gamma_B$ the structure of a totally ordered group. Clearly, if $V = Z(R)$, $\Gamma_V$ is an ordered subgroup of $\Gamma_B$. One finds that, unlike the commutative or invariant case, $\Gamma_B$ does not always classify the two-sided ideals of B. Nonetheless, $\Gamma_B$ is an interesting invariant that carries much information about the ideals of B, and does behave in many respects like the value group of an invariant valuation ring (cf. Th.3 and Cor.2 below). If B is a total valuation ring and $[S:Z(S)] < \infty$, then Gräter has shown in [$G_2$, Th.3.3] that $\Gamma_B$ is isomorphic to the center of Mathiak's value group (4) above.

Here are some examples of Dubrovin valuation rings:

(a) Clearly every invariant or total valuation ring is a Dubrovin valuation ring.

(b) If V is a valuation ring of a (commutative) field, then every Azumaya algebra A over V is a Dubrovin valuation ring [$D_2$, §2, Prop.1] and $\Gamma_A = \Gamma_V$ [$W_2$].

(c) Let V be a discrete (rank 1) valuation ring with quotient field F, and let S be an F-central simple algebra (with $[S:F] < \infty$). Then a subring B of S with $B \cap F = V$ is a Dubrovin valuation ring iff B is a maximal order of V in S. ("If" follows from [$D_1$, §1, Th.4] and [Re, (18.7)], and "only if" from [$D_2$, §3, Prop.2], [F, Th.2] and property (vi) above.)

For example, let $S = \left(\frac{-1,-1}{\mathbf{Q}}\right)$, the Hamiltonian quaternion division algebra over $\mathbf{Q}$, with its standard base {1, i, j, k}. Let p be a prime number, and $V = \mathbf{Z}_{(p)}$, the p-adic discrete valuation ring of $\mathbf{Q}$. If $p \neq 2$, $B = V + Vi + Vj + Vk$ is a Dubrovin valuation ring and an Azumaya algebra over V, but not a total valuation ring. In fact, $\overline{B} \cong M_2(\mathbf{F}_p)$, where $\mathbf{F}_n$ denotes the field with n elements, and $\Gamma_B = \Gamma_V \cong \mathbf{Z}$. Each such B has infinitely many conjugates, and these are all the Dubrovin valuation rings B of S with $B \cap \mathbf{Q} = \mathbf{Z}_{(p)}$ (cf. Th.2 in §2 below). For $p = 2$, let $\alpha = \frac{1}{2}(1 + i + j + k)$. Then for $V = \mathbf{Z}_{(2)}$, the ring $B = V\alpha + Vi + Vj + Vk$ is an invariant valuation ring of S, and is the unique Dubrovin valuation ring B of S with $B \cap \mathbf{Q} = \mathbf{Z}_{(2)}$. In this case, $\overline{B} \cong \mathbf{F}_4$ and $\Gamma_B = \frac{1}{2}\Gamma_V$.

(d) Let $V$ be a commutative valuation ring of rank 1 (i.e., Krull dimension 1) with quotient field $F$, and let $S$ be a central simple $F$-algebra. Let $(\hat{F}, \hat{V})$ be the completion of $(F,V)$ with respect to the topology induced by the ideals of $V$. Write $S \otimes_F \hat{F} \cong M_n(D)$ where $D$ is an $\hat{F}$-central division ring (with $[D:\hat{F}] < \infty$), and view $S \subseteq M_n(D)$. Because the complete rank 1 valuation ring $\hat{V}$ is Henselian (cf. §2 below) there is a unique invariant valuation ring $R$ of $D$ with $R \cap \hat{F} = \hat{V}$. Then every Dubrovin valuation ring of $M_n(D)$ contracting to $\hat{V}$ has the form $u M_n(R) u^{-1}$ for some $u \in M_n(D)^*$, and $B = (u M_n(R) u^{-1}) \cap S$ is a Dubrovin valuation ring of $S$ with $B \cap F = V$, by [$D_2$, §3, proof of Lemma 1] or [$BG_2$, Lemma 3.4, Th.5.2]. Moreover, every Dubrovin valuation ring of $S$ contracting to $V$ in $F$ is obtainable this way.

## 2. EXTENDING CENTRAL VALUATION RINGS AND HENSELIZATION

Let $E \subseteq D$ be division rings and $V$ some type of valuation ring of $E$. It is a basic problem to describe the extensions of $V$ to $D$, i.e., those valuation rings $W$ of $D$ with $W \cap E = V$. When $D$ is a field there is a well-developed extension theory, which begins with the observation that $V$ has at least one, often several, extensions to $D$. However, when $D$ is noncommutative it can be very hard to determine whether $V$ has any extensions to $D$ at all. There is a general extendibility criterion for invariant valuation rings with abelian value group, due to Krasner and rediscovered by P. M. Cohn (cf. [VG, p.31] and [CM, Th.2.3]), which generalizes Chevalley's criterion in the commutative case. However, this criterion is very difficult to apply in many cases.

We will here consider the extendibility problem for $D$ a division ring, where $E = Z(D)$ and $[D:E] < \infty$ ; we look first at extensions of a valuation ring of $E$ to invariant valuation rings, then to Dubrovin valuation rings. In the invariant case, the question of extendibility is closely tied to Hensel's Lemma. We will see that Henselization also yields much information about the structure of Dubrovin valuation rings.

We first recall a few basic facts about Henselian valuation rings. A valuation ring $V$ of a field $F$ is said to be *Henselian* if it satisfies the following equivalent conditions:

(a) For any monic polynomial $f \in V[X]$, if its image

$\bar{f} \in \bar{V}[X]$ has a simple root $a \in \bar{V}$, then $f$ has a simple root $\alpha \in V$ with $\bar{\alpha} = a$.

(b) For any field $K \supseteq F$, $K$ algebraic over $F$, $V$ has a unique extension to a valuation ring of $K$.

A nice account of the equivalence of (a) and (b) and many other characterizations of Henselian valuation rings can be found in [$R_2$]. Hensel's Lemma says that any complete rank 1 valuation ring is Henselian (cf. [En, pp.120-121]). Indeed, Henselian valuation rings play much the same rôle in general commutative valuation theory that complete valuation rings play in the rank 1 theory. In particular, every field with valuation ring (F,V) has a *Henselization* (F',V') (cf. [$R_1$] or [En]). The field $F'$ is a separable algebraic extension of $F$ usually of infinite degree, $V'$ is a Henselian valuation ring of $F'$ which extends $V$, and $\overline{V'} \cong \bar{V}$ and $\Gamma_{V'} = \Gamma_V$. Further, there is a universal mapping property for the Henselization that determines (F',V') uniquely up to isomorphism from (F,V).

Let $D$ be a division ring and $F = Z(D)$, with $[D:F] < \infty$, and let $V$ be a valuation ring of $F$. It is easy to construct examples of such $D$, $F$, $V$ for which $V$ has no extension to an invariant valuation ring of $D$ (e.g., the odd $p$ case of the example in (c) of §1 above). On the other hand, it has long been known that if $V$ is Henselian then $V$ has a unique extension to an invariant valuation ring $R$ of $D$ (cf. [S, Th.9, p.53] -- Schilling's term for Henselian is "relatively complete"). Indeed, $R$ consists of the (unique) extensions of $V$ to all the subfields of $D$ containing $F$. This extension property is a major reason why most work on invariant valuation rings on division algebras has been carried out over Henselian valued fields. But invariant valuation rings are also of interest in the non-Henselian case, as with constructions involving more than one valuation ring of the center (cf. [$JW_1$]), or when considering valuation rings on the generic abelian crossed product algebras of Amitsur and Saltman (cf. [$JW_2$,(5.14)(b)]). For $V$ not Henselian, P. Morandi has recently proved in [M] the following extendibility criterion:*

THEOREM 1. *Let* $V$ *be a valuation ring of a field* $F$, *and let* $D$ *be an* $F$-*central division ring with* $[D:F] < \infty$. *Let* (F',V') *be the Henselization of* (F,V). *Then,*

* Part (a) of Theorem 1 is also given in [$E_2$]; however, there is a gap in Ershov's proof which I do not know how to fill.

*(a)* $V$ *extends to an invariant valuation ring* $R$ *of* $D$ *if and only if* $D \otimes_F F'$ *is a division ring.*

*(b)* *Suppose* $D \otimes_F F'$ *is a division ring. Let* $R'$ *be the (unique) invariant valuation ring of* $D \otimes_F F'$ *with* $R' \cap F' = V'$ . *Then* $R = R' \cap D$ , $\overline{R} \cong \overline{R'}$, *and* $\Gamma_R = \Gamma_{R'}$ .

P. M. Cohn proved part (a) of this theorem for rank $V = 1$ using the completion instead of the Henselization of $V$ [C, Th.1]. An easy consequence of Theorem 1 is the criterion proved in [$W_1$] and earlier in [$E_2$]: $V$ extends to an invariant valuation ring of $D$ iff $V$ has a unique extension to each (commutative) field $K$ with $F \subseteq K \subseteq D$ . Furthermore, $V$ has at most one invariant extension to $D$ . Theorem 1 is a useful tool for reducing problems on invariant valuation rings to the better-understood Henselian case.

We return now to Dubrovin valuation rings, for which the problem of extending central valuation rings has the best solution one could hope for. Note that in view of property (vii) in §1 there is no loss in generality in working with Dubrovin valuation rings in division rings, rather than in simple Artinian rings.

THEOREM 2. *Let* $D$, $F$, *and* $V$ *be as in Theorem 1. Then,*

*(a)* $V$ *extends to a Dubrovin valuation ring* $B$ *of* $D$ .

*(b)* *If* $B'$ *is another Dubrovin valuation ring of* $D$ *extending* $V$ , *then* $B' = dBd^{-1}$ *for some* $d \in D^*$ .

Part (a) of Theorem 2 was proved in [$D_2$, §3, Th.2] and reproved, more convincingly, in [$BG_2$, Th.3.8]. Part (b), the conjugacy condition, was proved for $V$ of finite rank in [$BG_2$, Th.5.4], and the infinite rank case will be settled in [$W_2$].

Theorem 2 strikes me as a particularly significant result. To every valuation ring $V$ of $Z(D)$ there is a *unique* (up to isomorphism) associated Dubrovin valuation ring $B$ of $D$ . It is reasonable to expect that $B$ will carry much information about the arithmetic of $D$ in relation to $V$ . Given this theorem and the properties listed in §1 one might also hope that a generalization of the theory of maximal orders over discrete valuation rings could be developed for Dubrovin valuation rings over arbitrary central valuation rings. Some steps in this direction have already been taken in [$D_1$, §2].

To every valuation ring $V$ of $Z(D)$ there is an essentially uniquely determined Dubrovin valuation ring $B$ of $D$ and a uniquely associated invariant valuation ring $R$ of the corresponding division ring over a Henselization of $Z(D)$ with respect to $V$. Morandi's Theorem 1 showed how $B$ and $R$ are related in case $B$ is invariant. There is a corresponding result when $B$ is not assumed invariant:

THEOREM 3. *Let $D$ be a division ring and let $F = Z(D)$, with $[D:F] < \infty$. Let $B$ be a Dubrovin valuation ring of $D$, and let $V = B \cap F$. Let $(F',V')$ be the Henselization of $(F,V)$, and write*

$$D \otimes_F F' \cong M_n(D')$$

*where $D'$ is an $F'$-central division algebra. Let $R$ be the invariant valuation ring of $D'$ extending $V'$ in $F'$. Then,*

*(a)* $\overline{B}$ ( $= B/J(B)$ ) $\cong M_t(\overline{R})$, *where $t$ is an integer dividing $n$ which will be described in (7) below (and $\overline{R} = R/J(R)$ is a division ring).*

*(b)* $\Gamma_B \cong \Gamma_R$ *by an order-preserving isomorphism which is the identity on $\Gamma_V$.*

*(c) Define the* defect *of $B$, $\delta(B)$,* by

$$[D:F] = [\overline{B}:\overline{V}]\,|\Gamma_B:\Gamma_V|\,(n/t)^2\,\delta(B) .$$

*Then $\delta(B) = \delta(R)$, which by Draxl's "Ostrowski Theorem" [Dr, Th.2] equals 1 if $\mathrm{char}(\overline{V}) = 0$ and equals $p^a$ for some integer $a \geq 0$ if $\mathrm{char}(\overline{V}) = p \neq 0$.*

To give a formula for $t$ we consider other Henselizations. Let $\{P_j\}_{j\in J}$ be the linearly ordered set of nonzero prime ideals of $V$. For each $j \in J$, let $(F_j, V_j)$ be the Henselization of $(F, V_{P_j})$, where $V_{P_j}$ is the localization of $V$ at $P_j$; let $D \otimes_F F_j \cong M_{m_j}(D_j)$, where $D_j$ is a division algebra. So $1 \leq m_j \leq \sqrt{[D:F]}$. For $j,s \in J$ with $P_j \subseteq P_s$, $V_{P_s}$ is a refinement of $V_{P_j}$, so $F_j$ embeds in $F_s$ by the universal mapping property for Henselization (cf. [$R_1$, p.210, Cor.1]; hence $m_j | m_s$. It follows easily that for each $m \in \{m_j \mid j\in J\}$ there is a $P_j$ maximal such that $m_j = m$. Call such a $P_j$ a *jump prime ideal* of $V$ with respect to $D$. Let $Q_1 \subsetneq Q_2 \subsetneq \ldots \subsetneq Q_k$ be all the jump prime ideals of $V$ (re $D$). The number $k$ is called the *jump rank* of $V$ with respect to $D$. Note that $k \leq \sqrt{[D:F]} < \infty$.

Now, for each $i$, $1 \le i \le k-1$, let $B_{Q_i}$ denote the central localization of $B$ with respect to $Q_i$; $B_{Q_i}$ is a Dubrovin valuation ring of $D$ with $B_{Q_i} \cap F = V_{Q_i}$. We have $V_{Q_{i+1}}/Q_i$ is a valuation ring of the residue field $\overline{V_{Q_i}} = V_{Q_i}/Q_i$, and $V_{Q_{i+1}}/Q_i$ embeds in the center of $\overline{B_{Q_i}} = B_{Q_i}/J(B_{Q_i})$. Let

$\ell_{i+1}$ = the number of extensions of $V_{Q_{i+1}}/Q_i$ to valuation rings of $Z(\overline{B_{Q_i}})$.

Then, the integer $t$ in Theorem 3(a) is given by

$$t = n/\ell_2\ell_3 \ldots \ell_k . \tag{7}$$

The proofs of Theorem 3 and formula (7) will be given in [W2]. The proofs are by induction on the jump rank, which is a convenient invariant since it is always finite even when $V$ has infinite rank. We now point out some consequences of Theorem 3 and related theorems which will also be proved in [W2]. We retain throughout the notation and hypotheses of Theorem 3. For the first corollary note that for any $d \in st(B)$ conjugation by $d$ is a $V$-automorphism of $B$, so it induces a $\overline{V}$-automorphism of $\overline{B}$. Thus, there is a well-defined homomorphism $\varphi_B\colon st(B) \to \mathrm{Aut}_{\overline{V}}\, \overline{B}$. Since $\varphi_B(d)$ must send $Z(\overline{B})$ to itself, and is the identity on $Z(\overline{B})$ if $d \in B^* \cdot F^*$, $\varphi_B$ induces a homomorphism $\theta_B\colon \Gamma_B/\Gamma_V \to \mathrm{Aut}_{\overline{V}}\, Z(\overline{B})$.

COROLLARY 1. $Z(\overline{B})$ *is a normal (though not necessarily separable) finite degree field extension of* $\overline{V}$. *Also,* $\Gamma_B$ *is an abelian group, and the maps* $\varphi_B\colon st(B) \to \mathrm{Aut}_{\overline{V}}\, \overline{B}$ *and* $\theta_B\colon \Gamma_B/\Gamma_V \to \mathrm{Aut}_{\overline{V}}\, Z(\overline{B})$ *are surjective. (Hence, if* $Z(\overline{B})$ *is separable over* $\overline{V}$ *it is abelian Galois over* $\overline{V}$*.)*

COROLLARY 2. *Let* $A$ *be a ring,* $B \subseteq A \subseteq D$, *and let* $W = A \cap F$. *(So* $A$ *is a Dubrovin valuation ring of* $D$.*) Let* $\tilde{B} = B/J(A)$, *which is a Dubrovin valuation ring of* $\overline{A}$. *Then there is an exact sequence*

$$0 \to \Gamma_{\tilde{B}} \to \Gamma_B \to \Gamma_A \to \mathrm{Aut}_{\overline{V}}\, L \to 0,$$

*where* $L \subseteq Z(\overline{A})$ *is the decomposition field of the valuation ring* $\tilde{B} \cap Z(\overline{A})$ *relative to the field extension from* $\overline{W}$ *to* $Z(\overline{A})$.

Two positive integer invariants from $B$ are singled out in Theorem 3, namely

$n$ = matrix size of $D \otimes_F F'$ ,

$t$ = matrix size of $\overline{B}$ .

We have $t \mid n$ , and the extreme cases of the values of $t$ have interesting interpretations:

THEOREM 4. *The following are equivalent:*

*(a)* $t = n$ .

*(b)* $B$ *is integral over* $V$ .

*(c)* $B \otimes_V V'$ *is a Dubrovin valuation ring of* $D \otimes_F F'$

*(d)* *There is a Dubrovin valuation ring* $B'$ *of* $D \otimes_F F'$ *with* $B' \cap D = B$ .

*(e)* *Every principal two-sided ideal of* $B$ *is principal as a left ideal and as a right ideal of* $B$.

*(f)* *For every ring* $A$ *with* $B \subseteq A \subseteq D$ , *setting* $W = A \cap F$ , *we have* $V/J(W)$ *(a valuation ring of* $\overline{W}$ *) extends uniquely to a valuation ring of* $Z(\overline{A})$.

Note that formula (7) shows that Theorem 4 applies whenever $V$ has rank 1, or even jump rank 1.

THEOREM 5. *The following are equivalent:*

*(a)* $t = 1$ .

*(b)* $B$ *is a total valuation ring.*

*(c)* $B$ *has only finitely many different conjugates.*

*(d)* *The set* $T$ *of elements of* $D$ *integral over* $V$ *is a ring. (In fact,* $T = \bigcap_{d \in D^*} d B d^{-1}$ *.)*

*When these equivalent conditions hold, the number of conjugates of* $B$ *is exactly* $n$ *.*

In Theorem 5, (b) $\Rightarrow$ (c) and (b) $\Leftrightarrow$ (d) were proved in [BG1], where it was shown that the number of conjugates of $B$ is bounded by $\sqrt{[D:F]}$ . The exact value for the number of conjugates was rather a surprise.

Here is an example to illustrate these theorems. Let $k$ be any field with $\operatorname{char}(k) \neq 2$ ; let $F = k(x,y)$ where $x$ and $y$ are commuting indeterminates. Let $D = \left(\frac{1+x,\,y}{F}\right)$, a quaternion algebra over $F$, with its standard base $\{1,i,j,k\}$. Let $W = k(x)[y]_{(y)}$ the localization of $k(x)[y]$ at its prime ideal $(y)$. Then $W$ is a discrete valuation ring of $F$ with $\overline{W} \cong k(x)$ and $\Gamma_W \cong \mathbf{Z}$. It is easy to check (or apply [$JW_2$, (4.3)]) that $W$ extends to an invariant valuation ring $A$ of $D$; in fact, $A = W + Wi + Wj + Wk$ and $J(A) = J(W) + J(W)i + Wj + Wk$. Since $D$ has an invariant valuation ring, $D$ must be a division ring. Since $A$ is invariant it is the only Dubrovin valuation ring of $D$ extending $W$. For $A$ we have $n = t = 1$, $\overline{A} \cong \overline{W}(i) \cong k(x)(\sqrt{1+x})$, so the residue degree $[\overline{A}:\overline{W}]$ is 2, and $\Gamma_A = \frac{1}{2}\Gamma_W$, so the ramification index $|\Gamma_A:\Gamma_B| = 2$. Necessarily $\delta(A) = 1$ as $\text{char}(\overline{A}) \nmid [D:F]$. Let $w: D^* \to \Gamma_A$ be the valuation associated to $A$.

Now, let $V_0 = k[x]_{(x)}$, a discrete valuation ring of $\overline{W}$ which has two different extensions $Y_1$ and $Y_2$ to $Z(\overline{A}) = \overline{A}$. Let $V$ be the "composite" of $V_0$ with $W$, i.e., $V = \pi_W^{-1}(V_0)$, where $\pi_W: W \to \overline{W}$ is the natural projection. $V$ is a valuation ring of $F$. Let $B = \pi_A^{-1}(Y_1)$ (where $\pi_A: A \to \overline{A}$). This $B$ is a total valuation ring of $D$ which is not invariant. The conjugates $B$ and $\pi_A^{-1}(Y_2)$ are the only Dubrovin valuation rings of $D$ extending $V$. We have $\overline{B} \cong \overline{V} \cong k$ and $\text{st}(B) = \{a+bi+cj+dk \in D^* \mid w(a+bi) \leq w(c+di)\}$ so that $|D^*:\text{st}(B)| = 2$. One can check that $\text{st}(B) = F^* \cdot B^*$, hence $\Gamma_B = \Gamma_V \cong \mathbf{Z} \times \mathbf{Z}$. Thus, for $B$, $[\overline{B}:\overline{V}] = 1$, $|\Gamma_B:\Gamma_V| = 1$, $t = 1$, $\ell_2 = 2$, so $n = t\ell_2 = 2$, and $\delta(B) = 1$. Since $n^2 = [D:F]$, the Henselization of $F$ with respect to $V$ must split $D$. In fact, this Henselization contains the maximal subfield $F(\sqrt{1+x})$ of $D$. $V$ has rank and jump rank (re $D$) 2. $B$ is evidently not integral over $V$, as $Y_1$ is not integral over $V_0$.

Clearly many other examples of Dubrovin valuation rings can easily be constructed by this same process of composing valuation rings (i.e., composing their places).

The theory of Dubrovin valuation rings is still very new, but I think the results stated here provide strong motivation for further development of this theory. There are many questions that remain to be explored, and we close by mentioning a few of them.

Question 1. Suppose $B$ is a Dubrovin valuation ring of a division ring $D$, with $F = Z(D)$ and $[D:F] < \infty$. If $E$ is an F-subalgebra of $D$, under what conditions is $B \cap E$

a Dubrovin valuation ring of $E$ ? If $B \cap E$ is not Dubrovin, how is it related to the Dubrovin valuation rings of $E$ ?

For example, consider $D = \left(\frac{-1,-1}{\mathbf{Q}}\right)$, $V = \mathbf{Z}_{(5)}$ as in example (c) in §1, and let $E = \mathbf{Q}(i) \cong \mathbf{Q}(\sqrt{-1})$. Since $-1$ is a square mod 5 $V$ has two different extensions $V_1$ and $V_2$ to $E$. For any Dubrovin valuation ring $B$ of $D$ extending $V$, $B \cap E \subseteq V_1 \cap V_2$ since by Th. 4 $B$ is integral over $V$. Hence, $B \cap E$ is not a valuation ring. Likewise, neither $V_i$ can extend to a Dubrovin valuation ring of $D$ since $V_i$ is not integral over $V$.

Question 2. Suppose $B_i$ is a Dubrovin valuation ring of $D_i$, $i = 1,2$, where the $D_i$ are finite-dimensional F-central division algebras and $B_1 \cap F = B_2 \cap F = V$. Can one determine from $B_1$ and $B_2$ a Dubrovin valuation ring of $D_1 \otimes_F D_2$ extending $V$?

Note that in most cases $B_1 \otimes_V B_2$ is not a Dubrovin valuation ring (though this does occur if $B_1$ or $B_2$ is an Azumaya algebra). Some partial results on the corresponding problem for invariant valuation rings are given in [TW], [$JW_2$], and [M], but the proofs were usually hampered by having to get back to the underlying division algebra of $D_1 \otimes_F D_2$. Because Dubrovin valuation rings are defined over simple Artinian rings, one might be able to work directly in $D_1 \otimes_F D_2$ to obtain stronger results, with more natural proofs, than in the invariant case. The analogue to Question 2 can also be asked for Dubrovin valuation rings of $D \otimes_F K$ where $D$ is an F-central division algebra and $K$ is a field algebraic over $F$.

## REFERENCES


[$BG_1$] H.H.Brungs and J.Gräter, Valuation rings in finite dimensional division algebras, to appear in *J. Algebra*.

[$BG_2$] H.H.Brungs and J. Gräter, Extensions of valuation rings in central simple algebras, preprint, University of Alberta and Technische Universität Braunschweig, 1987.

[C] P.M. Cohn, On extending valuations in division algebras, *Studia Scient. Math. Hung., 16* (1981), pp. 65-70.

[CM] P.M. Cohn and M. Mahdavi-Hezavehi, Extensions of valuations on skew fields, pp. 28-41, in *Ring Theory Antwerp 1980. Proceedings*, ed. F. Van Oystaeyen, Lecture Notes in Math., No.825, Springer-Verlag, Berlin, 1980.

[Dr] P. Draxl, Ostrowski's theorem for Henselian valued skew fields, *J. Reine Angew. Math., 354* (1984), pp. 213-218.

[DK] P. Draxl and M. Kneser, eds., *$SK_1$ von Schiefkörpern*, Lecture Notes in Math., No.778, Springer-Verlag, Berlin, 1980.

[$D_1$] N.I. Dubrovin, Noncommutative valuation rings, *Trudy Moskov. Mat. Obshch., 45* (1982), pp.265-289; English transl: *Trans. Moscow Math. Soc., 45* (1984), pp.273-287.

[$D_2$] N.I. Dubrovin, Noncommutative valuation rings in simple finite-dimensional algebras over a field, *Mat. Sb., 123 (165)* (1984), pp.496-509; English transl: *Math. USSR Sbornik, 51*(1985), pp.493-505.

[En] O.Endler, *Valuation Theory*, Springer-Verlag, New York, 1972.

[$E_1$] Yu.L. Ershov, Henselian valuations of division rings and the group $SK_1$, *Mat. Sb., 117 (159)* (1982), pp.60-68; English transl: *Math. USSR Sbornik, 45* (1983), pp.63-71.

[$E_2$] Yu.L. Ershov, Valued division rings, pp.53-55, in *Fifth All Union Symposium, Theory of Rings, Algebras, and Modules*, Akad. Nauk SSSR Sibirsk. Otdel, Inst. Mat., Novosibirsk, 1982 (in Russian).

[F] E. Formanek, Noetherian P.I. rings, *Comm. Algebra, 1* (1974), pp.79-86.

[$G_1$] J.Gräter, Über Bewertungen endlich dimensionaler Divisionsalgebren, *Resultate der Math.,* 7 (1984), pp. 54-57.

[$G_2$] J.Gräter, Valuations on finite-dimensional division algebras and their value groups, preprint, Technische Universität Braunschweig, 1987.

[H] H. Hasse, Über p-adische Schiefkörper und ihre Bedeutung für die Arithmetik hyperkomplexer Zahlsysteme, *Math. Ann., 104* (1931), pp.495-534.

[$JW_1$] B. Jacob and A. Wadsworth, A new construction of noncrossed product algebras, *Trans. Amer. Math. Soc., 293* (1986), pp.693-721.

[$JW_2$] B. Jacob and A. Wadsworth, Division algebras over Henselian fields, preprint, Math. Sci. Research Inst., Berkeley, Calif., 1987.

[K] A. Kupferoth, Valuated division algebras and crossed products, *J. Algebra, 108* (1987), pp.139-150.

[Ma] K. Mathiak, *Valuations of Skew Fields and Projective Hjelmslev Spaces*, Lecture Notes in Math., No.1175, Springer-Verlag, Berlin, 1986.

[M] P. Morandi, The Henselization of a valued division algebra, preprint, Univ. of California at San Diego, 1987.

[P] V.P. Platonov, The Tannaka-Artin problem and reduced K-theory, *Izv. Akad. Nauk SSSR, Ser. Mat., 40* (1976), pp.227-261; English transl: *Math. USSR Izv., 10* (1976), pp.211-243.

[PY] V.P. Platonov and V.I. Yanchevskiĭ, Dieudonné's conjecture on the structure of unitary groups over a division ring, and Hermitian K-theory, *Izv. Akad. Nauk SSSR, Ser. Mat., 48* (1984), pp.1266-1294; English transl: *Math. USSR Izv., 25* (1985), pp.573-599.

[Re] I. Reiner, *Maximal Orders*, Academic Press, London, 1975.

[$R_1$] P. Ribenboim, *Théorie des Valuations*, Presses Univ. Montréal, Montréal, 1968.

[$R_2$] P.Ribenboim, Equivalent forms of Hensel's lemma, *Expo. Math., 3* (1985), pp.3-24.

[Sa] D. Saltman, Noncrossed products of small exponent, *Proc. Amer. Math. Soc., 68* (1978), pp.165-168.

[S] O.F.G. Schilling, *The Theory of Valuations,* Math. Surveys, No.4, Amer. Math. Soc., Providence, R.I., 1950.

[T] J.-P. Tignol, Cyclic and elementary abelian subfields of Malcev-Neumann division algebras, *J. Pure Appl. Algebra, 42* (1986), pp.199-220.

[TA] J.-P. Tignol and S.A. Amitsur, Kummer subfields of Malcev-Neumann division algebras, *Israel J. Math., 50* (1985), pp.114-144.

[TW] J.-P. Tignol and A.R. Wadsworth, Totally ramified valuations on finite-dimensional division algebras, *Trans. Amer. Math. Soc., 302* (1987), pp.223-250.

[VG] J. Van Geel, *Places and Valuations in Noncommutative Ring Theory,* Lecture Notes in Pure and Appl. Math., No.71, Marcel Dekker, New York, 1981.

[$W_1$] A.R. Wadsworth, Extending valuations to finite-dimensional division algebras, *Proc. Amer. Math. Soc., 98* (1986), pp.20-22.

[$W_2$] A.R. Wadsworth, Dubrovin valuation rings and Henselization, in preparation.

SPECIAL MONOMIAL ALGEBRAS OF FINITE GLOBAL DIMENSION

Dan Zacharia
Department of Mathematics
Syracuse University
Syracuse, New York U.S.A. 13244

ABSTRACT. We find bounds for the Loewy length and for the global dimension in terms of the number of nonisomorphic simple modules over a certain class of finite dimensional algebras of finite global dimension.

1. W. Gustafson proved in [GU] that if $\Lambda$ is a serial artinian ring of finite global dimension, then its Loewy length is less than 2n where n represents the number of nonisomorphic simple $\Lambda$-modules and, that its global dimension is bounded by 2n-2. His proof used the properties of the Kupisch series.

We show in this article that the same bounds can be obtained for a larger class of finite dimensional algebras which is a subclass of the "monomial" algebras. We will assume throughout this paper that $\Lambda$ is a finite dimensional k algebra, k being an algebraically closed field. Moreover we may assume that $\Lambda$ is basic, and then it is known that $\Lambda$ is given by a quiver with relations [GA]. Furthermore, if Q is the quiver of $\Lambda$, we will assume that Q has the property that for every oriented cycle C of Q, there is no arrow leaving the cycle. For instance Q may have the following shape:

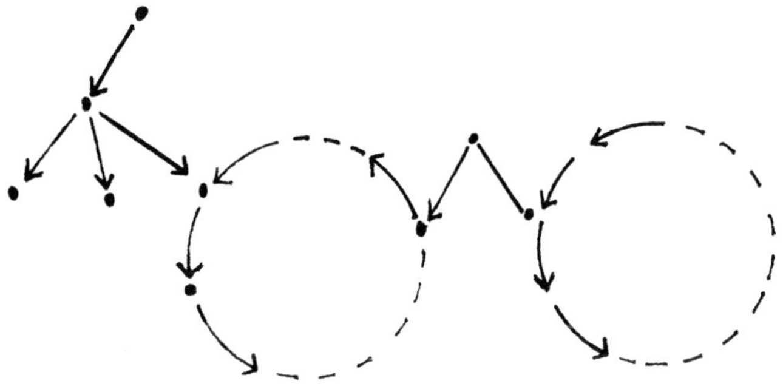



F. van Oystaeyen and L. Le Bruyn (eds.), Perspectives in Ring Theory, 375–378.

Then, we write $\Lambda \simeq kQ/I$ where kQ is the path algebra of the quiver Q and I is a two sided ideal of kQ generated by monomials (or paths) in Q. For the remainder we call such finite dimensional algebras special monomial algebras (or special zero relations algebras). We note that every left or right serial algebra, as well as every tree algebra is special.

Next we briefly review the algorithm for determining the minimal projective resolutions of the simple modules over a monomial algebra. If $\Lambda=kQ/I$ where I is generated by monomials, fix a minimal set of generators for I so that $I=\langle\rho\rangle=\langle\rho_1,\ldots,\rho_k\rangle$ where the $\rho_i$-s are oriented paths in Q of length at least 2. Next we look at the topological universal cover $(\tilde{Q},\tilde{\rho})$ of the quiver with relations $(Q,\rho)$. Clearly $\tilde{Q}$ is a tree (usually infinite) and $\tilde{\rho}$ is the set of all the paths in $\tilde{Q}$ which are liftings of the paths in $\rho$. As usual, we identify the finitely generated $\Lambda$-modules with the representations of $(Q,\rho)$, and the finite dimensional representations of $(\tilde{Q},\tilde{\rho})$ with the finitely generated graded $\Lambda$-modules where $\Lambda$ is group graded by $\pi(Q)$-the fundamental group of Q. Then, as showed in [GHZ] we reduce the problem of computing the minimal resolutions of the simple $\Lambda$-modules, to computing the minimal projective resolutions of the simple representations of the (possibly infinite) paths with relations of $(\tilde{Q},\tilde{\rho})$.

2. Lemma 2.1. Let $\Lambda$ be an arbitrary monomial algebra and let S be a simple $\Lambda$ module of finite projective dimension. Then $\mathrm{Ext}^1_\Lambda(S,S)=0$.

Proof: Let $\Lambda=kQ/\langle\rho\rangle$ where Q is the quiver of $\Lambda$ and $\rho$ is the ideal of relations. Let i be the vertex corresponding to S. It is clearly enough to show that there is no loop $i\xrightarrow{x}i$. If such an x exists then $x^n=0$ for some $n\geq 2$, since $\Lambda$ is finite dimensional, and if n is minimal with such property then $x^n\in\langle\rho\rangle$ is one of the generators. Now, on the infinite path $i\xrightarrow{x}i\xrightarrow{x}i\xrightarrow{x}\ldots$ in $(\tilde{Q},\tilde{\rho})$ we construct the associated sequence of relations as in [GHZ] and we observe that this sequence is infinite so that $\mathrm{pd}S=\infty$. ▲

We note that if   is an arbitrary artin ring of finite global dimension it is not known that $\mathrm{Ext}^1(S,S)=0$ for every simple module S.

Lemma 2.2. Let $\Lambda$ be a special monomial algebra of finite global dimension. Then, there exists a simple $\Lambda$-module S such that $\mathrm{pd}_\Lambda S=1$.

Proof: Clearly we may assume that Q has an oriented cycle, so let C be such a cycle with say m vertices. Assume that for every vertex j of C $\mathrm{pd}S(j)>1$. Then, to each such j, there is a relation $\rho_j\in\langle\rho\rangle$ starting at j. This implies that the associated sequence of relations corresponding to the vertices of C are infinite thus $\mathrm{pd}\,S(j)=\infty$, for every vertex in C which gives a contradiction. ▲

Lemma 2.3. Let $\Lambda$ be a special monomial algebra of finite global dimension and let $P_1,\ldots,P_n$ be a complete set of indecomposable projective

$\Lambda$-modules where pd $P_1/\underline{r}P_1=1$. Let $\Gamma=\mathrm{End}_\Lambda(P_2 \sqcup \ldots \sqcup P_n)^{op}$. Then $\Gamma$ is also a special monomial algebra and gldim $\Gamma \leq$ gldim $\Lambda$.

Proof: pd $P_1/\underline{r}P_1$ means that $\underline{r}P_1$ is either projective or zero. Thus by applying $\mathrm{Hom}_\Lambda(P_2 \sqcup \ldots \sqcup P_n, \cdot)$: mod $\Lambda \rightarrow$ mod $\Gamma$ to the minimal projective resolutions of the simple $\Lambda$-modules we obtain projective resolutions, not necessarily minimal of a complete set of nonisomorphic simple $\Gamma$-modules, therefore gldim $\Gamma \leq$ gldim $\Lambda$.

The fact that $\Gamma$ is also a special monomial algebra is trivial. ▲

We are now in position to prove the results about the promised bounds.

Proposition 2.4. Let $\Lambda$ be a special monomial algebra of finite global dimension having n nonisomorphic simple modules. Then

a) $LL(\Lambda) < 2n$
b) gldim $\Lambda \leq 2n-2$.

Proof. a) Clearly we may assume that Q contains a closed circuit and assume that there exists a vertex i in Q whose corresponding projective cover P(i) has Loewy length exceeding 2n. Then, there is a closed circuit $C \subset Q$ with say m vertices and a vertex j in C with $LLP(j) \geq 2m$. By induction we will arrive at a contradiction. Pick $i_0$ in C with pd $S(i_0)=1$ which is possible by Lemma 2.2 and consider

$$\Gamma = \mathrm{End}_\Lambda\Big(\coprod_{k \neq i_0} P(k)\Big)^{op}.$$

$\Gamma$ is again a special monomial algebra of finite global dimension, its quiver has n-1 vertices and also $\Gamma$ has a closed circuit with m-1 vertices, and the Loewy length of at least 2(m-1) for the projective cover corresponding to a vertex on that circuit. Now apply induction and the fact that special algebras have no loops.

b) If S is a simple $\Lambda$-module, [GHZ] shows that a minimal projective resolution of S has the form

$$0 \rightarrow \coprod \mathcal{K}_i \rightarrow P(s) \rightarrow S \rightarrow 0$$

where $\mathcal{K}_i$ are acyclic complexes whose terms are indecomposable projective $\Lambda$-modules. Then Gustafson's argument applies here as well. ▲

Corollary 2.5. Let $\Lambda$ be a special monomial algebra of finite global dimension and $P_1, \ldots, P_n$ be a complete set of nonisomorphic indecomposable projective $\Lambda$-modules. Then, for each i, $\dim_k \mathrm{Hom}_\Lambda(P_i, P_i) \leq 2$ and there exists a j such that $\dim_k \mathrm{Hom}_\Lambda(P_j, P_j)=1$.

Proof: Follows from the proposition and using induction on n. ▲

References:

[GA] P. Gabriel. Auslander-Reiten sequences and representation finite algebras. Lecture Notes in Mathematics. Vol. 831, Springer-Verlag, New York 1980.

[GHZ] E. L. Green, D. Happel and D. Zacharia. Projective resolutions over artin algebras with zero relations. Illinois Journal of Mathematics, vol. 29, No. 1, Spring 1985.

[GU] W. Gustafson. Global Dimension in Serial Rings. Journal of Algebra, vol. 97, 1985.

## INDEX OF DEFINITIONS.

**E**



**F**



**G**



**H**



**I, J**

## L



## M



## N



## P



## Q



## R

GPSR Compliance

*The European Union's (EU) General Product Safety Regulation (GPSR) is a set of rules that requires consumer products to be safe and our obligations to ensure this.*

*If you have any concerns about our products, you can contact us on ProductSafety@springernature.com*

In case Publisher is established outside the EU, the EU authorized representative is:

Springer Nature Customer Service Center GmbH
Europaplatz 3
69115 Heidelberg, Germany

**Batch number: 10815974**

Printed by Printforce, the Netherlands